T0329584

6G Frontiers

IEEE Press
445 Hoes Lane
Piscataway, NJ 08854

IEEE Press Editorial Board
Sarah Spurgeon, *Editor in Chief*

Jón Atli Benediktsson
Anjan Bose
Adam Drobot
Peter (Yong) Lian
Andreas Molisch
Saeid Nahavandi
Jeffrey Reed
Thomas Robertazzi
Diomidis Spinellis
Ahmet Murat Tekalp

6G Frontiers

Towards Future Wireless Systems

Chamitha de Alwis
University of Bedfordshire
Luton, United Kingdom

and

University of Sri Jayewardenepura
Nugegoda, Sri Lanka

Quoc-Viet Pham
Pusan National University
Busan, Republic of Korea

Madhusanka Liyanage
University College Dublin
Dublin, Ireland

and

University of Oulu
Oulu, Finland

Copyright © 2023 by The Institute of Electrical and Electronics Engineers, Inc. All rights reserved.

Published by John Wiley & Sons, Inc., Hoboken, New Jersey.
Published simultaneously in Canada.

No part of this publication may be reproduced, stored in a retrieval system, or transmitted in any form or by any means, electronic, mechanical, photocopying, recording, scanning, or otherwise, except as permitted under Section 107 or 108 of the 1976 United States Copyright Act, without either the prior written permission of the Publisher, or authorization through payment of the appropriate per-copy fee to the Copyright Clearance Center, Inc., 222 Rosewood Drive, Danvers, MA 01923, (978) 750-8400, fax (978) 750-4470, or on the web at www.copyright.com. Requests to the Publisher for permission should be addressed to the Permissions Department, John Wiley & Sons, Inc., 111 River Street, Hoboken, NJ 07030, (201) 748-6011, fax (201) 748-6008, or online at http://www.wiley.com/go/permission.

Trademarks: Wiley and the Wiley logo are trademarks or registered trademarks of John Wiley & Sons, Inc. and/or its affiliates in the United States and other countries and may not be used without written permission. All other trademarks are the property of their respective owners. John Wiley & Sons, Inc. is not associated with any product or vendor mentioned in this book.

Limit of Liability/Disclaimer of Warranty: While the publisher and author have used their best efforts in preparing this book, they make no representations or warranties with respect to the accuracy or completeness of the contents of this book and specifically disclaim any implied warranties of merchantability or fitness for a particular purpose. No warranty may be created or extended by sales representatives or written sales materials. The advice and strategies contained herein may not be suitable for your situation. You should consult with a professional where appropriate. Neither the publisher nor author shall be liable for any loss of profit or any other commercial damages, including but not limited to special, incidental, consequential, or other damages.

For general information on our other products and services or for technical support, please contact our Customer Care Department within the United States at (800) 762-2974, outside the United States at (317) 572-3993 or fax (317) 572-4002.
Wiley also publishes its books in a variety of electronic formats. Some content that appears in print may not be available in electronic formats. For more information about Wiley products, visit our web site at www.wiley.com.

Library of Congress Cataloging-in-Publication Data Applied for:

Hardback ISBN: 9781119862345

Cover Design: Wiley
Cover Image: © Rob Nazh/Shutterstock

Set in 9.5/12.5pt STIXTwoText by Straive, Chennai, India

To my parents

Contents

About the Authors

Chamitha de Alwis University of Sri Jayewardenepura
Chamitha de Alwis (Senior Member, IEEE) is a Lecturer, Researcher and Consultant in Cybersecurity. Presently he works as a Lecturer in Cybersecurity in the School of Computer Science and Technology, University of Bedfordshire, United Kingdom. He is the founder Head of the Department of Electrical and Electronic Engineering, University of Sri Jayewardenepura, Sri Lanka. He also provides consultancy services for telecommunication-related projects and activities. He received the BSc degree (First Class Hons.) in Electronic and Telecommunication Engineering from the University of Moratuwa, Sri Lanka, in 2009, and the PhD degree in Electronic Engineering from the University of Surrey, United Kingdom, in 2014. He has published peer-reviewed journal articles, conference papers, and book chapters, and delivered tutorials and presentations in international conferences. He also has contributed to various projects related to ICT, and served as a guest editor, reviewer, and TPC member in international journals and conferences. He has also worked as a Consultant to the Telecommunication Regulatory Commission of Sri Lanka, an Advisor in IT Services in the University of Surrey, United Kingdom, and a Radio Network Planning and Optimization Engineer in Mobitel, Sri Lanka. His research interests include network security, 5G/6G, blockchain, and IoT.

Quoc-Viet Pham Pusan National University
Quoc-Viet Pham (Member, IEEE) received the BS degree in Electronics and Telecommunications Engineering from the Hanoi University of Science and Technology, Vietnam, in 2013, and the MS and PhD degrees in Telecommunications Engineering from Inje University, South Korea, in 2015 and 2017, respectively. From September 2017 to December 2019, he was with Kyung Hee University, Changwon National University, and Inje University, in various academic positions. He is currently a Research Professor with Pusan National University, South Korea. He has been granted the Korea NRF Funding for

outstanding young researchers for the term 2019–2023. His research interests include convex optimization, game theory, and machine learning to analyze and optimize edge/cloud computing systems and future wireless systems. He received the Best PhD Dissertation Award in Engineering from Inje University, in 2017. He received the top reviewer award from the *IEEE Transactions on Vehicular Technology* in 2020. He is an editor of the *Journal of Network and Computer Applications* (Elsevier), an associate editor of the *Frontiers in Communications and Networks*, and the lead guest editor of the *IEEE Internet of Things Journal*.

Madhusanka Liyanage is currently an Assistant Professor/Ad Astra Fellow and Director of Graduate Research at the School of Computer Science, University College Dublin, Ireland. He is also acting as a Docent/Adjunct Professor at the Center for Wireless Communications, University of Oulu, Finland, and Honorary Adjunct Professor of Network Security, The Department of Electrical and Information Engineering, University of Ruhuna, Sri Lanka. He received his Doctor of Technology degree from the University of Oulu, Finland, in 2016. He was also a recipient of the prestigious Marie Skłodowska-Curie Actions Individual Fellowship during 2018–2020. During 2015–2018, he had been a Visiting Research Fellow at the CSIRO, Australia, the Infolabs21, Lancaster University, United Kingdom, Computer Science and Engineering, The University of New South Wales, Australia, School of IT, University of Sydney, Australia, LIP6, Sorbonne University, France and Computer Science and Engineering, The University of Oxford, United Kingdom. He is also a senior member of IEEE. In 2020, he received the "2020 IEEE ComSoc Outstanding Young Researcher" award by IEEE ComSoc EMEA. Dr. Liyanage is an expert consultant at the European Union Agency for Cybersecurity (ENISA). In 2021, Liyanage was elevated as Funded Investigator of Science Foundation Ireland CONNECT Research Centre, Ireland. He was ranked among the World's Top 2% Scientists (2020) in the List prepared by Elsevier BV, Stanford University, USA. Also, he was awarded an Irish Research Council (IRC) Research Ally Prize as part of the IRC Researcher of the Year 2021 awards for the positive impact he has made as a supervisor. Moreover, he is an expert reviewer at different funding agencies in France, Qatar, UAE, Sri Lanka, and Kazakhstan. More info: www.madhusanka.com

Preface

While the fifth-generation (5G) mobile communication networks are deployed worldwide, multitude of new applications and use cases driven by current trends are already being conceived, which challenges the capabilities of 5G. This has motivated academic and industrial researchers to rethink and work toward the next generation of mobile communication networks called 6G hereafter. 6G networks are expected to mark a disruptive transformation to the mobile networking paradigm by reaching extreme network capabilities to cater to the demands of the future data-driven society.

Recent developments in communications have introduced many new concepts such as edge intelligence, beyond sub 6, GHz to THz communication, nonorthogonal multiple access, large intelligent surfaces, and self-sustaining networks. These concepts are evolving to become full-fledged technologies that can power future generations of communication networks. On the other hand, applications such as holographic telepresence, extended reality, smart grid 2.0, and Industry 5.0 are expected to emerge as mainstream applications of future communication networks. However, requirements of these applications such as ultra-high data rates, real-time access to powerful computing resources, extremely low latency, precision localization and sensing, and extremely high reliability and availability surpass the network capabilities promised by 5G. IoT, which is enabled by 5G, is even growing to become Internet of Everything (IoE) that intends to connect massive numbers of sensors, devices, and cyber-physical systems beyond the capabilities of 5G. This has inspired the research community to envision 6G mobile communication networks. 6G is expected to harness the developments of new communication technologies, fully support emerging applications, connect a massive number of devices, and provide real-time access to powerful computational and storage resources.

6G mobile networks are expected to provide extreme peak data rates over 1, Tbps. The end-to-end delays will be imperceptible and lie even beneath 0.1, ms. 6G networks will provide access to powerful edge intelligence that has processing delays falling below 10, ns. Network availability and reliability are expected to go beyond 99.99999%. An extremely high connection density of over 10^7 devices/ per km^2 is expected to be supported to facilitate IoE. The spectrum efficiency of 6G will be over 5× than 5G, while support for extreme mobility up to 1000, kmph is expected. It is also envisioned that the evolution of 6G will focus around a myriad of new requirements such as Further-enhanced Mobile Broadband (FeMBB), ultra-massive Machine-Type Communication (umMTC), Mobile BroadBand and Low-Latency (MBBLL), and massive Low-Latency Machine Type communication (mLLMT). These requirements will be enabled through emerging technologies such as THz spectrum, federated learning, edge artificial intelligence (AI), compressive sensing, blockchain, and 3D networking. Moreover, 6G will facilitate emerging applications such as unmanned aerial vehices (UAVs), holographic telepresence, IoE, Industry 5.0, and collaborative autonomous driving. In light of this vision, many new research work and projects are themed toward developing the 6G vision, technologies, use cases, applications, and standards. The vision for 6G is envisaged to be framed by 2022–2023 to set forth the 6G requirements and evaluate the 6G development, technologies, standards, etc.

In order to further provide a full understanding of 6G frontiers and boost the research and development of 6G, we are motivated to provide an authored book on 6G, future wireless systems. To the best of our knowledge, this book covers all the aspects of 6G. In the first part of this book, we present the evolution of mobile networks, from 0G to 6G, which are followed by the introduction to driving trends, requirements, and key-enabling technologies. In the second part, we present potential architectural directions of 6G, including zero-touch network and service management, intent-based networking, edge AI, intelligent network softwarization, and radio access networks. Then, the technical aspects of 6G are discussed in detail. In particular, we focus on (i) hyper-intelligent networking, (ii) security and privacy, trust, (iii) energy management and resource allocation, (iv) harmonized mobile networks, and (v) legal aspects and standardization. In the final part of this work, we focus on vertical applications which are expected to emerge in future 6G network systems. More specifically, we focus on four main kinds of applications, including (i) healthcare/well-being, (ii) smart cities, (iii) industrial automation (e.g. Industry 5.0. collaborative robots, and digital twin), and (iv) wild applications (e.g. space tourism and deep sea tourism).

Intended Audience

This books will be of key interest for

- *Researchers*: Developing 6G enabling technologies is already at the forefront of today's communications research. This book will provide a clear idea on how different technologies will mature toward developing the 6g framework.

- *Academics*: Academics who are teaching and performing research work in the area of emerging communication technologies are in need of a textbook on the envisaged 6G technologies and framework, which is provided through this book.
- *Technology Architects*: Technology architects need to envision 6G and develop and align technologies toward realizing 6G.
- *Mobile Network Operators* (MNOs): MNOs require knowledge on 6G in order to plan their future work considering 6G technologies, framework, applications, and use cases as discussed in this book.
- *Industry Experts*: Industry experts are expected to envision future applications and use cases and develop businesses and invest accordingly.
- *Regulators and Standards*: Regulators and Standards institutions are required to be aware of the forthcoming technologies and applications in order to set regulations and standards.

Book Organization

This book begins with introducing the concept of 6G mobile communication networks in Chapter 1. Subsequently, the key driving trends toward 6G mobile networks are explained in Chapter 2. Then 6G requirements, including the vision for 6G together with enabling 6G applications and technologies, are discussed in Chapter 3. Chapter 4 explains the key 6G technologies, while Chapter 5 introduces 6G architectural visions. Zero-Touch Network and Service Management is explained in Chapter 6. Chapter 7 elaborates Edge AI, while Chapter 8 discusses intelligent network softwarization with 6G. Chapter 9 explains 6G radio access technologies. Security and privacy aspects of 6G are discussed in Chapter 10, while Chapter 11 discusses about resource efficient 6G networks. Chapter 12 elaborates how 6G will be deployed as harmonized mobile networks to provide extreme global coverage. Chapter 13 discusses 6G standardization efforts and legal aspects. Chapters 14, 15, and 16 explain emerging directions for 6G applications in healthcare, smart cities, and industrial automation. Chapter 17 provides insights on some wild 6G applications that are expected to emerge in the coming decade. Chapter 18 concludes this book.

Nugegoda, Sri Lanka — *Chamitha de Alwis*
Dublin, Ireland — *Madhusanka Liyanage*
South Korea — *Quoc- Viet Pham*

Acknowledgments

This book would not have been possible without the great help and support of many. The concept of publishing this book to facilitate 6G-related studies, research, development, and standardization came to light during our research work in projects such as STHRD R1/SJ/01 Project, University of Sri Jayewardenepura ASP/01/RE/ENG/2022/85 Research Project, Korea NRF-2019 R1C1C1006143, and Science Foundation Ireland under Connect Center (13/RC/2077_P2) Project, the Academy of Finland under 6Genesis Flagship (Grant 318927) project and European Commission under H2020 SPATIAL project (Grant 101021808). We would also like to acknowledge all the partners of those projects. Furthermore, we would like to thank our universities, University of Sri Jayewardenepura, Pusan National University, and University College Dublin, for all the support extended toward the successful completion of this book. We would also like to thank chapter contributors, including Dr. Pardeep Kumar, Dr. Thippa Reddy Gadekallu, Dr. Sweta Bhattacharya, Dr. Praveen Kumar Reddy Maddikunta for their invaluable contribution to complete this book. We also thank all the reviewers for helping us select suitable chapters for this book. We are also grateful to Sandra Grayson, Teresa Netzler, and the whole John Wiley & Sons team for their support toward getting this book published.

Last but not least, we would offer our heartiest gratitude to our families, who gladly allowed us to share our time with them toward the completion of this book.

Chamitha de Alwis
Quoc-Viet Pham
Madhusanka Liyanage
C. V. M.

Acronyms

ASTA	Arrivals See Time Averages
BHCA	Busy Hour Call Attempts
BR	Bandwidth Reservation
b.u.	bandwidth unit(s)
CAC	Call / Connection Admission Control
CBP	Call Blocking Probability(-ies)
CCS	Centum Call Seconds
CDTM	Connection Dependent Threshold Model
CS	Complete Sharing
DiffServ	Differentiated Services
EMLM	Erlang Multirate Loss Model
erl	The Erlang unit of traffic-load
FIFO	First in - First out
GB	Global balance
GoS	Grade of Service
ICT	Information and Communication Technology
IntServ	Integrated Services
IP	Internet Protocol
ITU-T	International Telecommunication Unit–Standardization sector
LB	Local balance
LHS	Left hand side
LIFO	Last in - First out
MMPP	Markov Modulated Poisson Process
MPLS	Multiple Protocol Labeling Switching
MRM	Multiretry Model
MTM	Multithreshold Model
PASTA	Poisson Arrivals See Time Averages
PDF	Probability Distribution Function
pdf	probability density function

PFS	Product Form Solution
QoS	Quality of Service
r.v.	random variable(s)
RED	random early detection
RHS	Right hand side
RLA	Reduced Load Approximation
SIRO	service in random order
SRM	Single-Retry Model
STM	Single-Threshold Model
TCP	Transport Control Protocol
TH	Threshold(s)
UDP	User Datagram Protocol
3GPP	3rd Generation Partnership Project
A2G	Air-to-Ground
AEC	AI-assistive Extreme Communications
AI	Artificial Intelligence
AR	Augmented Reality
AV	Autonomous Vehicles
BAN	Body Area Network
BCI	Brain Computer Interface
CAV	Connected Autonomous Vehicles
CPS	Cyber-Physical Systems
CS	Compressive Sensing
D2D	Device-to-Device
DLT	Distributed Ledger Technologies
DRL	Deep Reinforcement Learning
EI	Edge Intelligence
ELPC	Extremely Low-Power Communication
ETSI	European Telecommunications Standards Institute
eMBB	enhanced-Mobile Broadband
eMTC	enhanced Machine Type Communication
eRLLC	extremely Reliable Low-Latency Communication
FeMBB	Further-enhanced Mobile Broadband
FL	Federated Learning
H2H	Hospital-to-Home
HCS	Human-Centric Services
HT	Holographic Telepresence
IoBNT	Internet of Bio-NanoThings
IIoMT	Intelligent Internet of Medical Things
IIosT	Internet of Industrial smart Things
IIoT	Industrial Internet of Things

IoE	Internet of Everything
IoH	Internet of Healthcare
IoNT	Internet of Nano-Things
IoT	Internet of Things
IoV	Internet of Vehicles
IP	Internet Protocol
ITU	International Telecommunication Union
IRS	Intelligent Reflecting Surface
ITS	Intelligent Transport System
IWD	Intelligent Wearable Devices
KPI	Key Performance Indicator
LDHMC	Long Distance and High Mobility Communications
LED	Light Emitting Diodes
LIS	Large Intelligent Surfaces
LSTM	Long Short Term Memory
LTE	Long-Term Evolution
MBBLL	Mobile BroadBand and Low-Latency
mBBMT	massive Broadband Machine Type
MEC	Multiaccess Edge Computing
mHealth	mobile Health
MIMO	Multiple-Input and Multiple-Output
ML	Machine Learning
mLLMT	massive Low-Latency Machine Type
MMS	Multimedia Message Services
mMTC	massive Machine Type Communication
MR	Mixed Reality
MTC	Machine Type Communication
MTP	Motion-To-Photon
NB-IoT	Narrowband Internet of Things
NOMA	Non-Orthogonal Multiple Access
NTN	Nonterrestrial Networks
QoL	Quality of Life
QoPE	Quality-of-Physical-Experience
RF	Radio Frequency
RIS	Reconfigurable Intelligent Surface
SAGINs	Snetworks, and space-Air-Ground Interconnected Networks
SDN	Software Defined Networking
SDO	Standards Developing Organizations
SMS	Short Message Services
SSN	Self-Sustaining Networks
U2X	UAV-to-Everything

UAV	Unmanned Aerial Vehicles
UHD	Ultra High Definition
umMTC	ultra-massive Machine-Type Communication
uRLLC	ultra-Reliable Low Latency Communication
VANET	Vehicular Ad Hoc Networks
VoIP	Voice Over IP
VR	Virtual Reality
VLC	Visible Light Communincaiton
XR	Extended Reality
RPL	Low-Power and Lossy Networks
ZSM	Zero touch network and Service Management

Part I

Introduction

1

Evolution of Mobile Networks

Mobile networks have been evolving since the 1980s, resulting in a new generation of mobile network every decade. Presently, fifth-generation (5G) mobile networks are being deployed. However, mobile communication research and development work suggest that we can expect to see sixth-generation (6G) mobile networks by 2030. After reading this chapter, you should be able to

---

- Explain the evolution of mobile networks from 0G to 6G.
- Understand the present context of 6G development.

1.1 Introduction

While fifth-generation (5G) mobile communication networks are deployed worldwide, multitude of new applications and use-cases driven by current trends are already being conceived, which challenges the capabilities of 5G. This has motivated researchers to rethink and work toward the next-generation mobile communication networks "hereafter 6G" [1, 2]. The sixth-generation (6G) mobile communication networks are expected to mark a disruptive transformation to the mobile networking paradigm by reaching extreme network capabilities to cater to the demands of the future data-driven society.

So far mobile networks have evolved through five generations during the last four decades. A new generation of mobile networks emerges every ten years, packing more technologies and capabilities to empower humans to enhance their work and lifestyle. The precellphone era before the 1980s is marked as the zeroth-generation (0G) of mobile communication networks that provided simple radio communication functionality with devices such as walkie-talkies [3]. The first-generation (1G) introduced publicly and commercially available

6G Frontiers: Towards Future Wireless Systems, First Edition.
Chamitha de Alwis, Quoc-Viet Pham, and Madhusanka Liyanage.
© 2023 The Institute of Electrical and Electronics Engineers, Inc. Published 2023 by John Wiley & Sons, Inc.

cellular networks in the 1980s. These networks provided voice communication using analog mobile technology [4]. The second-generation (2G) of mobile communication networks marked the transition of mobile networks from analog to digital. It supported basic data services such as short message services in addition to voice communication [5]. The third-generation (3G) introduced improved mobile broadband services and enabled new applications such as multimedia message services, video calls, and mobile TV [6]. Further improved mobile broadband services, all-IP communication, Voice Over IP (VoIP), ultrahigh definition video streaming, and online gaming were introduced in the fourth-generation (4G) [7].

The 5G mobile communication networks are already being deployed worldwide. 5G supports enhanced Mobile Broadband (eMBB) to deliver peak data rates up to 10 Gbps. Furthermore, ultra-Reliable Low Latency Communication (uRLLC) minimizes the delays up to 1 ms while massive Machine Type Communication (mMTC) supports over 100× more devices per unit area compared to 4G. The expected network reliability and availability is over 99.999% [8]. Network softwarization is a prominent 5G technology that enables dynamicity, programmability, and abstraction of networks [9]. Capabilities of 5G have enabled novel applications such as Virtual Reality (VR), Augmented Reality (AR), Mixed Reality (MR), autonomous vehicles, Internet of Things (IoT), and Industry 4.0 [10, 11].

Recent developments in communications have introduced many new concepts such as Edge Intelligence (EI), beyond sub 6 GHz to THz communication, Nonorthogonal Multiple Access (NOMA), Large Intelligent Surfaces (LIS), swarm networks, and Self-Sustaining Networks (SSN) [12, 13]. These concepts are evolving to become fully fledged technologies that can power future generations of communication networks. On the other hand, applications such as Holographic Telepresence (HT), Unmanned Aerial Vehicles (UAV), Extended Reality (XR), smart grid 2.0, Industry 5.0, and space and deep-sea tourism are expected to emerge as mainstream applications of future communication networks. However, requirements of these applications such as ultrahigh data rates, real-time access to powerful computing resources, extremely low latency, precision localization and sensing, and extremely high reliability and availability surpass the network capabilities promised by 5G [14, 15]. IoT, which is enabled by 5G, is even growing to become Internet of Everything (IoE) that intends to connect massive numbers of sensors, devices, and Cyber-Physical Systems (CPS) beyond the capabilities of 5G. This has inspired the research community to envision 6G mobile communication networks. The 6G is expected to harness the developments of new communication technologies, fully support emerging applications, connect a massive number of devices, and provide real-time access to powerful computational and storage resources.

1.2 6G Mobile Communication Networks

The 6G networks are expected to be more capable, intelligent, reliable, scalable, and power-efficient to satisfy all the expectations that cannot be realized with 5G. The 6G is also required to meet any new requirements, such as support for new technologies, applications, and regulations, raised in the coming decade. Figure 1.1 illustrates the evolution of mobile networks, elaborating key features of each mobile network generation. Envisaged 6G requirements, vision, enablers, and applications are also highlighted to formulate an overview of the present understanding of 6G.

1.2.1 6G as Envisioned Today

The 6G mobile communication networks, as envisioned today, are expected to provide extreme peak data rates over 1 Tbps. The end-to-end delays will be imperceptible and lie even beneath 0.1 ms. The 6G networks will provide access to powerful edge intelligence that has processing delays falling below 10 ns. Network availability and reliability are expected to go beyond 99.99999%. An extremely high connection density of over 10^7 devices/km^2 is expected to be supported to facilitate IoE. The spectrum efficiency of 6G will be over 5× than 5G, while support for extreme mobility up to 1000 kmph is expected [12].

It is envisioned that the evolution of 6G will focus around a myriad of new requirements such as further-enhanced mobile broadband (FeMBB), ultramassive Machine-Type Communication (umMTC), Mobile BroadBand and Low-Latency

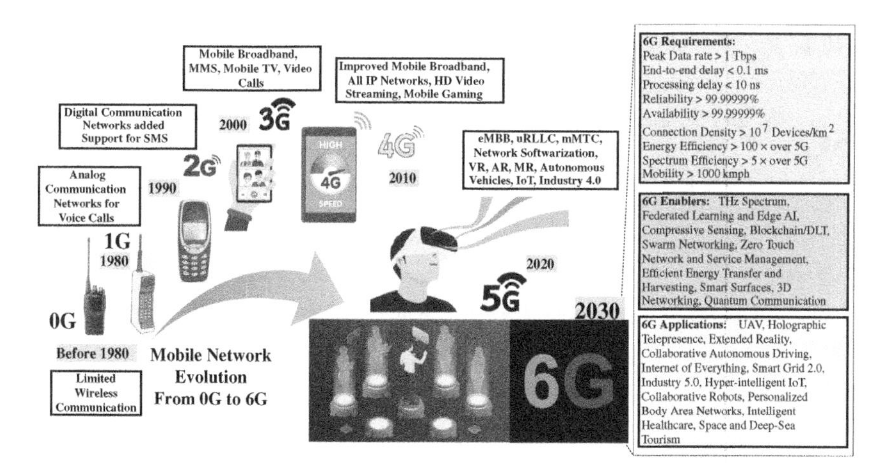

Figure 1.1 Evolution of Mobile Networks from 0G to 6G. Source: vectorplus / Adobe Stock.

(MBBLL), and massive Low-Latency Machine Type communication (mLLMT). These requirements will be enabled through emerging technologies such as THz spectrum, Federated Learning (FL), edge Artificial Intelligence (AI), Compressive Sensing (CS), blockchain/Distributed Ledger Technologies (DLT), and 3D networking. Moreover, 6G will facilitate emerging applications such as UAVs, HT, IoE, Industry 5.0, and collaborative autonomous driving. In light of this vision, many new research work and projects are themed toward developing 6G vision, technologies, use-cases, applications, and standards [1, 2].

1.2.2 6G Development Timeline

The 6G developments are expected to progress along with the deployment and commercialization of 5G networks, and the final developments of 4G Long-Term Evolution (LTE), being LTE-C, which followed LTE-Advanced and LTE-B [16]. The vision for 6G is envisaged to be framed by 2022–2023 to set forth the 6G requirements and evaluate the 6G development, technologies, standards, etc. Standardization bodies such as the International Telecommunication Union (ITU) and Third-Generation Partnership Project (3GPP) are expected to develop the specifications to develop 6G by 2026–2027 [16]. Network operators will start 6G research and development (R&D) work by this time to do 6G network trials by 2028–2029, to launch 6G communication networks by 2030 [14, 16–18]. Global 6G development initiatives are illustrated in Figure 1.3, while the expected timeline for 6G development, standardization, and launch is presented in Figure 1.2.

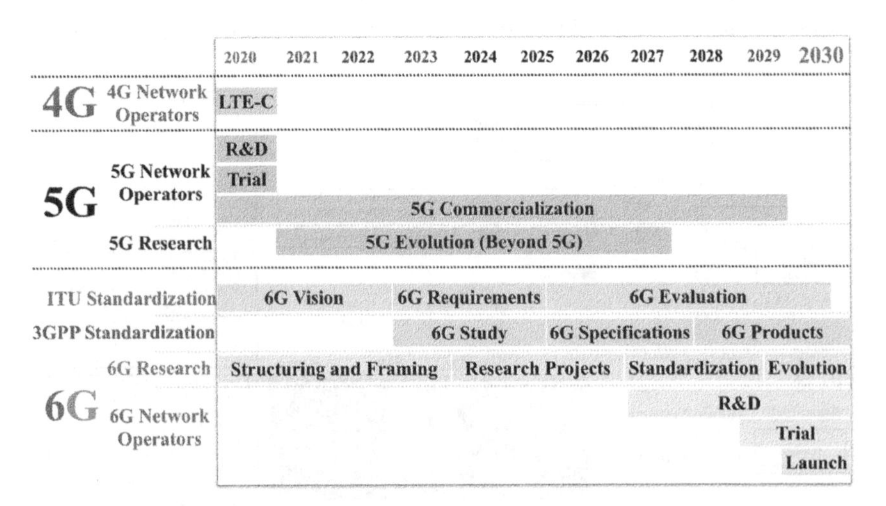

Figure 1.2 Expected Timeline of 6G Development, Standardization and Launch. Source: Adapted from [14, 16–18].

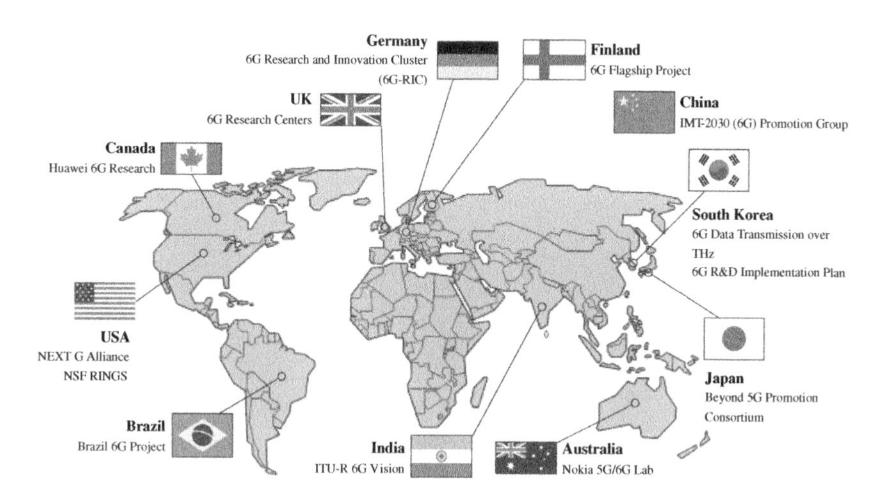

Figure 1.3 Global 6G Development Initiatives.

2

Key Driving Trends Toward 6G

Emerging applications and use-cases of the future society demands mobile networks to be more dense and capable. After reading this chapter, you should be able to

- Understand the future trends of 6G networks.
- Importance of the driving trend toward the development and definition of the requirements of 6G network.
- Identify components related to environmental and energy infrastructure.

2.1 Introduction

A new generation of mobile communication has emerged every 10 years over the last four decades to cater to society's growing technological and societal needs. This trend is expected to continue, and 6G is seen on the horizon to meet the requirements of the 2030 society [19, 20]. The technologies, trends, requirements, and expectations that force the shift from 5G toward the next generation of networks are identified as 6G driving trends. These driving trends will shape 6G into the key enabler of a more connected and capable 2030 society.

This chapter discusses the key 6G driving trends elaborating why and how each trend demands a new generation of communication networks. Figure 2.1 illustrates the 6G driving trends that are discussed in this section.

- *Expansion of IoTs*: It is expected that the number of IoT devices in the world will grow up to 24 billion by 2030. Moreover, the revenue related to IoT will hit the market capitalization of USD 1.5 trillion by 2030 [23].

6G Frontiers: Towards Future Wireless Systems, First Edition.
Chamitha de Alwis, Quoc-Viet Pham, and Madhusanka Liyanage.
© 2023 The Institute of Electrical and Electronics Engineers, Inc. Published 2023 by John Wiley & Sons, Inc.

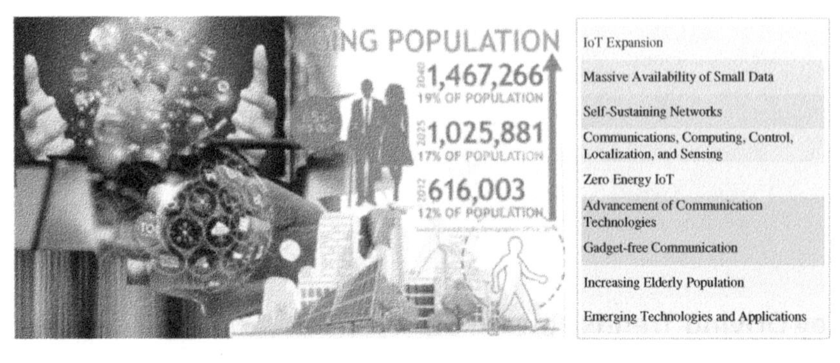

Figure 2.1 6G Driving Trends. Source: [21, 22]/IEEE.

- *Massive Availability of Small Data*: Due to the anticipated popularity of 6G-based IoT devices and new 6G-IoT services, 6G networks will trend to generate an increasingly high volume of data. Most of such data will be small, dynamic, and heterogeneous in nature [12, 24].
- *Availability of Self-Sustained Networks*: 6G mobile systems need to be energy self-sustainable, both at the infrastructure side and at the device side, to provide uninterrupted connectivity in every corner of the world. The development of energy harvesting capabilities will extend the life cycle of both network infrastructure devices and end devices such as IoE devices [25, 26].
- *Convergence of Communication, Sensing, Control, Localization, and Computing*: Development of sensor technologies and direct integration of them with mobile networks accompanied by low-energy communication capabilities will lead to advanced 6G networks [12, 27]. Such a network will be able to provide sensing and localization services in addition to the exciting communication and computing features [12, 27, 28].
- *Zero Energy IoT*: Generally, IoT devices will consume significantly more energy for communication than sensing and processing [29]. The development of ultralow-power communication mechanisms and efficient energy harvesting mechanisms will lead to self-energy sustainable or zero-energy IoT devices [29].
- *More Bits, Spectrum, and Reliability*: The advancement of wireless communication technologies, including coding schemes and antenna technologies, will allow to utilize new spectrum as well as reliably send more information bits over existing wireless channels [12, 19].
- *Gadget-Free Communication*: The integration of an increasing number of smart and intelligent devices and digital interfaces in the environment will lead to a change from gadget-centric to user-centric or gadget-free communication model. The hyperconnected digital surroundings will form an "omnipotential"atmosphere around the user, providing all the information, tools, and services that a user needs in his or her everyday life [30–32].

- *Increasing Elderly Population*: Due to factors such as advanced healthcare facilities and the development of new medicines, the world's older population continues to grow at an unprecedented rate. According to the "An Aging World: 2015" Report, nearly 17 percent (1.6 billion) of the world's population will be aged 65 and over by 2050 [22].
- *Emergence of New Technologies*: By 2030, the world will experience new technological advancements such as stand-alone cars, Artificial Intelligence (AI)-powered automated devices, smart clothes, printed bodies in 3D, humanoid robots, smart grid 2.0, industry 5.0, and space travel [12, 19]. The 6G will be the main underline communication infrastructure to realize these technologies.

2.2 Expansion of IoT toward IoE

IoT envisions to weave a global network of machines and devices that are capable of interacting with each other [33]. The number of IoT devices is on the rise and is expected to grow up to 24 billion by 2030 due to the growth of applications such as the Industrial Internet of Things (IIoT). The total IoT market is also expected to rise to USD 1.5 trillion in 2030 [23]. IoE is expected to expand the scope of IoT to form a hyperconnected world connecting people, data, and things to streamline the processes of businesses and industries while enriching human lives [34]. IoE will connect many ecosystems involving heterogeneous sensors, actuators, user equipment, data types, services, and applications [35].

The importance of this driving trend is discussed, considering the challenges in overcoming the limitations of existing networks to facilitate IoT development toward IoE. One of the key challenges in this development is the integration of AI and Machine Learning (ML) technologies into mobile communication networks [36]. These technologies are essential to process massive amounts of data collected from heterogeneous IoE devices to obtain meaningful information and enable new applications and use-cases envisioned with 6G [37]. Processing massive amounts of data using AI and ML requires future communication networks to provide real-time access to powerful computational facilities (Figure 2.2). The communication between IoE devices and mobile networks should also be power-efficient to minimize the carbon footprint. For instance, intelligent traffic control and transportation systems in future smart cities are expected to utilize future 6G communication networks to massively exploit data-driven methods for real-time optimization [38]. Such systems will require AI and ML to efficiently process large amounts of data collected from heterogeneous sensors in real-time to provide insights that will minimize traffic.

Preserving data security and privacy in existing IoT networks is yet another important requirement. Since everything in IoE is connected to the Internet,

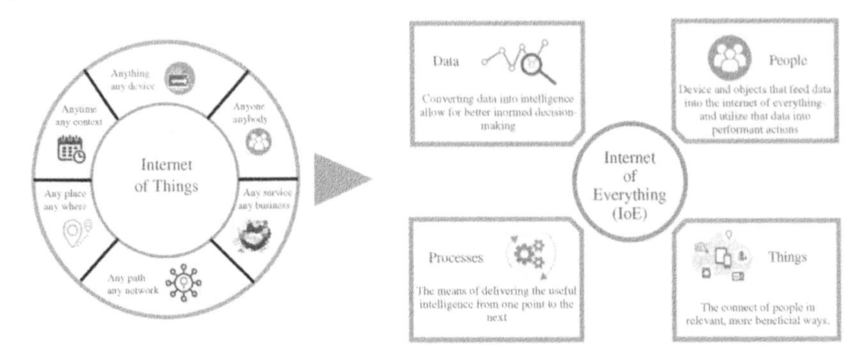

Figure 2.2 IoT to IoE Transition.

distributed AI technologies will be required for training data sets spread unevenly across multiple edge devices. This exposes IoE networks to security vulnerabilities associated with distributed AI, such as poisoning attacks and authentication issues [39]. Solving these issues requires AI and DLT-based adaptive security solutions that should be integrated with future communication networks [40]. DLTs and blockchain, in particular, are key enablers of IoE. The decentralized operation, immutability, and enhanced security of blockchain are instrumental in overcoming the challenges concerned with the exponential expansion of IoT [41].

Moreover, traditional Orthogonal Multiple Access (OMA)-based schemes cannot provide access to a massive number of IoE devices due to limitations in the radio spectrum. This requires new technologies such as NOMA to be applied to cellular IoT to provide access to a massive number of IoT devices [13, 36]. In addition, providing seamless connectivity to IoE devices that lie beyond the coverage of terrestrial cellular networks requires unmanned aerial vehicle (UAV) and satellites to work in coordination to form a cognitive satellite-UAV network [42]. Such technologies are expected to be integrated with the next generation of mobile communication networks to facilitate the smooth progression from IoT toward IoE.

2.3 Massive Availability of Small Data

The widespread heterogeneous IoT sensor nodes that continuously acquire massive amounts of diverse data are expected to generate over 30 exabytes of data per month by 2020 [43]. Collecting, storing, and processing this type of data through widespread communication networks will be one of the challenging requirements that should be met by future communication networks. The term "Small Data" refers to small data sets representing a limited pool of data in a niche area of interest [44]. Such data sets can provide meaningful insights to manage massive amounts of IoT devices. Unlike Big Data that are concerned with large sets of historical data, Small Data are concerned about either real-time data or statistical data of a limited time. Small Data will be instrumental in many applications, including the real-time equipment operation and the maintenance of massive numbers of machines connected in IIoT. IIoT is expected to grow connecting billions of CPS, devices, and sensors in the coming decade as discussed in Section 2.2.

Another example is the growing demand for Small Data-based analytics in the retail industry that collects data from various sensors, personal wearables, and IoT devices [45]. Such analytics are helpful to provide real-time personalized services for customers. These applications give rise to the generation of massive amounts of small data sets that should be efficiently collected and processed using AI and ML [12].

The importance of this driving trend is discussed, considering the processing and communication limitations of existing mobile communication networks. Processing massive amounts of Small Data sets is not efficient in existing cloud computing and edge computing infrastructure that is designed to process large data sets [46]. This requires new means of efficiently processing massive amounts of Small Data sets in the Edge AI infrastructure in future communication networks. This will also require new ML techniques beyond classical, big data analytics to enhance network functions and provide new services envisaged in future communication networks [12]. Furthermore, future networks should maximize the energy efficiency of offloading massive amounts of Small Data to edge computing facilities. This requires the optimization of joint radio and computation resources while satisfying the maximum tolerable delay constraints [47].

On the other hand, communication networks will need to support massive amounts of Small Data transmission from heterogeneous IoT devices. Overheads of this type of communication can be significant compared to the size of data that is being transmitted, making this type of data communication less efficient [24, 48]. This requires new methods to reduce transmission and contention overheads in future communication networks.

2.4 Availability of Self-Sustaining Networks

Self-Sustaining Networks (SSNs) can perform tasks such as self-managing, self-planning, self-organizing, self-optimizing, self-healing, and self-protecting network resources to continuously maintain its Key Performance Indicators (KPIs) [12]. This is performed by adapting network operation and functionalities considering various facts, including environmental status, network usage, and energy constraints [49]. These types of intelligent and real-time network operations in SSNs are facilitated using machine learning/deep learning/quantum machine learning techniques that enable fast learning of rapid network changes and dynamic user requirements [50]. Using SSNs, future networks are expected to enable seamless access to emerging application domains under highly dynamic and complex environments [12].

The importance of this driving trend is discussed, considering the incapability of existing mobile networks to function as SSNs. SSNs require the ability to obtain network statistics in real time to automatically manage resources and adapt functionalities to maintain high KPIs [12]. Therefore, SSNs require a novel self-sustaining network architecture that can adapt to rapid changes in the environment and user requirements. These operations should be facilitated through real-time analysis of massive amounts of Small Data obtained by network nodes. Small Data analysis can be performed using edge intelligence capabilities envisaged in future networks, as explained in Section 2.3.

Furthermore, self-optimization of radio resources needs to bank on software-defined cognitive radios through operations such as radio scene analysis [50, 51]. In addition, SSNs should facilitate energy self-sustainability at the infrastructure side as well as the device side to provide uninterrupted and seamless connectivity. Therefore, energy harvesting in network infrastructure should play a pivotal role to extend the range and stand-by times [25, 26]. This also requires future communication networks to be designed in an energy-aware fashion to enable devices to harvest energy, be self-powered, share power, and last long [17, 52]. Furthermore, handling massive numbers of IoT devices in an energy-efficient manner under various channel conditions and diverse applications requires self-learning through context-aware operation to minimize the energy per bit for a given communication requirement [43].

2.5 Convergence of Communications, Computing, Control, Localization, and Sensing (3CLS)

Future communication networks are expected to converge computing resources, controlling architecture, and other infrastructure used for precise localization and sensing [12]. This convergence is essential to facilitate highly personalized and time-critical future applications. For instance, Human-Centric Services (HCS) are expected to bank on 3CLS services to facilitate efficient communication and real-time processing of a large number of data streams gathered through sensors that are centered around humans [53, 54].

The development of 3CLS services is an important driving trend toward the next generation of mobile communication networks as existing 5G technologies have not fully explored the interdependence between computing, communication, control, localization, and sensing in an end-to-end manner [55]. Realizing 3CLS services will require future mobile communication networks to possess collective network intelligence at the edge of the network to run AI and ML algorithms in real time [12, 56]. Moreover, the network architecture should also be open, scalable, and elastic to facilitate AI orchestrated end-to-end 3CLS design services [12, 57]. Precise localization and sensing should also coexist with communication networks by sharing network resources in time, frequency, and space to facilitate emerging applications such as extended reality, connected robotics, connected and automated vehicles (CAVs), sensing, and 3D mapping [12, 28].

2.6 Zero Energy IoT

Zero energy IoT devices can harvest energy from the environment to obtain infinite power [58]. For instance, radio frequency (RF) energy harvesting can harvest energy from RF waves to extend the network lifetime. Nodes that harvest more energy can share their energy with other nodes using energy cooperation. Presently, only about 0.6% of the 1.5 trillion objects in the real world is connected to the Internet [29]. The remaining devices are also expected to be connected in an energy-efficient fashion together with the growth of future communication technologies and applications.

Zero energy IoT is an important driving trend toward future communication networks to enable maintenance-free and battery-less operation of a massive number of IoT devices. This requires mobile networks to be able to support ultralow-power communication and efficient energy harvesting [29]. However, existing 5G network infrastructures do not support energy harvesting, especially as the electronic circuitry cannot efficiently convert the harvested energy into electric current [17]. Therefore, electronic circuitry in future communication networks should be designed and developed to support efficient energy harvesting. Furthermore, circuits that harvest energy should allow devices to be self-powered to enable off-grid operations, long-lasting IoT devices, and longer stand-by times [17]. Wireless power transfer is also expected to play a key role in the next generation of mobile communication networks considering the feasibility of doing so due to much shorter communication distances in denser communication networks [59]. Furthermore, data communication stacks can also be optimized in an energy-aware fashion to minimize energy usage.

2.7 Advancement of Communication Technologies

Mobile communication has seen significant technological advances recently. For instance, electromagnetically active Large Intelligent Surfaces (LIS) made using meta-materials placed in walls, roads, buildings, and other smart environments with integrated electronics will provide massive surfaces for wireless communication [60]. Furthermore, novel channel access schemes such as NOMA have offered many advantages, such as being more spectral efficient than prevailing schemes. Beyond-millimeter Wave (mmWave) communication at THz frequency bands is also being exploited to provide uninterrupted connectivity in local and wide-area networks [61]. Key advancements of communication technologies are illustrated in Figure 2.3.

The emergence of new communication technologies that cannot be integrated with existing 5G networks is discussed to highlight the importance of this driving trend. For instance, future communication networks will need to shift from existing small cells toward tiny cells to support high-frequency bands in the THz spectrum. This requires a new architectural design supporting denser network deployments and mobility management at higher frequencies [61]. Furthermore, multimode base stations will be necessary to facilitate networks to operate in a

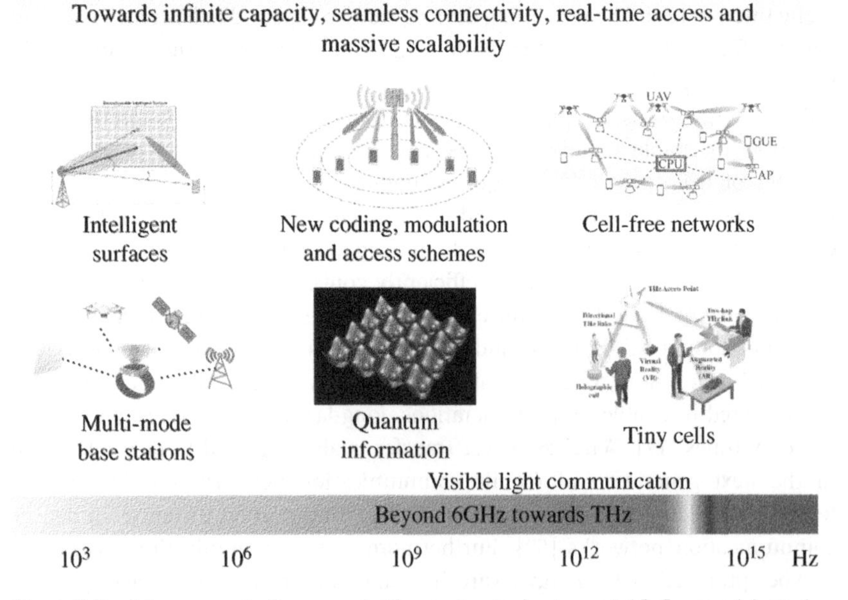

Figure 2.3 Advancement of communication technologies toward 6G. Source: Adapted from [12, 60–62]

wide range of spectra ranging from microwave to THz to provide uninterrupted connectivity. Furthermore, utilizing LISs as transceivers requires low-complexity channel demodulation banking on techniques such as joint compressive sensing (CS) and deep learning, which is not feasible with 5G [63]. Also, none of the recent advancements in communication technologies such as providing AI-powered network functionalities using collective network intelligence, Visible Light Communication (VLC), NOMA, cell-free networks, and quantum computing and communications are realized in 5G [12, 62, 64]. Therefore, the integration of these advanced communication technologies demands a new paradigm of mobile communication networks.

2.8 Gadget-Free Communication

Gadget-free communication eliminates the requirement for a user to hold physical communication devices. It is envisaged that the digital services centered around smart and connected gadgets will move toward a user-centric, gadget-free communication model as more and more digital interfaces, intelligent devices, and sensors get integrated to the environment [30, 32, 65]. Since most of our data and services are already based on cloud platforms, the move toward a ubiquitous gadget-free environment seems to be the natural progression. The hyperconnected smart digital surroundings will provide an *omnipotential* environment around the user to provide all the digital services needed in their everyday life. Hence, in the future, any user can live *naked*, i.e. users can access Internet-based services without any personal devices, gadgets, or wearables [31].

The limitations of present 5G network technologies to facilitate gadget-free communication also highlights it as an essential driving trend toward the next generation of networks. Gadget-free communication requires users to stay connected seamlessly with high availability, high-network performance, increased energy efficiency, and lower costs (Figure 2.4). Future communication networks need to be highly automated, context-aware, adaptable, flexible, secure, and self-configurable to provide users with a satisfactory service [31]. Facilitating such requirements demands future communication networks be equipped with powerful distributed computing with edge intelligence, which is lacking in present 5G implementation [56]. Future networks are also required to facilitate extreme data rates, negligible latencies, and extreme reliability to facilitate holographic communication that will enable users to fully utilize the potential of gadget-free communication [66]. Furthermore, existing network security measures and privacy also need to be improved. For instance efficient, secure, and privacy, ensuring authentication mechanisms using lightweight operations are required

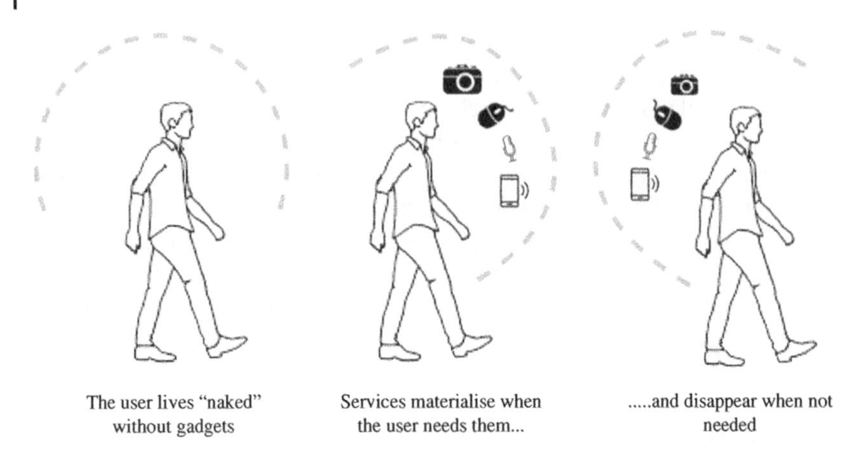

Figure 2.4 Gadget-free Communication.

to be integrated with future communication networks to facilitate gadget-free communication [67].

2.9 Increasing Elderly Population

The world's older population continues to grow exponentially due to advance healthcare facilities, life prospects, and access to new medicine and healthcare facilities. Presently, there are more 60-year-olds than children under the age of five, and this trend is expected to grow [21]. The World Health Organization (WHO) in 2015 has also predicted that the elderly populations will double from 12% to 22% by 2050 [22]. The elderly population is prone to old-age diseases. Thus, they need continuous health monitoring to ensure well-being. However, frequent hospital visits might not be feasible due to costs, transportation difficulties, and body movement restrictions. This requires technologies to aid physicians to manage their patients in real time while measuring parameters such as heart rate, body and skin temperature, blood pressure, respiration rate, and physical activity using multiple wearable devices and environmental sensors [68]. Concepts such as Human Bond Communication (HBC) are developed to detect and transmit information using all five human senses (sight, smell, sound, touch, and taste) [21]. Ambient-Assisted Living (AAL) is another developing concept that will allow remote monitoring of health as well as other hazards such as smoke or fire [69].

The importance of increasing the elderly population as an emerging trend is identified considering the limitations of existing network infrastructure to provide smart healthcare and other related facilities to the increasing elderly population.

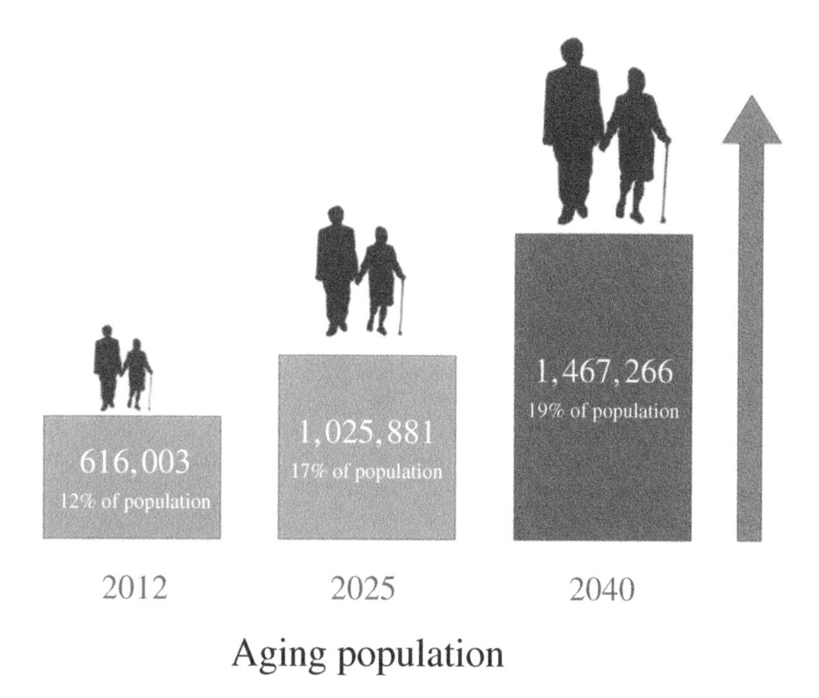

Figure 2.5 Increasing elderly population.

It is observed that the requirements of future healthcare applications can extend up to very high data rates, extremely high reliability (99.99999%), and extremely low end-to-end delays (≤1 ms) [17, 70]. Furthermore, emerging applications such as the Intelligent Internet of Medical Things (IIoMT) require powerful edge intelligence to process massive amounts of data in real time for the early detection of adverse medical conditions such as cancers (Figure 2.5). Similarly, Hospital-to-Home (H2H) services that can provide urgent treatments for patients will also require seamless connectivity with extreme reliability [70]. In addition, VLC is expected to be integrated with mobile networks to facilitate in-body sensors to provide vital information for patient monitoring [21]. Moreover, the massive amounts of health information that will be gathered should be protected by future networks with powerful and intelligent measures to ensure data security and user privacy [70, 71]. These requirements are beyond 5G capabilities and demand a new generation of mobile communication networks.

3

6G Requirements

The 6G networks are required to develop over existing 5G networks to support emerging technologies and applications. After reading this chapter, you should be able to

---

- Understand how the 6G requirements develop over existing networks capabilities.
- Obtain an insight on how 6G requirements can enable future technologies and applications.

3.1 6G Requirements/Vision

To realize new applications, 6G networks have to provide extended network capabilities beyond 5G networks. Figure 3.1 depicts such requirements which need to be satisfied by 6G networks to enable future applications.

As adopted from the various studies [37, 72–75], 6G networking requirements can be divided into different categories as follows:

- *Further Enhanced Mobile Broadband (FeMBB)*: The mobile broadband speed has to be further improved beyond the limits of 5G and provide the peak data rate at Terabits per second (Tbps) level. Moreover, the user-experienced data rate should also be improved up to Gigabits per second (Gbps) level [70].
- *Ultramassive Machine Type Communication (umMTC)*: Connection density will further increase in 6G due to the popularity of Internet of Things (IoT) devices and the novel concept of IoE. These devices communicate with each other and offer collaborative services in an autonomous manner [76, 77].
- *Enhanced Ultrareliable, Low-Latency Communication (ERLLC/eURLLC)*: The E2E latency in 6G should be further reduced up to μs level to enable new high-end, real-time 6G applications [20].

6G Frontiers: Towards Future Wireless Systems, First Edition.
Chamitha de Alwis, Quoc-Viet Pham, and Madhusanka Liyanage.
© 2023 The Institute of Electrical and Electronics Engineers, Inc. Published 2023 by John Wiley & Sons, Inc.

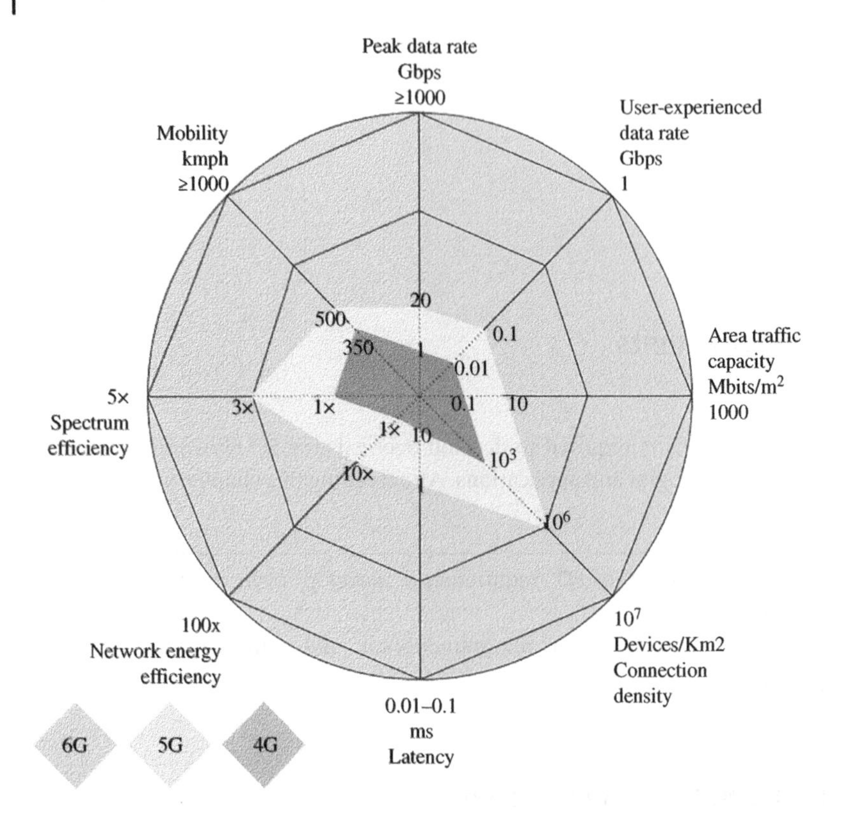

Figure 3.1 6G Requirements.

- *Extremely Low-Power Communications (ELPC)*: The network energy efficiency of 6G will be improved by 10× than 5G and 100× than 4G. It will enable ELPC channels for resource constrained IoT devices [20, 78].
- *Long Distance and High Mobility Communications (LDHMC)*: With the support of fully integrated satellite technologies, 6G will provide communication for extreme places such as space and the deep sea. Moreover, AI-based automated mobility management systems and proactive migration systems will be able to support seamless mobility at speed beyond 1000 kmph [70].
- *High Spectrum Efficiency*: The spectrum efficiency will be further improved in 6G up to five times as in 4G and nearly two times as in 5G networks [20].
- *High Area Traffic Capacity*: The exponential growth of IoT will demand the improvement of the area traffic capacity by 100 times than 5G networks. It will lead up to 1 Gbps traffic per square meter in 6G networks.
- *Mobile Broadband and Low-Latency (MBBLL)*: MBBLL will enable high data rates (> 1 Tbps) and low response time (< 0.1 ms), even in high mobility use-cases (> 1000 km/h) to support applications such as XR.

- *Massive Broad Bandwidth Machine Type (mBBMT)*: The 6G networks are expected to support large number of sensors, devices, equipment, and other machines (e.g. $100/m^3$).
- *Massive Low Latency Machine Type (mLLMT)*: In 6G, URLLC and massive machine-type communication (mMTC) services should be linked and novel unified solutions are needed to meet the challenge of offering efficient and fast massive connectivity.
- *AI-Assistive Extreme Communications*: AI is expected to immerse in every aspect of future communication networks.

Rest of the chapter discusses how the 6G each requirement can be improved as compared to the existing networks. Each requirement is then followed by several enabling applications and their key enabling technologies.

3.2 Further-Enhanced Mobile Broadband (FeMBB)

eMBB represents a continuing evolution from traditional LTE, which enables mobile broadband in limited applications. The speed of eMBB is the gigabits in 4G. In 5G, eMBB is being enhanced greatly [72]. In addition, it has been predicted that a series of exciting immersive applications including 3D extended reality features, 3D multimedia, IoE will be enabled by high quality of services that would need the peak of tens of Gbps [37]. Therefore, 6G mobile broadband speed has to be further improved beyond the limits of 5G to provide the peak of mobile broadband data rate at Tbps level. Moreover, as the end users will be using more high-definition contents, their mobile data rates should also be improved up to Gbps level.

3.2.1 Enabling 6G Applications

It will enable many use-cases, for instance FeMBB can dramatically enhance broadband in highly dense or populated regions including public transportation (e.g. high-speed trains and smart cities) to provide super-fast hot spot. Furthermore, FeMBB would enable many exciting ultrahigh definition media applications (e.g. 4D video gaming and mobile TV [79]) to facilitate enhanced multimedia applications. Other areas of FeMBB growth in the technical work include autonomous manufacturing and growth of connected wearables and sensors. As a result, 6G will enable broadband everywhere on the planet.

3.2.2 Enabling 6G Technologies

The 6G will be deployed across large number of areas through the fixed wireless access, which will leverage 6G technologies to deliver wireless broadband to everywhere on the planet [80]. Another popular technology is THz band, which is one of the main frontiers in beyond 5G communications as of today. The THz band would offer virtually unbounded capacity for supporting wide-channels and extremely high data rates [61]. In addition, the work in [81] deliberately supported that the THz communications can attain high data rates through VLC.

Recently, AI/ML has been proposed to be used at the physical and MAC layers [82]. ML can optimize synchronization, manage power allocation, and modulation and coding schemes. Furthermore, ML would assist with efficient spectrum sharing, channel estimation, and enable adaptive and real-time massive Multiple-Input and Multiple-Output (MIMO) beamforming. However, such AI/ML-based solutions are still under research. Therefore, it becomes viable that we would need more intelligent algorithms that can determine in which domains two systems can share the spectrum with high coexistence efficiency [82].

3.3 Ultramassive, Machine-Type Communication

In the IoE revolution, 5G communications are expected to support massive machine-type communication for billions of devices. The ability to connect and transfer data up to 1 million sensors per km^2. In addition, the work in [73] suggested that the scale of machine-type communications will be turned upside down by IoT devices and their connectivity in IoE world. In the IoE architecture surprisingly, a trillion of sensors and actuators will be automated to send their data back and forth. In such massive scale networks, the current machine-type communication architecture would not be able to cater to effective and efficient connectivity. However, beyond-5G and/or 6G networks potentially require umMTC architecture that can support reliable connectivity to massive scales of networks, e.g. trillion of devices [83]. Thus, connection density will be further improved in 6G due to the popularity of novel concept of IoE.

3.3.1 Enabling 6G Applications

In 6G, umMTC will enable several key applications including Internet of Industrial smart Things (IoIsT), smart buildings, Internet-enabled supply-chain, logistics, and fleet management, as well as air and water quality monitoring [73, 84]. In addition, other applications will be ultradense cellular IoT networks, container tracking, nature/wildlife sensing, mines/road and/or forest works monitoring [85].

3.3.2 Enabling 6G Technologies

Technologies, such as SigFoX and LoRa [73], will be the potential candidates for network connectivity and coverage toward the 6G network. In another vein,

Machine Type Communication (MTC) architecture and its features can also be realized in 6G using a licensed spectrum that would overlay on existing communication infrastructure (e.g. RAN). In addition, such an architecture can provide guaranteed reliability to the devices in the 6G communications. In order to achieve this, mainly two technologies can be utilized as enabling technologies: (i) enhanced MTC and, (ii) Narrow Band Internet of Things (NB-IoT), among others [83]. The enhanced Machine-Type Communication (eMTC) can provide high bandwidth data rate (e.g. up to 1 Mbps) and can support high mobility to many of 6G-enabled applications such as Internet of Vehicles (IoV) [84]. Whereas the NB-IoT can enable and/or support many applications which required low data bandwidth (e.g. in the order of Kbps) [83].

In another vein, Massive MIMO is one of the substantial candidates that can enhance efficiency of spectrum in multiuser environment [86]. As a result, it can enhance the channel capacity in 5G and beyond networks. In order to simulate massive MIMO, the authors in [86], applied MIMO to solve the issues of perfect channel allocation (PCA) issue. The authors used bit-error rate for PCA. Moreover, they utilized a narrowband (shared) communication to collect network data traffic from MTC devices (i.e. connected devices in the network). To enhance the broadband efficiency, they used clustering technique where heterogeneous devices share own cluster's resources [86].

3.4 Extremely Reliable Low Latency Communication

The 5G is backed by uRLLC and its reliability is 99.999% [74]. However, by 2030, innovations would need extremely super high ultrareliability not only in MTC but also in several other communications as well, such as device-to-device communication, Wi-Fi, device-to-cloud, and so on. In a practical scenarios, consider a medical surgeon is sitting in front of a telecommunications-based smart surface or console in Chicago city, while the patient lies on an operation bed 4000 miles away in Dublin, Ireland, as shown in Figure 3.2. Using the smart surface and other communication technologies, the medical surgeon can remotely control the movement of a multiarmed surgical robot to remove the 70-year-old patient's diseased gallbladder. However, such time-critical robotic surgery application will demand ultrahigh reliability and low latency communication. Therefore, in 6G, the researchers must explore and develop new or enhanced techniques that can enable eRLLC to provide the high reliability rate (99.99999%) than the 5G.

3.4.1 Enabling 6G Applications

eRLLC and (enhanced Ultrareliable Low-Latency Communication (eURLLC) will enable several applications such as telemedicine, XR, Internet of Healthcare (IoH)

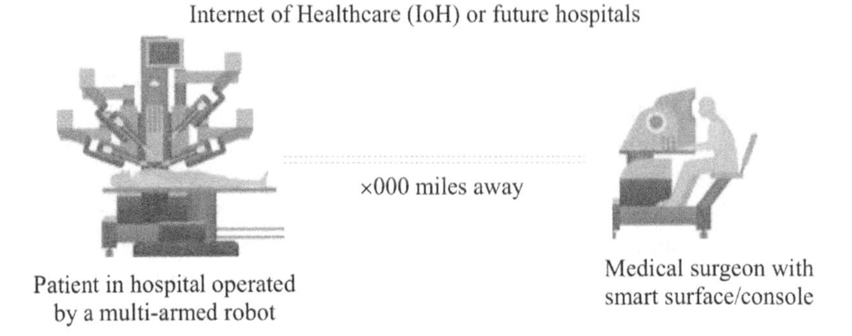

Figure 3.2 An application example (IoT Healthcare, Source: Adapted from [87]) for eURLLC.

revolution includes HT, and AI-Healthcare. All of these applications demands superior-ultrareliable communication [12].

3.4.2 Enabling 6G Technologies

Designing of eRLLC systems (i.e. high reliability and low latency), demand various parameters, such as end-to-end fast turnaround time, intelligent framing and coding, efficient resource management, intelligent up-link and down-link communication, and so on. For more details, please refer to [74]. Madyan *et al.* [79] discussed another enabling technology that would be useful in uplink grant-free structures and in reducing the transmission latency. In [79], the authors proposed to not use a middle-man cognitive operation that would typically require a committed scheduling grant technique.

3.5 Extremely Low Power Communication

Internet-enabled resource-constrained objects are increasing rapidly, and these objects required highly efficient hardware that can be self-powered. However, research revealed that the traditional devices are integrated with large-scale antenna arrays (e.g. MIMO) that will inevitably bring high power consumption [13]. In 5G network, several technologies are available to facilitate low power communication with 5G communication networks, for instance back-scatter communication, hybrid analog/digital hybrid precoding, lens-based beam domain transmission technology, sparse array, and sparse RF link designing. Nevertheless, such approaches may not be fully able to control the environmental issues such as the nature of the wireless communication that may consume more power [13]. Therefore, 6G communication must focus on maintaining high-speed transmission while reducing energy consumption.

3.5.1 Enabling 6G Applications

Several applications will require ELPCs; however, currently the promising applications are smart homes, smart cars, UAV, etc.

3.5.2 Enabling 6G Technologies

In order to achieve ELPC, the researchers propose to deploy the Intelligent Reflecting Surface (IRS), which is known as Reconfigurable Intelligent Surface (RIS) [13]. With the use of few antennas, the IRS may help to reduce the dependencies of hardware complexities in transmitters and receivers. In addition, utilizing passive artificial arrays to greatly reduce energy consumption. In another vein, the IRS technology is being quite attractive from an energy consumption point of view. In IRS, no power amplifier is being used to amplify and forward the incoming signal. As a result, since no amplifier is used, an IRS will consume much less energy than a regular amplify-and-forward relay transceiver [88].

3.6 Long Distance and High Mobility Communication

In large dimension networks, LDHMC are indispensable requirements in 6G [89]. In 5G, LDHMC services are undeniable as they can support up to 500 km/h. Consider a science-fiction example, a high-speed rail will operate over 500 km/h in the next few decade and that will require long distance and high mobility to support communications and services for on-board crew and passengers. Therefore, 5G-based LDHMC may not be enough for the future applications, as they may require long-distance communication for many thousands of kilometers and may require seamless mobility. In addition, the research reveals that the node mobility is highly challenging in different environments (such as deep waters) [90]. Therefore, 6G must require long distance and high mobility communication (e.g. >1000 km/h)-based seamless services for future applications.

3.6.1 Enabling 6G Applications

Following the [90], 6G will enable many exciting applications, few of the examples are as space sightseeing, deep-sea tourism, high-speed transportation, as shown in Figure 3.3.

3.6.2 Enabling 6G Technologies

There are several enabling technologies such as accurate channel estimation. Due to severe time and frequency spreading in high-mobility wireless communications, channel estimation is very challenging. Filter-based alternative

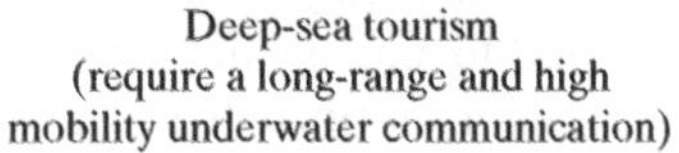

Figure 3.3 Application examples (such as space tourism [91] (Studiostoks/Adobe Stock), deep-sea tourism [92], high-speed trains [93]) for LDHMC.

waveforms [94] (alternatives to orthogonal frequency division multiplexing), such as filter-bank multicarrier and universal filtered multicarrier are good candidates for 6G high mobility communications.

3.7 High Spectrum Efficiency

As projected in [95, 96], a high magnitude of devices or smart objects is anticipated to grow many times in 6G network. Specifically in the region of thousand(s) of smart devices including machines, equipment, sensors, and many more, in a given cubic meter [97]. Nevertheless, ultrahigh definition video streams such as holographic contents need will require high bandwidth spectrum that may not be supported by the spectrum of the millimeter-wave. This will pose an unmanageable disturbance related to the area efficiency where high number of devices may not be able to connect appropriately to the current network. This will lead to deploying of the new technology in the 6G domain such as sub-THz and THz bands, that can bridge the requirements for the high spectrum efficiency network-based applications [97].

3.7.1 Enabling 6G Applications

6G network will enable data-hungry applications, such as augmented reality/mixed reality. This is going to enable new smart services in smart cities, smart agriculture, retail, supply chain, and much more to function seamlessly.

3.7.2 Key Enabling 6G Technologies

In [98], two open loop beamforming methods with the help of location information (i.e. the location-based MIMO precoding and the location-assisted MIMO precoder cycling) are being proposed. These techniques can be attributed to the early enabling technologies for high spectrum efficiency.

3.8 High Area Traffic Capacity

Area traffic capacity corresponds to the total traffic throughput served per geographic area (bit/s/m^2) [99]. In this regard, it is widely anticipated that 5G may enable a traffic capacity of 10 Mbps per square meter in dedicated hotspot areas. However, the applications such as 3/4-D multimedia would require high traffic capacity and that may be not supported by current 5G communications. Therefore, 6G must provide ten times the area traffic capacity of 5G, as suggested in [89]. More importantly, this will reach up to 1 Gb/s/m^2 for the real-world applications [89].

3.8.1 Enabling 6G Applications

Enabling applications are followings: urban networks, weather forecasts, automated vehicles, high-density rail networks.

3.8.2 Enabling 6G Technologies

To attain high traffic capacity in automated vehicles, the work in [100] proposed a novel technique. The authors assumed that each smart vehicle comprises of two distinct modules, a leader and a follower. However, the proposed technique ensures a high traffic capacity and vehicle density in a dense geographical location. In addition, the authors claimed that their proposal can avoid the traffic congestion significantly. For more details, the interested readers may refer to [100].

3.9 Mobile Broadband and Low Latency (MBBLL)

MBBLL will be the enabling necessity in 6G communications. In order to understand the concept of MBBLL, consider an example of a VR application [101]. In VR environments, requiring high latency is the utmost demand for a pleasant experience of an immersive VR headset to its users [102]. The human eye typically requires free and perfect smooth movement, i.e. low Motion-To-Photon (MTP) response time without any interruption. Here, the MTP is the time within a moment and the pixels of a picture frame that represents to the new field of view (FoV) which has shown to the human eye [102]. However, a high MTP response time may direct contradictory signal values to the vestibulo-ocular reflex (VoR),i.e. between the head movement and eye. In addition, a high MTP response time might lead to an apparent motion illness. More precisely, practically, in the simple setting, the value for MTP's upper bound is $<15-20$ ms. At the same time, the loop back response time of 4/5G is 25 ms in the given ideal functioning conditions. However, such VR-based applications require high data bandwidth rates (e.g. downlink peak data rate >1 Tbps, and user experienced data rate >10 Gbps) including ultrahigh definition pictures, videos, and other immersive instructions such as human gestures. Moreover, it demands low response time for real-time voice-based commands (<0.1 ms) and prompt control receptions (<1 ms). In addition, these requirements must also be assured in high-mobility use-cases (>1000 km/h), for example space tourism, deep-sea tourism, high speed transportation, and so on.

3.9.1 Enabling 6G Applications

Typical MBBLL applications include mobile AR, VR, and HT [101].

3.9.2 Enabling 6G Technologies

The work in [103] introduced a proposal where multiedge computing is being used to attain an end-to-end guaranteed low latency for VR video streaming using the immersive technologies. The authors designed a low-complexity mechanism that can offload the high computation tasks to MEC and can achieve energy efficiency. Such a proposal can be attributed as the enabling techniques for 6G communications.

3.10 Massive Broadband Machine-Type Communications

The 5G promises to address the performance-related issues as well as enabling entirely new use cases. The ability to connect and transfer data up to 1 million sensors per km^2, allowing continuous collection of data from vast numbers of sensors, will enable remote monitoring and predictive maintenance of manufacturing assets. Low latency together with edge cloud capabilities will underpin real-time processes such as collaborative robots for process automation, and high reliability will support mission-critical operations. For, the tactile IIoE will enable many of paramount application use-cases in the next decade or so, that will require massive data rate to experience high quality of service and everything. However, such IIoE will also expect monolithic connections for densely employed (e.g. $100/m^3$) sensors, devices, equipment, and other machines to capture environment data and translate the sensed data into the digital data.

3.10.1 Enabling 6G Applications

Motivating applications are industrial smart networks, ultradense cellular IoT networks, container tracking, nature/wildlife sensing, mines/road/forest works monitoring [85].

3.10.2 Enabling 6G Technologies

Inspired by 5G technologies, there are several key enabling technologies which can enhance the mBBMT capabilities in 6G. Massive MIMO is one of the substantial candidates that can enhance efficiency of spectrum in multiuser environment [86]. As a result, it can enhance the channel capacity in 5G and beyond networks. In order to simulate massive MIMO, the authors in [86], applied MIMO to solve the issues of PCA issue. The authors used bit-error rate for PCA. Moreover, they utilized a narrowband (shared) communication to collect network data traffic from MTC devices (i.e. connected devices in the network). To enhance the broadband efficiency, they used clustering technique where heterogeneous devices share own cluster's resources [86].

3.11 Massive Low Latency Machine-type Communications (mLLMT)

The purpose of MTC in the 6G automation is associated with the many services, such as data availability, ultrascalability, and more importantly, low latency in the 6G-enabled applications. Such low latency services are highly paramount for time-critical applications where decision-making will happen on a scale of fraction of milliseconds. However, such requirements (e.g. low latency, high availability) may not be met by the existing wireless network including 4/5G network due to several challenges, such as confined resources of communication technologies, lack of automation of operations, human-centric devices, and so on. Hence, 6G has to increase its mission-critical mLLMT communication to support future applications.

3.11.1 Enabling 6G Applications

Multiple application domains include such as home and building automation, integration of distributed energy resources with the energy plants, unmanned vehicle systems, IoT-enabled healthcare infrastructures, and controlling and monitoring industrial 4.0 use-cases. To boost such innovative and immersive IoE applications and many others (e.g. space tourism), there is a pressing demand for new wireless technologies that can enable and support a massive number of connections among IoE devices and ground to space and vice versa.

3.11.2 Enabling 6G Technologies

There are several existing technologies that can directly be adopted to enable the mLLMT services in 6G ecosystem. For instance, Park *et al.* [104], proposed a mechanism that may enable low latency, machine-type communication where the resources of IoE devices can be shared within a fraction of response time (in ms). The authors designed a novel finite memory multistate sequential learning framework that will suitably fulfill the requirements in several scenarios, such as delay-tolerant applications, periodic messages delivery, and urgent and critical messages exchanges. Park *et al.* claimed that their suggested learning framework will enhance the latency of IoE devices or MTC to learn the several number of critical messages and to redistribute the network and communication resources for the delivery of periodic messages that are to be used for the critical messages in many of 6G applications.

3.12 AI-Assistive Extreme Communications

In the next two to three decades, AI will immerse in every aspect of communication and will be heavily used for communication purposes. Here, we propose a term *AI-assistive extreme communications (AEC)*. However, as of today, 4/5G network would not be able to deal with such massive scale of devices, applications, heterogeneous standard, and nonstandard practices, different stakeholders, etc. Therefore, 6G must require such AEC.

3.12.1 Enabling 6G Applications

The AEC network may control and monitor trillions of devices in various verticals (e.g. IoT manufacturing and supply chain) across the globe. These devices keep track of several intelligent parameters – bandwidth allocation, decision-making in data routing and aggregation, knowledge sharing, are few examples. Other application opportunities are AI-empowered, data-driven network planning, operation, intelligent mobility, handover management, and smart spectrum management to achieve many of real-world traditional environments to dynamic communication environments in 6G.

3.12.2 Enabling 6G Technologies

There are various machine learning techniques that can enable dynamic communication, networking, and security and trust elements in vehicular-to-infrastructure networks and envision the ways of supporting AI-centric futuristics 6G smart or driverless vehicular networks. In [105], the authors surveyed several ML techniques including supervised multilayer perceptron to equalize channels by intelligently and creatively redefining the communication symbols. A support vector machine-based nonlinear equalization mechanism can be used in the wireless network and communication of 6G, where the temporal correlation exists in the collected intersymbol interference data. For the more details, interested readers may refer to [105].

4

Key 6G Technologies

The previous chapters have focused on the evolution, driving trends, and key requirements of future 6G wireless systems. Several key technologies have been proposed to realize 6G, and these technologies are discussed in this chapter. After reading this chapter, you should be able to:

- Gain an overview of key technologies in future 6G wireless systems.
- Explore each key technology with a preliminary, the role in 6G, and a review of representative studies.

4.1 Radio Network Technologies

In this section, we present important 6G radio network technologies, including THz communications and nonterrestrial networks toward 3D networking.

4.1.1 Beyond Sub-6 GHz toward THz Communication

The rapid increase in wireless data traffic has been estimated to be seven-fold in mobile data traffic from 2016 to 2021 [106]. Wide radio bands such as millimeter waves (up to 300 GHz) are expected to fulfill the demand for data in 5G networks. However, applications such as holographic telepresence, brain–computer interfaces, and XR are expected to require data rates in the range of Tbps, which would be difficult with mmWave systems [107]. This requires exploring the terahertz (THz) frequency band (0.1–10 THz). This type of communication will especially be useful for ultrahigh data rate communication with zero error rates within short distances.

6G is expected to deliver over a 1000x increase in the data rates compared to 5G to meet the target requirement of 1 Tbps. More spectrum resources beyond

6G Frontiers: Towards Future Wireless Systems, First Edition.
Chamitha de Alwis, Quoc-Viet Pham, and Madhusanka Liyanage.
© 2023 The Institute of Electrical and Electronics Engineers, Inc. Published 2023 by John Wiley & Sons, Inc.

sub-6 GHz are explored by researchers to cater to this significant increase in data rates. Early 6G systems are expected to bank on sub-6 GHz mmWave wireless networks. However, 6G is expected to progress by exploiting frequencies beyond mmWave, at the THz band [61]. The size of 6G cells is expected to shrink further from small cells in 5G toward tiny cells that will have a radius of only a few tens of meters. Thus, 6G networks will require to have a new architectural design and mobility management techniques that can meet denser network deployments than 5G [12]. 6G transceivers will also be required to support integrated frequency bands ranging from microwave to THz spectra. The applications of THz for 6G networks are illustrated in Figure 4.1, where THz communication is used for high speed transmissions between radio towers and mobile devices, integrated access and backhaul networks, and high speed satellite communication links.

THz waves are located between mmWave and optical frequency bands. This allows the usage of electronics-based and photonics-based technologies in future networks. As for electronic devices, nano-fabrication technologies can facilitate the progress of semiconductor devices that operate in the THz frequency band. The electronics in these devices are made from indium gallium arsenide

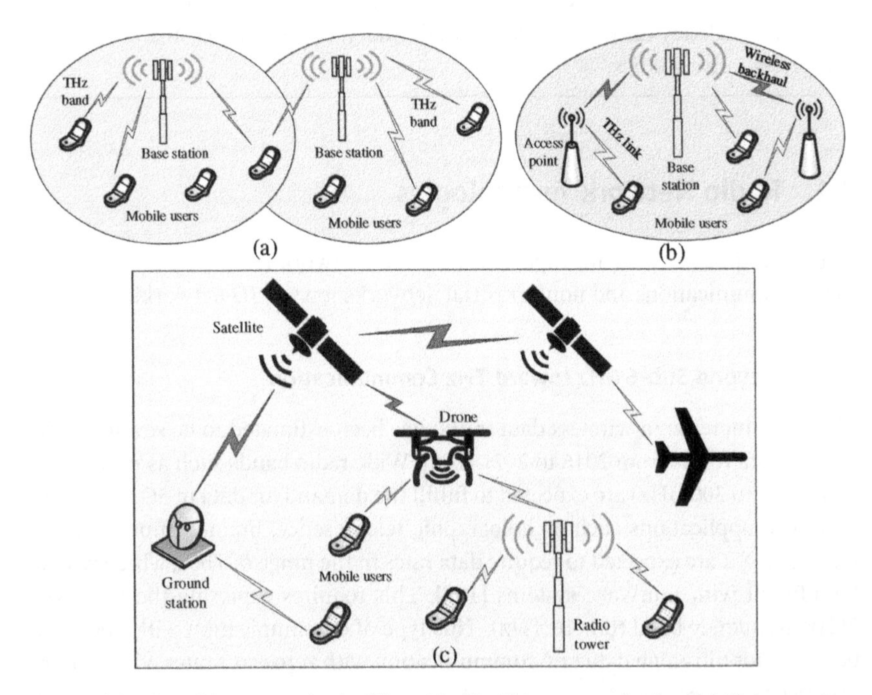

Figure 4.1 Promising scenarios in 6G enabled by THz communication: (a) high speed transmission, (b) integrated access and backhaul networks, and (c) high speed satellite communication links.

phosphide and various silicon-based technologies [108]. A scalable silicon architecture allows synthesis and shaping of THz wave signals in a single microchip [109]. The feeding mechanism of optical fibers to THz circuits is prominent to achieve higher data rates in photonic devices. Conventional materials used at lower frequencies in the microwave and mmWave ranges are not efficient enough for high frequency wireless communication. Devices made from such materials exhibit large losses at the THz frequency range. THz waves require electromagnetically reconfigurable materials. In this context, graphene has been identified as a suitable candidate to reform THz electromagnetic waves by using thin graphene layers [110, 111]. Graphene-based THz wireless communication components have exhibited promising results in terms of generating, modulating, and detecting THz waves [112]. THz wireless communication allows small antenna sizes to achieve both diversity gain and antenna directivity gain using MIMO. For example, 1024x1024 ultramassive MIMO is introduced in [113] as an approach to increase the communication distance in THz wireless communication systems.

THz band channel is highly frequency-selective [114]. These channels suffer from high atmospheric absorption, atmospheric attenuation, and free-space path loss. This requires the development of new channel models to mimic the behavior of THz communications [106]. The first statistical model for THz channels is proposed in [115], which depends on performing extensive ray-tracing simulations to obtain statistical parameters of the channel. Some recent studies [116, 117] provide more accurate channel models. Various research works have also focused on applications of THz communication. A hybrid radio frequency and free-space optical system is presented in [118], where a THz/optical link is envisaged as a suitable method for future wireless communication. In addition, THz links can be used in data centers to improve performance while achieving massive savings in minimizing the cable usage [119].

Summary: THz communications are expected to pave the way for Tbps data rate to meet the demands of future applications and have the potential to strengthen backhaul networks. Nevertheless, they suffer from high propagation losses and demand line of sight (LoS) for communications. More efforts are required to understand the behavior of THz signals, and better channel models are required.

4.1.2 Nonterrestrial Networks Toward 3D Networking

In conventional ground-centric mobile networks, the functioning of base stations is optimized to primarily cater to the needs of ground uses. Moreover, the elevation angle provided to the antennas at ground base stations focuses on the ground user for better directivity and hence cannot support aerial users [120].

Such a mobile network allows marginal vertical movement (i.e. above and below the ground surface), thus predominantly offering two-dimensional (2D) connectivity. Nonterrestrial networks expand the 2D connectivity by adding altitude as the third dimension [121]. Nonterrestrial networks are capable of providing coverage, trunking, backhauling, and supporting high speed mobility in unserved or underserved areas through the integration of UAVs, satellites (in particular very low Earth orbit (VLEO)), tethered balloons, and high-altitude platform (HAP) stations [12, 121]. The development of protocols and architectural solutions for new radio (NR) operations in nonterrestrial networks is promoted in 3GPP Rel-17 and is expected to continue in Rel-18 and Rel-19 [121]. The 3D networking further extends the nonterrestrial network paradigm, allowing 6G to emerge as a global communication system by extending its coverage from ground to air toward space, underground, and underwater [122]. Interestingly, aerial base stations powered by UAV technology can offer on-demand, broadband, and reliable wireless coverage in a cost-effective and agile way. Some of the promising characteristics of UAV-enabled aerial base stations [120, 122] are as follows:

- Intelligent 3D mobility and ease of maneuvering.
- Varying capabilities in terms of computation, storage, power backup, etc., to meet heterogeneous demands.
- LoS communication links that allow effective beamforming in 3D.
- High flexibility in terms of the number of antenna elements when UAVs are used to create antenna arrays for 3D MIMO.

There has been an exponential increase in the number of connected devices, and the trend will continue with a higher rate of increase in the future. In particular, the future is expected to see a significant increase in aerial users or aerial-connected devices. Technological advancements in various fields, such as electronics and sensor technology, high speed links, data communication networking, and aviation technology, provide a necessary ecosystem for the robust growth of UAVs (also known as drones), which have, in turn, extended the horizon of UAVs' applications. By 2022, the fleet of small model UAVs (primarily used for recreational purposes by hobbyists) is expected to reach a mark of 1.38 million units, whereas small nonmodel UAVs (primarily used for commercial purposes) are forecast to be 789000 million units as per the Federal Aviation Administration (FAA)'s report [123]. Moreover, by the same year, i.e. 2022, the global market of UAVs is estimated to value at US$ 68.6 billion [124]. Hence, 6G mobile networks are expected to provide the required connectivity to such an increasing number of aerial users. To fulfill this expectation, the 3D networking paradigm is going to play a key enabling role in 6G.

A framework for UAV-based 3D cellular network for beyond 5G provides solutions for placement of UAV-enabled aerial base stations in 3D (using the truncated octahedron approach) as well as latency-sensitive association of UAV-based users to UAV-enabled aerial base stations [120]. In [125], a 3D nonstationary geometry-based stochastic model (GBSM) is investigated for UAV-to-ground channels that are envisioned in UAV-integrated 6G mobile networks. The main purpose of GBSM is to work well for mmWave and massive MIMO configurations. Moreover, intelligence can be extended to edge 3D networks (i.e. beyond the premises of 2D networks) by leveraging edge computing (e.g. MEC and fog computing) [126]. In particular, the work envisioned to integrate flying base stations with terrestrial stations by using emerging technologies, such as mobile edge computing (MEC), software-defined networking, and AI.

Summary: Nonterrestrial networks are evolved toward 3D networking to enable global radio coverage and capacity in 3D for future 6G networks. Nonterrestrial networks represent a gamut of technologies such as UAVs, HAPs, satellites, and other flying gadgets that are anticipated to work in harmony to offer seamless coverage over space, air, ground, underwater, and underground. AI/ML-based solutions are expected to play an important role in overcoming the limitations posed by the physical absence of human beings.

4.2 AI/ML/FL

Thanks to distinctive features and remarkable abilities, AI has various applications in wireless and mobile networking. Massive data generated by massive IoT devices can be exploited by AI approaches to extract valuable information, thus improving the network operation and performance. Recently, FL has emerged as a new AI concept that leverages on-device processing power and improves user data privacy [127, 128]. The rationale is to collaboratively train a shared model such that participating devices train the local models and only share the updates (instead of data) with the centralized parameter server [15, 129]. Mobile and IoT devices, such as mobile phones, IoT devices, and autonomous, are getting increasingly powerful in terms of storage and computational capabilities, and this has paved the growing interest in FL. According to [130], FL can be classified into horizontal FL, vertical FL, and federated transfer learning:

- *Horizontal FL*: the overall feature spaces of various datasets are the same, but FL users have different sample distributions, as shown in Figure 4.2(a).
- *Vertical FL*: the feature spaces of various datasets are different, but FL users share the same sample distribution, as illustrated in Figure 4.2(b).
- *Federated transfer learning*: FL users have datasets with different feature spaces and sample space distributions, as shown in Figure 4.2(c).

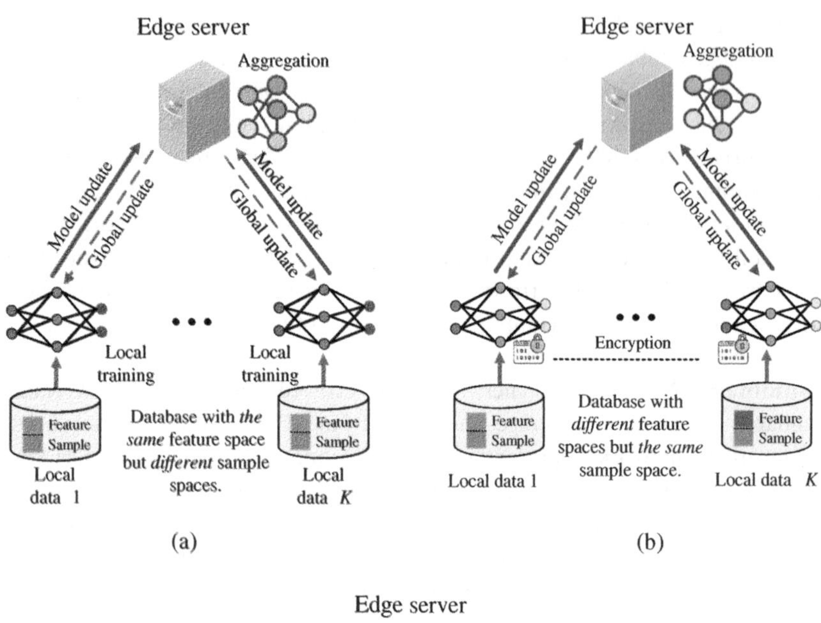

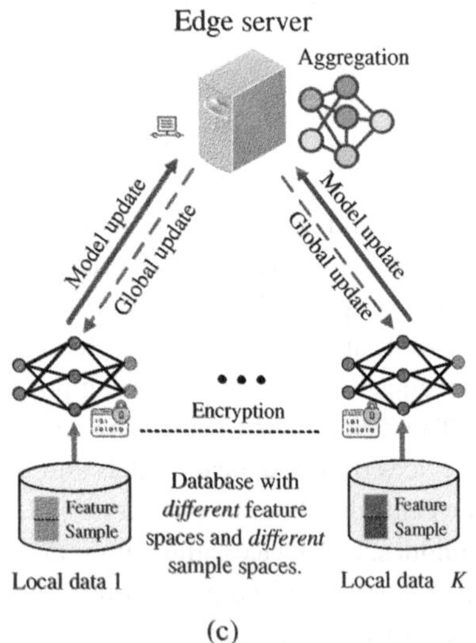

Figure 4.2 Three categories of FL: (a) horizontal FL, (b) vertical FL, and (c) federated transfer learning.

With 6G being envisioned to have AI/ML at its core, the role of AI/FL becomes important to 6G. The use of conventional centralized ML approaches is suitable for network scenarios, where centralized data collection and processing are available. As the amount of mobile data and advancements in computing hardware and learning advancements are increasing, numerous problems in future 6G networks can be effectively resolved by AI approaches such as modulation classification, waveform detection, signal processing, and physical layer design [79, 131, 132]. However, due to the centralized nature of such AI/ML-based systems, they suffer from single-point-of-failure as well as security vulnerabilities. Thus, FL is gaining popularity and is emerging as a viable distributed AI solution that enables "ubiquitous AI" vision of 6G communications [133]. The FL offers a multitude of benefits to 6G, as summarized by authors in [134], such as communication-efficient distributed AI, support for heterogeneous data originating from different devices pertaining to different services that can lead to nonidentically distributed (non-IID) dataset, privacy-protection since data remains locally and is not uploaded anywhere, and enabling large-scale deployment.

The last few years have witnessed the use of different AI techniques for numerous problems in wireless networks. For example, the work in [135] reviews applications of deep reinforcement learning (DRL) for three important topics, including network access and rate control, data caching and computation offloading, security and connectivity preservation, and for a number of miscellaneous issues such as resource allocation, traffic routing, signal detection, and load balancing. The applications of ML for wireless networks are reviewed in [136], where AI-enabled resource management, networking, mobility, and localization solutions are discussed. The use of transfer learning, deep learning, and swarm intelligence for future wireless networks can be found in [137, 138], and [139], respectively. Some of the related work pertain to the use of FL in wireless networks. The work in [140] proposed a Stackelberg game-based approach to incentivize the interaction between the global server and devices participating in training model. The work in [133] proposed an FL framework for IoT networks with the aim to simultaneously maximize the utilization efficiency of edge resources and minimize the cost for IoT networks. An incentive FL mechanism using contract theory is proposed in [141] to allow highly reputed devices to engage in the training task. Moreover, a reputation scheme based on a multiweight model is proposed for selection devices, and blockchain is used for managing this reputation system. Despite the hype, there are numerous challenges that need to be addressed to gain the maximum benefits of FL in the 6G realm. Some challenges of FL are highlighted in [15, 127], including the cost of communications, significant hardware heterogeneity, high device churn, privacy leakage through model update, and security issues.

Summary: AI and FL techniques enable the design of intelligent mechanisms for future 6G networks via exploiting massive mobile data and increasingly computing resources available at the network edge. Participating devices in FL locally train ML models (in use) by leveraging their on-board resources. By doing so, these devices will not have to share the raw data, which may be private and prone to security attacks. Therefore, what is shared is the updated model at the centralized servers, thereby making the learning to be federated. Nevertheless, in order to effectively realize AI, the very first challenge is the existence of high-quality training datasets. Another challenge is the inclusion of AI in beyond 5G and future 6G networks [16]. Moreover, the bottlenecks of AI caused by 6G wireless networks with many new advanced technologies, device heterogeneity, and emerging intelligent applications should be further studied. Regarding FL, the devices need to be incentivized to participate in the training process, and also, the rouge devices need to be detected as early as possible. Moreover, the challenges such as privacy leakage via model updates and hardware heterogeneity also need to be met [142].

4.3 DLT/Blockchain

The last couple of years have witnessed the rise of DLT, in particular, blockchain technology. DLT is envisioned to unlock the doors to the decentralized future by overcoming the well-known impediments of centralized systems [143]. Blockchain is a type of DLT, which maintains a digital ledger in a secure and distributed way. This ledger holds all the transactions in a chronological order and is cryptographically sealed [144]. The illustration of a blockchain is depicted in Figure 4.3. If a hacker intends to edit the transaction in a blockchain, he or she has to modify hash of not only that block but also every other hashes, which is nearly impossible. This security property of blockchain makes it an ideal choice for usage in many sectors such as banking, insurance, government services, and supply chain management. Moreover, blockchain offers numerous advantages such as disintermediation, immutability, nonrepudiation, proof of provenance, integrity, and pseudonymity; therefore, blockchain has received all-around attention equally by industry and academia [128].

Many sectors have already acknowledged the pragmatic use of blockchain technology and its efficacy as they are running/offering blockchain-based technological solutions [145]. Examples of some of these business sectors are finance and banking, industrial supply chain and manufacturing, shipping and transportation, medical health care and patient records, and educational processes and credentialing. Blockchain can play a cardinal role in improving the following aspects of future 6G wireless systems.

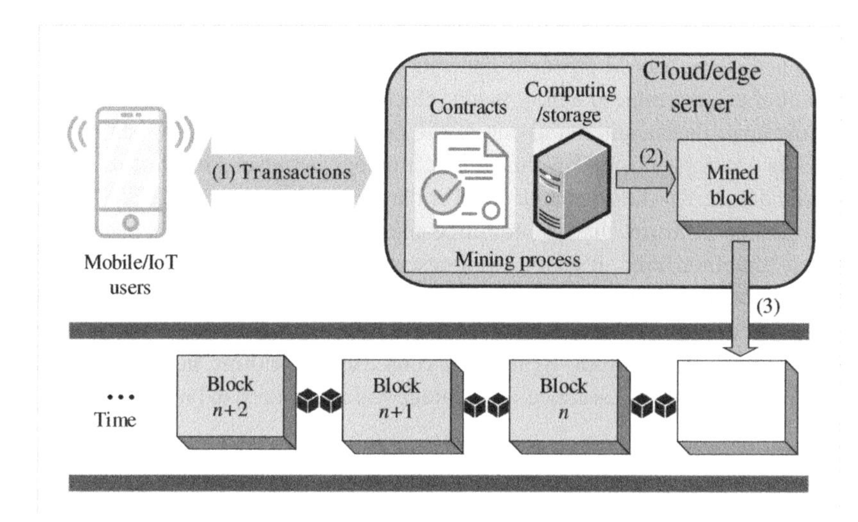

Figure 4.3 Illustration of the general transaction process of blockchain: (1) A mobile user or an IoT device sends a transaction to smart contracts to request data. (2) A new block is created to represent the verified transaction. (3) A mined block is appended to the blockchain.

- Management and orchestration in terms of interference mitigation, resource allocation, spectrum, and mobility management [146, 147].
- Operations in terms of cell-free communications and 3D networking.
- Business models in terms of decentralized and trustless digital markets involving various stakeholders, such as infrastructure providers (InPs), network tenants, industry verticals, over-the-top (OTT) providers, and edge providers [148, 149].

Further, blockchain has an immense potential to strengthen the existing service arena of mobile networks as well as to set the floor for futuristic applications and use cases of 6G.

Blockchain has been identified as one of the key enabling technologies for 6G. Numerous efforts are being made to leverage its potential to improve both the technical aspects of 6G and use cases of the 6G ecosystem. For instance, the work in [146] presented the use of blockchain for decentralized network management of 6G. In particular, the work showed the blockchain plus smart contract for spectrum trading. In [150], blockchain-based radio access network is proposed (B-RAN) to allow small subnetworks to collaborate in a trustless environment to create larger cooperative networks. The work in [147] presented how blockchain can be leveraged to remove the intermediate layer and develop a distributed mobility-as-a-service that is hosted as an application on edge computing facility. Improved transparency and trust among all the stakeholders are the manifested advantages of this work. The work in [151] advocated the use of blockchain for ensuring data security AI-powered applications for 6G. The two main applications are indoor positioning and autonomous vehicles.

Summary: Blockchain being one of the prominent types of DLT has turned out to be a very promising key enabling technology because of its built-in strong security nature. On the one hand, it can enhance the technical aspects of 6G such as dynamic spectrum sharing, resource management, mobility management, and on the other hand, it enables unforeseen applications such as holographic telepresence, XR, fully connected autonomous vehicle, Industry 5.0, and many more. Nevertheless, to harness the best use of blockchain for 6G challenges such as computational overheads, lightweight consensus algorithms, high transaction throughput, quantum resistance, and storage scalability need to be mitigated.

4.4 Edge Computing

One of the very first edge computing concepts, the so-called cloudlet, was proposed in 2009 [152]. Conceptually, a cloudlet is defined as a trusted and resource-rich computer or a cluster of computers located in a strategic location at the edge and well connected to the Internet. The main purpose of cloudlet is to extend cloud computing to the network edge and support resource-poor mobile users in running resource-intensive and interactive applications. Although the storage and computing capabilities of a cloudlet are relatively smaller than that of a data center in cloud computing, cloudlets have the advantages of low deployment cost and high scalability [153]. Mobile users exploit virtual machine (VM)–based virtualization on customizing service software on proximate cloudlets and then using those cloudlet services over a wireless local area network to offload intensive computations [152, 154]. There are two different VM approaches for computation offloading [152], including VM migration and VM synthesis. Since VM provisioning is used in cloudlets, cloudlets can operate in standalone mode without the intervention of the cloud [154, 155]. Mobile users can access cloudlet services through Wi-Fi. The idea of cloudlet to distribute the cloud computing in close proximity to mobile users is similar to the Wi-Fi concept in providing Internet access [156]. The Wi-Fi connection between users and cloudlets can be a serious drawback. In this way, mobile users are unable to access cloudlets in the long distance and use both Wi-Fi and cellular connection simultaneously [153], i.e. users have to switch between the mobile network and Wi-Fi if they use cloudlet services.

To address the inherent drawbacks of cloud computing, fog computing has emerged as a promising solution. Fog computing, a term put forward by Cisco in 2012, refers to the extension of the cloud computing from the core to the network edge, thus bringing computing resources closer to end users [157].

The main purpose of fog computing is to bring computational resources to end users as well as reduce the amount of data needed to transfer to the cloud for processing and storage. Therefore, instead of being transferred to the cloud, most intensive-computations from mobile users and data collected by end users (e.g. sensors and IoT devices) can be processed and analyzed by fog nodes at the network edge, thus reducing the execution latency and network congestion [158]. As a complement to cloud computing, fog computing stands out in the following features [155, 159]:

- Edge location, low latency, and location awareness.
- Wide-spread geographic distribution.
- Support for mobility and real-time applications.
- A very large number of nodes as a consequence of geographic distribution.
- Heterogeneity of fog nodes and predominance of wireless access.

Due to its characteristics, fog computing plays an important role in many use cases and applications [159]. In terms of the node type, a fog node can be built from heterogeneous elements (e.g. routers, switches, IoT gateways), and a cloudlet can be referred to as a cloud data center in a box [160]. In terms of node location, a fog node can be deployed at a strategic location ranging from end devices to centralized clouds and a cloudlet can be deployed indoors as well as outdoors. The close similarity between cloudlet and fog computing is that cloudlets and fog nodes are not integrated into the mobile network architecture; thus, fog nodes and cloudlets are commonly implemented and owned by private enterprises, and it is not easy to provide mobile users with the Quality of Service (QoS) and Quality of Experience (QoE) guarantees [155, 156].

The MEC concept was initiated in late 2014. As a complement of the C-RAN architecture, MEC aims to unite the telecommunication and information technology (IT) cloud services to provide the cloud-computing capabilities within radio access networks in the close vicinity of mobile users [161]. The general architecture of MEC can be illustrated in Figure 4.4. The main purposes of MEC are [162]: optimization of mobile resources by hosting compute-intensive applications, optimization of the large data before sending to the cloud, enabling cloud services within the close proximity of mobile subscribers, and providing context-aware services with the help of radio access network information. From the illustration in Figure 4.4, MEC enables a wide variety of applications, where the real-time response is strictly required, e.g. driverless vehicles, VR, AR, robotics, and immersive media. To reap the additional benefits of MEC with heterogeneous access technologies, e.g. 4G/5G/6G, Wi-Fi, and fixed connection, the name of mobile edge computing is changed to mean multiaccess edge computing in 2017 [163]. After this scope expansion, MEC servers can be deployed by the network operators at various locations within RAN and/or collocated with

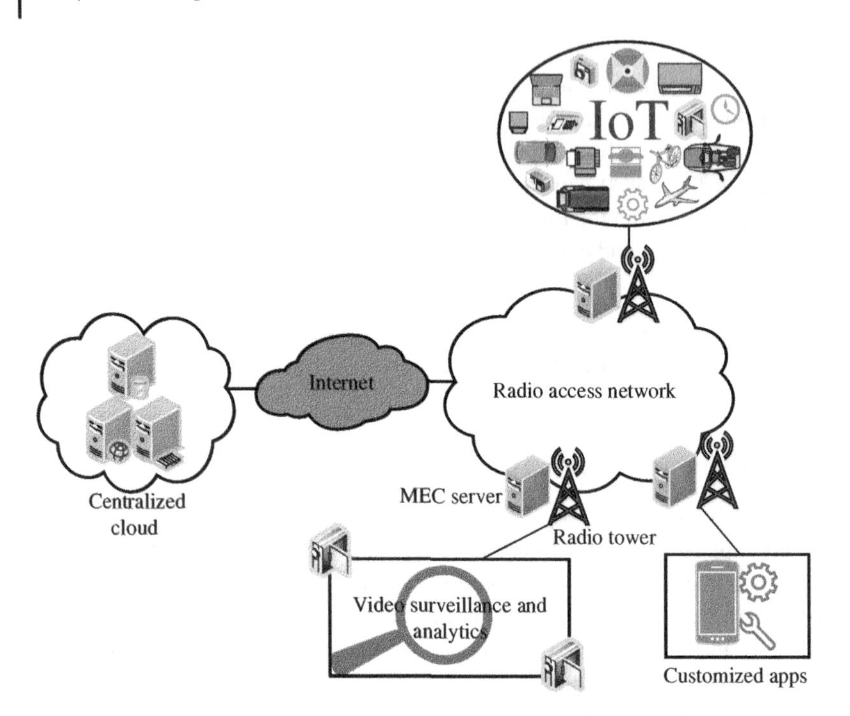

Figure 4.4 A general architecture of MEC.

different elements that establish the network edge. While the BSs in traditional cellular networks are mainly used for communication purposes, MEC enables them to collocate with MEC servers for facilitating mobile users with additional services. This transformation pushes intelligence toward the traditional BSs so that they can be used for not only communication purposes but also computation, caching, and control services [71].

Edge computing has been a key technology in 5G networks and will still be a key technology in beyond 5G and future 6G networks. The implementation of edge computing enables many new services and applications, for example autonomous vehicles, smart IoT, smart cities, smart healthcare, and more. As a result, there exists a strong need for the close integration of edge computing with enabling technologies, such as artificial intelligence, big data, intelligent surfaces, blockchain, IoT, and VR/AR/metaverse. Significantly, there will be more applications and services requiring the deployment of AI at the edge of the network and the deployment of intelligent edge, creating the concept of edge AI, as presented in Chapter 7. Moreover, there are several new concepts thanks to the combination of edge computing with emerging technologies. For example, in-network computing can further enhance conventional edge computing concepts. In particular, in-network computing enables in-network devices, such as routers and switches, to perform computing functions instead of forwarding the data transmission [164]. The amalgamation of aerial access networks and edge computing introduces a novel concept, referred to as aerial computing [165]. Aerial computing is expected to provide advanced services, e.g. communication, computing, caching, sensing, navigation, and control, at a global scale. Facilitated by advantages of high mobility, fast deployment, global availability, scalability, and flexibility, aerial computing complements conventional computing paradigms (e.g. cloud computing, fog computing, MEC) and is thus considered a pillar of the comprehensive computing infrastructure in future 6G networks.

Summary: Different edge computing paradigms are proposed to provide applications and services at the edge networks. The availability of edge computing resources and the explosion of IoT data enable the deployment of many new services with low latency and intelligent capabilities.

4.5 Quantum Communication

Quantum computing–enabled communications have derived a lot of research and development interests recently. Slowly but beyond mere science fiction, QC will be a transformative reality in the 6G paradigm in the next decade or so. As 6G must meet stringent requirements, such as massive data rates, fast computing, and strong security, QC will be a potential enabler. The driving force of QC is that it exploits traditional concepts of physics, i.e. photons are being used to process the computation to the quantum qubits. These qubits are then sent from a sender (or emitter) machine to a receiver machine. Using flying qubits in communications brings enormous advantages, such as weak interaction with an environment, faster computations and communications, quantum teleportation, communication security, and low transmission losses in communication.

As per [166], the role of quantum systems for telecommunication and networking fall under two different categories: quantum communication and quantum computing. Quantum communication is a way to transfer a quantum state from a sender to a receiver [167]. It can enable the execution of tasks that either cannot be performed or are inefficiently performed using classical techniques [166, 167]. Some of the interesting offerings of quantum communication are quantum key distribution (QKD), quantum secure direct communication (QSDC), quantum secret sharing (QSS), quantum teleportation, quantum network (quantum channel, quantum repeaters, quantum memory, and quantum server) [168]. One of the promising uses of quantum communication is the secure distribution of cryptographic keys referred to as QKD. The techniques that use quantum entanglement for this purpose are called entanglement-based QKD [169]. Any kind of man-in-middle attack while using QKD can be detected easily as the attacker disturbs the quantum (shared joint) state, and this disturbance can be known by examining the correlations between the communicating entities [169].

Quantum communication is foreseen to play a crucial role in realizing secure 6G communications. In particular, the underlying principles of quantum entanglement and its nonlocality, superposition, inalienable law, and noncloning theorem pave the way for strong security. The next generation of services that are going to be increasingly supported by quantum communication is HT, tactile Internet, BCI, and extremely massive and intelligent communications [166, 170]. These QKD protocols have shown most progress and numerous practical implementation of such protocols have been shown, which reflects their potential applicability in future 6G wireless networks [59, 166]. Yet another interesting use of quantum communication is its applicability for secure long-distance communication [59]. This, in particular, would be interesting since it is envisioned that 6G will have a special focus on LDHMC that would deal with extremely long-distance communications.

The work in [51] surveyed various paradigms such as quantum communication, quantum computing–assisted communication, and quantum-assisted machine learning–based communication that utilizes machine learning and quantum computing in a synergetic manner. The advent of quantum computing facilities poses a significant treat to traditional cryptographic techniques that are used to encrypt data in current wireless communication systems. Thus, the work in [171] exploited the QKD mechanism for key generation and management in 5G IoT scenarios, namely, quantum key GRID for authentication and key agreement (QKG-AKA), and analytically showed that security cannot be broken in polynomial time. The work in [172] proposed two hash functions (for 5G applications) that utilize quantum walks (QW), namely, QWHF-1 and QWHF-2 (Quantum Walk Hash Function 1 and 2). Further, QWHF-1 is in turn used to develop authentication key distribution (AKD) protocol and QWHF-2 is used to develop authenticated quantum direction communication (AQDC) protocol for device-to-device (D2D) communications. The work in [173] studied the applicability of QKD for securing fronthaul that offers low-latency connectivity to a myriad of 5G terminals. In particular, a fronthaul link integrates BB84 and QKD and supports AES encryption. Their work paves the way toward quantum secure fronthaul infrastructure for 5G/B5G networks.

Summary: Quantum communication is a very promising application of the principles of quantum physics in the world of communication. Some of the techniques provided by quantum communication are quantum teleportation, quantum network, QKD, QSDC, and QSS. Various upcoming applications of quantum communication are quantum optical twin, holographic telepresence, tactile Internet, brain–computer interface, and long-distance intelligent communication.

In spite of the hype, establishing a synergy between quantum communication and classical communication will be challenging. So far, most advancements are limited to QKD.

4.6 Other New Technologies

4.6.1 Visible Light Communications

Visible light communications (VLC), which uses visible light for short-range communication, is one of the promising optical wireless communication (OWC) technologies [174]. The frequency spectrum for VLC is between 430 THz to 790 THz. Further, in VLC, the most common devices used for transmission are light-emitting diode (LED) and light amplification by stimulated emission of radiation (LASER) diodes, whereas for reception photodetectors such as silicon photodiode, PIN photo-diode (PD) and PIN avalanche photo-diode (APD) are used [174–176]. Some of the advantages of using VLC technology based on [174, 176, 177] are as follows:

- The visible light spectrum is free to use since it falls under the unlicensed band.
- Very high bandwidth compared to RF signals (visible light spectrum is 10^4 times higher than radio waves [176]).
- High spatial reusability since visible light is blocked by objects like walls.
- Precise estimation of direction-of-arrival.
- Very high data rate, e.g. 10 Gbps using LED and 100 Gpbs using LASER diodes [178].
- Low energy consumption with the use of LEDs.
- Inherently secure due to unidirectional propagation, signal isolation, and non-penetrating nature of visible light.
- Less costly compared to radio communication especially in mmWave and THz range.
- Safe to be used for communication since it meets the eye and skin regulations [176].

Moreover, the existing radio frequency band and the visible light band are well separated; thus, there is no electromagnetic interference [179].

VLC came into existence with the emergence of white LEDs and has matured over the last two decades for it to be considered as an enabling technology for 6G [180]. The technology has been successfully used for various application scenarios, such as vehicular communications, underwater communications, indoor scenarios, visible light identification systems, wireless local area networks, and underground mines. Since VLC operates in the THz range, it offers ultrahigh bandwidth, which means it can well satisfy the capacity and data rate demands of 6G

[181]. From the 6G point of view, a hybrid communication infrastructure can be developed leveraging the best of visible light communications and other conventional communications such as RF, Wi-Fi, infrared (IR), and power line communication (PLC) [180, 182, 183]. For instance, to establish RF-VLC hybrid system, use of RIS has been suggested. Here, RIS can control the propagation environment and ensure the LoS communication between base stations and mobile devices equipped with a photodetector [184].

4.6.2 Large Intelligent Surfaces

The emerging paradigm for controlling the propagation environment via smart and intelligent surfaces has been referred to with numerous names such as LIS, IRS, RIS, software-defined surface (SDS), and many more [185]. LIS plays a vital role when direct LoS communication is not feasible or degrades in quality such that it hampers sensible communications. LIS provides a way to transform human-made structures (e.g. building, roads, indoor walls/ceilings) into an intelligent and electromagnetically active wireless environment [186]. This transformation is performed by augmenting these structures either with a large array of small, low-cost, and passive antennas or with meta-materials (aka meta-surfaces), such that they help in controlling the characteristics such as reflection, scattering, and refraction of propagation environment [185]. Although changes can be made to different characteristics of the incident electromagnetic signals by LIS, most of the work focuses on changes in the phase of the incident signal.

LIS has been identified as a key technological enabler for 6G since it is going to operate at higher frequencies and is expected to go beyond the massive MIMO [12, 187]. Various advantages of LIS over conventional massive MIMO, as highlighted in [186], include reduced noise, lower interuser interference, and reliable communication. Interestingly, LIS would allow the use of holographic RF as well as holographic MIMO [12]. Moreover, in the 6G realm, LIS can be used to better sense wireless environments by capturing CSI [188].

4.6.3 Compressive Sensing

Sampling is an integral part of modern digital signal processing and stands at the interface between analog (physical) and digital worlds. Traditionally, for efficient transmission, flexible processing, noise immunity, security inclusion (using encryption and decryption), low cost, etc., the Nyquist sampling theorem has been used. According to this, for a band-limited signal, if the samples are taken at a rate greater than or equal to twice the highest frequency of that signal, then the exact replica of the signal can be reconstructed using these samples.

Sampling is usually followed by a compression process where the sampled data is compressed to maintain some acceptable level of quality [189]. As pointed out in [190], with the increase in the transmission bandwidth with 5G and future 6G mobile networks, the continued use of the Nyquist sampling technique will result in numerous challenges such as significant overheads, large complexity, and higher power consumption. In this context, compressive sensing, also known as compressive sampling or sparse sampling, has been proposed as an intriguing solution that has the potential to overcome the limits imposed by traditional sampling. Compressive sensing is basically a sub-Nyquist sampling framework. Compressive sensing states that provided a signal is characterized by sparsity and incoherence, it can be sampled at a rate lower than the Nyquist rate and the resulting (smaller set of) samples are sufficient to reconstruct the original signal [190]. This is achieved in a computationally efficient manner and by finding a solution to underdetermined linear systems. Moreover, in compressive sensing, both sampling and compression are carried out at the same time. In a nutshell, the sampling rate in compressive sensing depends on sparsity and incoherence characteristic of the signal being sampled and does not depend on the bandwidth of the signal [14]. This property of compressive sensing opens the door for its applicability to 6G networks. In general, compressive sensing is proposed to be used for reducing the data generated by IoT devices for mMTC [191]. Another proposed use of compressive sensing is to enable NOMA at the transmitter in the landscape of mMTC scenario [192]. This is carried out by assigning nonorthogonal spreading codes to devices and applying compressive sensing-based multiuser detection techniques since very less percentage of total devices are active at any given time. The advantage is that this scheme incurs no control signaling overhead. Yet another use of compressive sensing along with deep learning techniques is to overcome the issues pertaining to the expected intensive usage of LISs in the next-generation networks [63].

4.6.4 Zero-touch Network and Service Management

Zero-touch network and service management (ZSM) is an evolving concept that aims to provide a framework for building a fully automated network management, primarily driven by the initiative of ETSI. The idea of ZSM is to empower networks so that they can perform self-configuration to carry out autonomous configuration without the need for explicit human intervention, self-optimization to better adapt as per the prevailing situation, self-healing to ensure correct functioning, self-monitoring to track the functioning, and self-scaling to dynamically engage or disengage resources as per need [193]. To stress the importance of the ZSM framework, ETSI in [194] identifies a list of scenarios that are grouped into seven different broad categories as follows:

- The end-to-end network and service management category talks about the automation of operational and functional tasks involved in the end-to-end lifecycle management of different types of network resources and services that are part of core network, transport network and radio access network.
- Network-as-a-Service (NaaS) presents the requirement of exposing some of the service capabilities from all the parts of the network to enable zero-touch automation.
- Analytics and ML scenarios emphasize the need for integration of ML and AI capabilities for realizing ZSM.
- Collaborative service management category emphasizes the need for collaborative management spanning domains of multiple operators.
- Security highlights the need for strong security and privacy mechanisms for ZSM framework.
- Testing scenario points out the need for automated testing of resources as well as services.
- Tracing scenario needs are driven by requirements for automated troubleshooting and root cause analysis.

In this direction, a framework, named self-evolving networks (SENs) [195], is proposed that aims to automate the network management with self-efficient resource utilization, coordination and conflict management, inherent security and trust, reduce cost, and high quality of experience.

Future 6G wireless networks will be heterogeneous with multitenancy, multioperator, and multi-(micro) services features. To make such networks work at their best and at a low cost, they are envisioned to be fully automated. Thus, ZSM becomes highly important in 6G. The use of AI/ML capabilities within the framework of ZSM has indeed the potential to add many new capabilities (as mentioned above) and set the floor for AI-enabled autonomous networks. However, security remains a great concern. This is because ML techniques are vulnerable to attacks such as poisoning attacks or evasion attacks [196]. Here, the use of blockchain as a common communication channel can do the required. Further, enabling automated service updates without affecting service interoperability as well as end-user experience is another challenge when using ZSM [166].

4.6.5 Efficient Energy Transfer and Harvesting

Energy harvesting has been the much-sought area of research when it comes to a future sustainable way of energizing the growing number of connected devices. The aim of energy harvesting is to replace the conventional ways of powering devices and sensors by tapping the energy from ambient environments.

The two broad classes of sources for energy harvesting are natural sources and human-made sources [197]. Natural sources include renewable energy sources such as solar, mechanical vibrations, wind, thermal, microbial fuel cells, and human activity powered [197]. Human-made energy harvesting happens through wireless energy transfer (WET), where a dedicated power beacon is used to transfer energy from source to destination [197]. Since the natural sources of energy harvesting suffer unpredictability and periodicity, they fail to offer guaranteed quality of service; thus, WET is the hot research area [198].

With the vision of a universal communication system and providing solid underpinning to IoE, the future 6G networks will be proliferated with a large number of connected devices. The conventional way of powering these devices with rechargeable or replaceable batteries might not efficiently scale in the 6G era. The reason being that such solutions are, in general, costly, inconvenient, risky, and have adverse effects when the devices are operating inside the body [198]. Thus, energy harvesting technologies are considered to be an efficient alternative solution for next-generation mobile networks [197]. In this context, much of the excitement revolves around the idea that radio signals can simultaneously transfer energy as well as information [199]. This is referred to as radio frequency energy harvesting (RF-EH) [200]. Nevertheless, to have the pragmatic use of such techniques and to gain the maximum benefit, the challenge lies in efficient integration of both wireless information transfer and wireless energy transfer provided that their hardware and operational requirements are different [197]. The other open issues are high mobility, multiuser energy and information scheduling, resource allocation and interference management, health issues, and security issues [198, 200].

Part II

Architectural Directions

5

6G Architectural Visions

6G mobile network architecture is envisaged to undergo a remarkable evolution from 5G to facilitate the flexible infrastructure demanded by future technologies and applications. After reading this chapter, you should be able to

- Understand the evolution of mobile network architecture toward 6G.
- Describe how 6G evolves to be an intelligent and energy efficient network.

5.1 Evolution of Network Architecture

Mobile networks in the era of 0G–4G provided one-fits-all network solutions for mobile applications. Thus, 0G–4G networks consisted of a fixed and inflexible network architecture. However, as mobile applications evolved demanding faster, more reliable, and real-time connectivity, the one-fits-all network solutions deemed to be insufficient. In response, with the advent of 5G, mobile network architecture evolved to be dynamic and agile through network softwarization [57]. Thus, 5G networks were able to provide customized networking and computing solutions to diverse mobile applications through enhanced mobile boradband (eMBB), ultra reliable low latency communication (uRLLC), and massive machine-type communication (mMTC).

The evolution of conventional mobile networks to softwarized networks is enabled through network virtualization and cloudification, service migration, orchestration of networks and services, and automation of network services [201]. Network virtualization is enabled through Software-Defined Networking (SDN), Network Function Virtualization (NFV), and cloud computing [202]. SDN separates the control plane and the data plane of networks. Therefore, the

6G Frontiers: Towards Future Wireless Systems, First Edition.
Chamitha de Alwis, Quoc-Viet Pham, and Madhusanka Liyanage.
© 2023 The Institute of Electrical and Electronics Engineers, Inc. Published 2023 by John Wiley & Sons, Inc.

network infrastructure can be abstracted, programmed, and controlled through softwarized network functions [203]. Hence, softwarized network functions are decoupled from proprietary networking hardware by NFV to run the network functions as software instances in virtual machines [204]. These virtual machines catering for network elements and network functions can be hosted in the cloud through network cloudification [205]. In addition, network functions are orchestrated to synchronize E2E network operations to ensure the smooth functionality of mobile networks. Furthermore, network operations and services are automated through technologies such as multiaccess edge computing (MEC) and network slicing [201]. MEC is able to extend cloud computing toward the Radio Access Network (RAN) edge to provide real-time access to radio network resources [206]. Network slicing configures network functions to allocate end-to-end logical networks identified as network slices to facilitate diverse network requirements of 5G applications.

However, emerging applications such as smart homes, smart cities, connected autonomous vehicles, advanced healthcare, Industry 5.0, deep sea, and space communication applications demand mobile networks to provide better connectivity and intelligence beyond the capabilities envisaged through 5G networks [25]. This requires beyond-5G/6G mobile network architecture to evolve beyond network softwarization, specifically toward supporting Artificial Intelligence (AI), Machine Learning (ML), and Deep Learning-(DL) based applications. Hence, the 6G architecture should support AI, ML, and DL naively to enable network intelligentization. This will allow 6G mobile networks to be more capable, agile, and efficient while learning dynamic network dynamics and user requirement [16]. Considering these requirements, several research works have proposed their vision and suggestions to formulate the architecture for 6G networks.

5.2 Intelligent Network of Subnetworks

The network architecture proposed by Letaief *et al.* [16] proposes the 6G networks to be more intelligent while allowing networks to be efficient and flexible by learning network and user characteristics and dynamics. Accordingly, 6G will also evolve to be a network of subnetworks, which will be more flexible, efficient, scalable, and upgradable, banking on AI capabilities. Furthermore, the separation of algorithm and hardware allows heterogeneous devices in the networks to support dynamic upgrades.

These subnetworks can upgrade themselves as single cells or neighboring cells. This will facilitate to support newer technological developments and protocols in the subnetworks with minimal testing. Therefore, ultimately, the whole network will be upgraded with minimal effort in contrast to setting up the whole network at once, which is not only time-consuming but also costly.

However, evolving existing networks toward an intelligent network of subnetworks requires overcoming several challenges. Three of those challenges are briefly described as follows:

- Subnetworks are required to collect data on the dynamic wireless environment and user requirements. These data can be analyzed through AI and ML to facilitate the efficient network operation and seamless upgradability.
- Manage any protocol changes within the subnetworks through game and learning techniques to ensure the coordination between nodes.
- Each upgrade in subnetworks should be evaluated by the control plane. This can be performed through AI/ML considering the network condition and user requirements.

Furthermore, rapid developments in 6G hardware such as radio networks and other infrastructure will require an architecture with algorithm-hardware separation. For instance, communication hardware and transceiver algorithms are jointly designed in 0G–5G networks. Therefore, hardware capabilities, such as number of antennas, sampling rate of analog to digital converters, and decoder computational capabilities remain fixed. However, in 6G, transceivers can be automatically configured by estimating their hardware capabilities, enabling the hardware capabilities in 6G base stations and mobile devices to be upgradable. This can be facilitated by having an operating system that runs in between network hardware and transceiver algorithms. The operating system will measure network parameters, estimate hardware capabilities, and configure the transceiver algorithm. Hence, this type of a framework can be identified as "intelligent radio," which develops on an architecture where algorithms and hardware are separated. This architecture also supports protocols above the network layer to support AI/ML-based services and applications to analyze network bottlenecks caused by hardware components. This type of analysis can support hardware manufacturers to produce more efficient and cost-effective network devices having less time to market. Figure 5.1 illustrates an overview of the 6G architectural changes proposed in this framework.

Accordingly, the evolution of intelligent subnetworks and intelligent radio in beyond-5G and 6G will facilitate the development of novel protocols and technologies considering the dynamic network conditions and network/user requirements.

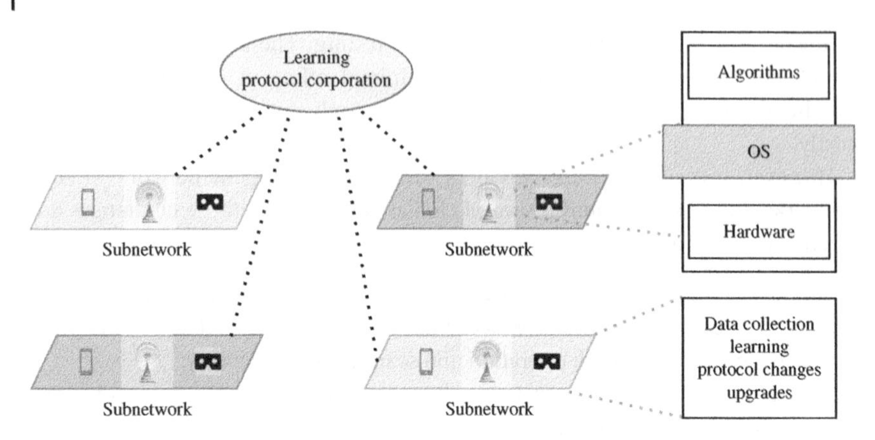

Figure 5.1 6G architectural changes.

5.3 A Greener Intelligent Network

6G networks are expected to be greener, i.e. more energy-efficient, while satisfying the extreme connectivity requirements expected by future communication networks. However, this requires network architecture to evolve in several directions [207].

6G networks are expected to shift from terrestrial networks toward providing 3D network coverage. Furthermore, legacy mobile networks are not able to facilitate emerging communication applications ranging from deep-sea communication to space communication. 6G is envisaged to integrate several nonterrestrial networks to a ubiquitous 3D network coverage. Once such prospective nonterrestrial communication mode is space networks. This consists of high throughput satellites, which are able to provide connectivity similar to terrestrial networks. The significant delays caused by geostationary orbit (GEO) satellites can be minimized by the use of nongeostationary orbit (NGSO) satellites that can be integrated with terrestrial networks facilitating space-backbone networks, intersatellite links, and space-based access networks to provide global coverage with low latency and high data rates. Some of the NGSO systems that are expected to be commercialized include Starlink, OneWeb, and Hongyan. In addition, aerial networks consisting of High-Altitude Platforms (HAP) and Low-Altitude Platforms (LPF) will provide flexible relay services for 6G networks. The HAPs will provide a wider coverage. On the other hand, LAPs consisting of Unmanned Aerial Vehicles (UAV) will provide a more flexible communication infrastructure for high bitrate and low-latency, short-range communication. UAVs will also enable communication in emergencies. However, utilizing UAVs requires energy-efficient measures, such as trajectory optimization, and computational offloading. Furthermore, under-sea

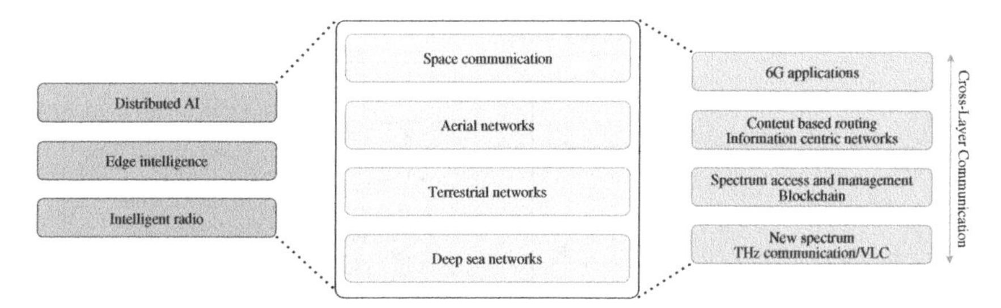

Figure 5.2 Toward greener intelligent networks.

networks in future 6G infrastructure will utilize RF, optical, and acoustic channels.

AI- and ML-based intelligent networks toward self-configuring, self-aware, and self-maintaining energy efficient networks also require AI and ML capabilities within the network structure. One of the key enablers of intelligent networks is edge intelligence, which utilizes computational resources in high-performance hardware in the environment. Unlike centralized cloud resources, edge intelligence provides real-time access to powerful computational resources for critical applications such as autonomous driving. Hence, network intelligence will be provided through decentralized resources, resulting in distributed AI for computation, communication, caching, and control of 6G networks. Also, a new network protocol stack is expected to be developed supporting features such as flexible packet architectures and cross-layer communication capabilities.

Moreover, technologies such as THz communication supporting Tbps bitrates with advanced beamforming technologies, energy-efficient modulation schemes, and energy-efficient channel coding algorithms should be supported by 6G network architecture. In addition, the architecture should support technological developments, such as Visible Light Communication (VLC), molecular communication, blockchain technologies, and intelligent energy harvesting and management.

The architecture proposed under this work is presented in Figure 5.2 [207].

5.4 Cybertwin-based Network Architecture

6G networks are envisaged to utilize a cybertwin-based network architecture toward realizing a better spectrum utilization, energy efficiency, and user experience. The cybertwin-based network architecture proposes a cybertwin-based model with cloud-centric Internet and a new radio-access network architecture [208].

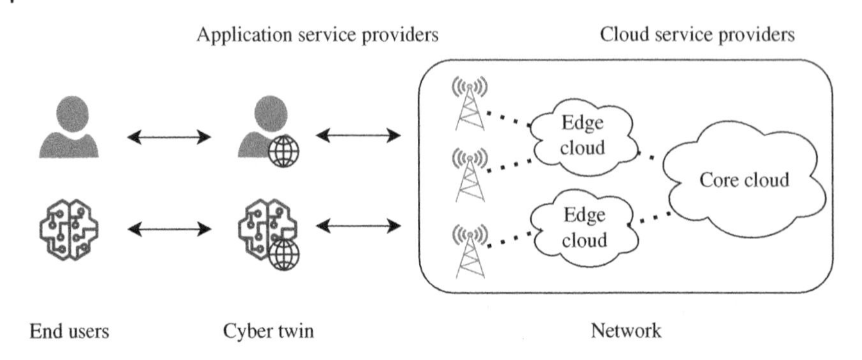

Figure 5.3 Toward a cloud-centric network architecture.

A cybertwin is a digital representation of a person or an object, that can be accessed through a network. A cybertwin provides the functionalities of communication assistant, network behavior logger, and digital assets to enhance the functionalities of network architecture. Communication assistant functionality obtains necessary services from the network, while the network behavior logger functionality obtains and logs data for users. Furthermore, the digital asset functionality removes personal information from user data to produce digital assets.

A cybertwin-based communication model can be developed on top of a cloud-centric architecture. The cloud architecture relies on two components in the network architecture, namely, the core cloud and the edge cloud, as illustrated in Figure 5.3. The core cloud will provide the computing, caching, and communication services. On the other hand, the edge cloud, which is in between the access network and the core cloud, caters to low latency and high reliability requests. A cloud operator can manage the network through a cloud network operating system. Application service providers are able to deploy their services on top of the services provided by cloud service providers.

In addition, cybertwin-enabled network functions will be utilized to provide communication and security services on 6G networks – a Fully Decoupled Radio Access Network (FD-RAN) architecture. This architecture will physically decouple the control plane and the user plane of the network to efficiently manage network resources. Furthermore, the uplink and downlink will be decoupled physically to enhance spectrum and energy efficiency. Furthermore, through centralized resource management and coordinated multipoint, the spectrum usage, energy usage, and cloud resource utilization in 6G networks can be optimized.

6

Zero-Touch Network and Service Management

Future 6G networks are expected to have fully automated (zero-touch) network services and management. After reading this chapter, you should be able to

---

- Understand the needs of automated network and service management for 6G
- Understand the zero-touch network and service management (ZSM) architecture and its components

---

6.1 Introduction

Owing to the emergence of new applications such as autonomous vehicles and virtual reality, as well as the proliferation of massive Internet-of-Things (IoT) services, numerous technologies have been developed for fifth-generation (5G) and beyond-5G networks (B5G). Among the potential technologies, Network Function Virtualization (NFV), software-defined networking (SDN), network slicing (NS), and multiaccess edge computing (MEC) have been considered as crucial enablers of 5G and B5G networks [71, 76, 209, 210]. In NFV, the network functions are decoupled from the underlying hardware which enables fast deployment of new services and also enables quick adaptability to scalable yet agile needs of the customers [211]. Furthermore, SDN allows network configuration and monitoring in a dynamic and programmed manner, helping to manage the entire network holistically and globally regardless of the underlying technologies [212]. MEC is another key technology that moves computing resources and IT functionalities from the central cloud to the network edge close to IoT and mobile users [71]. NS, on the other hand, is a critical technology that facilitates service customization and resource isolation, enabling multiple logical networks on the same physical infrastructure [213]. There exists three primary services that are supported in 5G networks which include enhanced mobile broadband

6G Frontiers: Towards Future Wireless Systems, First Edition.
Chamitha de Alwis, Quoc-Viet Pham, and Madhusanka Liyanage.
© 2023 The Institute of Electrical and Electronics Engineers, Inc. Published 2023 by John Wiley & Sons, Inc.

(eMBB), massive machine-type communication (mMTC), and ultrareliable and low latency Communication (URLLC). It is expected that these services would be supported to a great extent to help unforeseeable applications such as extended reality (XR), holographic telepresence, and collaborative robots which would be available in 2030. These applications demand joint optimization of computation communication, caching, control, and sensing [76].

Network infrastructures and supporting technologies have witnessed immense growth both horizontally and also vertically [76]. The conventional 5G cellular networks are composed of terrestrial infrastructures such as IoT and mobile devices, small cells, and macrocells. In order to support such massive connectivity ensuring global coverage, future sixth-generation (6G) wireless systems would comprise of underground, underwater, and aerial communications. In particular, an aerial radio access network consists of three main tiers, including low-altitude platforms (LAPs), high-altitude platforms (HAPs), and low earth orbit (LEO) satellites. LAP systems usually connect to users directly and support very high quality of services (QoS). The LEO satellite tier supports sparse-connectivity scenarios and global coverage with reasonable QoS, while HAP systems maintain a balance of LAP systems and LEO satellite communications. Along with a massive number of IoT and mobile devices, managing the network in a fully automated manner is a great challenge [76]. Although many solutions and concepts have been developed over the last decade, such as NFV, SDN, MEC, and NS, still manual processes are required for the operation and management of present network systems, i.e. human intervention is a must to ensure fully autonomous network and service management solutions [214–216]. These difficulties have motivated extra efforts from academic and industry communities.

6.2 Need of Zero-Touch Network and Service Management

The following limitations of existing network management and orchestration (MANO) solutions have motivated the adoption of the zero-touch network and service management (ZSM) concept [194, 217–225].

- *Network Complexity*: Massive IoT connectivity, many emerging services, and new 5G/6G technologies result in extremely heterogeneous and complex mobile networks, and thus significantly increase the overall complexity of the network orchestration and management.
- *New Business-Oriented Services*: Many new services will be available in future networks, which should be quickly implemented to meet business opportunities. Along with key-enabling technologies, such as NS, NFV, and MEC, the ZSM concept allows an agile and more straightforward deployment of new services.

- *Performance Improvement*: Diverse QoS requirements and the need to reduce the operational cost and improve network performance trigger robust solutions of network operation and service management.
- *Revolution for Future Networks*: Even 5G networks are not fully available worldwide, and numerous activities have been dedicated to the research and development of future 6G wireless systems. Many new technologies, services, applications, and IoT connections will be available, which will make the future network very complex and complicated to be efficiently managed by conventional MANO approaches [76].

The above limitations explain a strong need for the ZSM concept for the complete automation and management of future networks. In order to eliminate such limitations, enabling fully automated network operation and management solutions, the European Telecommunications Standards Institute (ETSI) ZSM group was established in December 2017.

6.3 Overview of Zero Touch Network and Service Management

One of the main design objectives of the ZSM reference architecture is its ability to achieve zero-touch-enabled network and service management, irrespective of the vendors. The ZSM reference architecture provides flexible management services, which align with the industry trend of alienating from the management systems [214]. A detailed discussion on the ZSM reference architecture, its key components, and interfaces is provided in this section.

6.3.1 ZSM Architecture Principles

ZSM is designed based on the principle of supporting self-contained, loosely coupled facilities. It allows the accommodation of new services and the modules to be scaled and deployed independently. ZSM architecture facilitates portability, reusability, vendor-neutral resource, and service management. The use of closed-loop management automation is to achieve and maintain a set of goals without any intervention. It allows management functions to be separated from the data storage and processing [226].

Management services are planned in such a way that they can resume their regular services after the issues have been resolved. Services and resources are managed based on these resources in a management domain (MD). Exterior to the MD, the domain resource's complexity can be abstracted from the service users. The E2E cross-domain service management coordinates the MD activities and manages

E2E services that span across MDs. The MDs expose the management resources. These management resources can be merged to form new management services. The intent-based interfaces are exposed to high-level abstractions, and the behavior of related entities is interpreted [227].

The architecture is simple and satisfies all the functional and nonfunctional specifications. The components and functionalities of the ZSM architecture assist in network and service management's automation.

6.3.2 ZSM Architecture Requirements

The ZSM reference architecture specifications are derived from ETSI GS ZSM 001 scenarios and requirements [214]. It also identifies functional and nonfunctional requirements that have to be satisfied by the architecture [228–231].

6.3.2.1 Nonfunctional Requirements

General Nonfunctional Requirements The ZSM reference architecture is expected to support the ability to achieve a defined degree of availability, wherein the management actions would be able to comply with relevant regulatory requirements accordingly. Furthermore, it shall be energy-efficient and remain independent from the vendor, the operator, and the service provider.

Nonfunctional Requirements for Cross-Domain Data Services The ZSM platform reference architecture is expected to accommodate QoS specifications for data services in and outside the ZSM framework reference architecture. It should achieve high data availability and the capabilities to process the data and possess the ability to complete management tasks in a predetermined amount of time.

Nonfunctional Requirements for Cross-Domain Service Integration The ZSM reference architecture provides new and legacy management functions. The changes in the management functions should not be required to integrate management resources into the ZSM framework. ZSM framework reference architecture is expected to allow management resources to be added or removed on demand and also enable multiple management service versions to coexist at the same time.

6.3.2.2 Functional Requirements

General Functional Requirements The ZSM framework reference architecture should provide frameworks for managing the services and the resources, including resource-facing service and customer-facing service that is provided by the MDs. In addition, adaptive closed-loop management and cross-domain management of E2E resources should be supported. It needs to enable the operator to constrain the automated decision-making processes with rules and policies.

Any ambiguity of MDs and E2E services is to be hidden. All the technology domains required to implement an E2E service should be supported.

Automation of operational life cycle management functions should be promoted by the ZSM MDs that apply to the services and the resources. The ZSM framework reference architecture should provide access control and open interfaces.

Functional Requirements for Data Collection The ZSM architecture needs to provide functionality that allows collecting live data, providing features for storing the collected data. It is expected that the up-to-date data collected across the management is allowed to be accessed that supports the capability of implementing data governance while providing shared access to the interdomain aggregation and (pre-)processing/filtering of the data collected. The reference architecture of the ZSM should allow various kinds and levels of data with cadence, velocity, and volume and control the distribution of collected data according to the needs and keep it consistent while allowing metadata to be attached to the collected data [231].

Functional Requirements for Cross-Domain Data Services The reference architecture of the ZSM needs to support data services across the domains to provide features that allow data storage to be separated from the data processing, where data have to be shared within the reference architecture. It is expected to offer features that will enable automatic data recovery, redundancy management in stored data, consistency, data service failure, and overload handling. It should also provide capabilities for logically centralized data storage processing based on the policies of multiple data resources with various data types [57].

Functional Requirements for Cross-Domain Service Integration and Access The ZSM framework reference architecture should provide functionality that allows management resources to be registered, discovered, and offer details regarding access to discovered services. Furthermore, it should enable asynchronous and synchronous communication between consumers and service producers and provide features that make it easier to invoke management resources indirectly, where discovered management resources' direct invocation by the service user is not prohibited ZSM [232].

Functional Requirements for Lawful Intercept The undetectability of lawful intercept should be assisted by the ZSM architecture endorsing the ability to prevent lawful interception from being interrupted [233].

6.3.2.3 Security Requirements

The ZSM framework reference architecture must include features that allow data protection in use, in transit, and at rest. An optimum level of security is expected in the management functions, managed services, and infrastructure resources. It should ensure the security of management data and the integrity of data, managed services, and management infrastructure resources and functions. Furthermore, it needs to ensure the availability of infrastructure resources, data, management functions, and managed services and provide the features that allow personal data privacy, such as privacy-by-design and privacy-by-default. Authenticated service users should approve service access using the ZSM framework reference architecture and endorse the ability to automatically implement acceptable security policies based on the individual management services' compliance status concerning security requirements. Automated attack/incident detection, recognition, prevention, and mitigation should also be supported. To avoid the spread of vulnerabilities and attacks, the ZSM platform reference architecture should enable capabilities to supervise/audit the decisions of the ML/AI on privacy and security issues [234].

6.4 ZSM Reference Architecture

The architecture of the ZSM was created to fully automate the network and service management in the environments with multidomains, where the operations span across the legal boundaries of the organizations [235, 236]. Cross-domain data services, multiple MDs, intra- and cross-domain integration fabrics, and an E2E service MD are all part of the system architecture, as shown in Figure 6.1.

Every MD is responsible for smart automated resource and service management within their scope. The E2E service MD is treated as an in-charge of E2E service management across various administrative domains. The differentiation of MDs and E2E service MDs encourages device modularity and helps them to grow independently. Each MD is made up of several management functions organized into logical groups through which service interfaces are exposed by management services [237].

The intradomain integration fabric is used to provide and consume resources that are local to the MD. The cross-domain integration fabric consumes resources that are spread across domains. Intelligence services within E2E service MDs and MDs may use data in cross-domain data services to support cross-domain and domain-level AI-based, closed-loop automation [238].

6.4.1 Components

The components of the ZSM reference architecture are discussed in brief below.

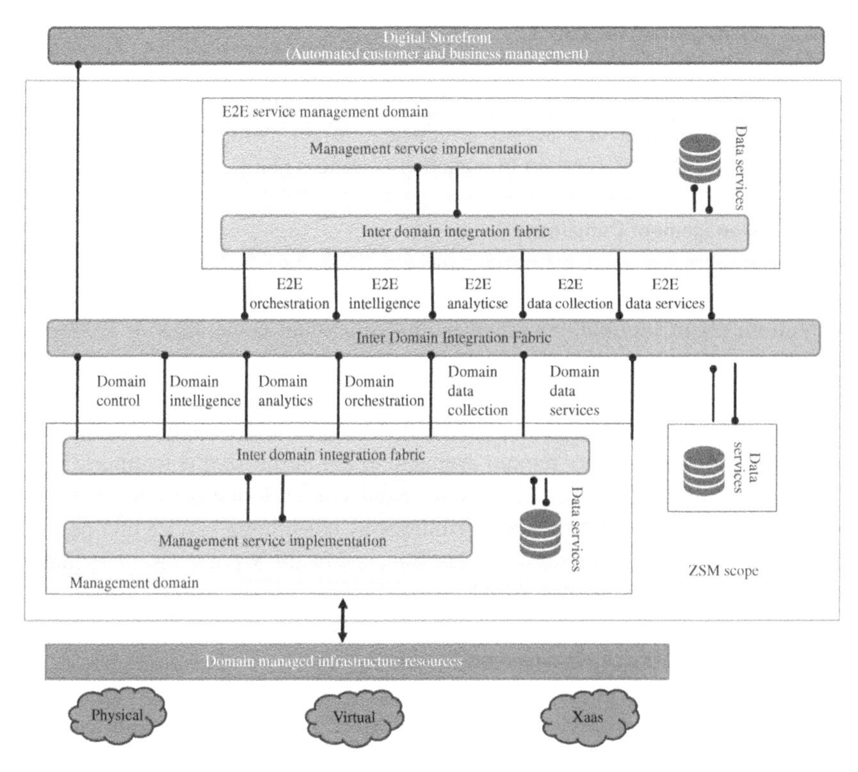

Figure 6.1 The ZSM framework reference architecture. Source: Adapted from [217].

6.4.1.1 Management Services

The fundamental building block in the ZSM architecture is its "management service." Management services provide capabilities for consumption by consumers with the help of standardized management service end-points. A management service's capabilities collectively describe its role in the organization it manages. Multiple service customers may use the same service capabilities. Several service end-points may be allocated to one or more service capabilities. For invocation, all management services have a consistent collection of capabilities. In the case of interactions between MDs, it allows a high degree of automation and consistency. Management services that are already available can be merged to create new management services. Management resources with higher abstraction and broader reach are supported by each higher layer in the composition hierarchy. The infrastructure resources communicate with the management service producers to provide their capabilities, either directly through their management interfaces or indirectly through the consumption of other management services through their service end-points [239].

6.4.1.2 Management Functions

Management functions produce and use management services, which can be both producer and consumer of the service. If the management function produces certain capabilities, it is a service producer. On the contrary, it is a service consumer if it consumes certain management services [240].

6.4.1.3 Management Domains

The administrative responsibilities are classified by the MDs to establish "separation of concerns" in a given ZSM framework, depending on several implementation, organizational, governance, and functional constraints. It federates management services with the capabilities required to control the resources/resource-facing services in a given domain. For example, some management services are constrained by approvals when the authorized consumers consume the MD. In contrast, others remain available to the authorized consumers, both within and outside the MD, at all times. Management domains manage one or more entities and provide service capabilities by consuming service end-points. Sometimes, the consuming service being managed by the MD can also potentially consume management services [241].

6.4.1.4 The E2E Service Management Domain

It is a unique MD that offers E2E management of customer-facing services, combining resource-facing services from several MDs. However, it does not control infrastructure resources directly [242].

6.4.1.5 Integration Fabric

It facilitates communication and interoperation between management functions that include the communication between management functions, discovery, registration, and invocation of management services. It also offers management service integration, interoperation, and communication capabilities, and consumes capabilities [243].

6.4.1.6 Data Services

Registered consumers can access and persist shared management data across management services using the data services. Data processing and data persistence are enabled by removing management functions to handle their data persistence [244].

6.4.2 ZSM Interfaces

6.4.2.1 Domain Data Collection

Domain data collection systems track managed entities and consumed managed services, providing fault data and live output to support closed-loop automation, requiring checking how the network responds to the changes. Domain data collection systems track managed services and managed entities, providing fault data and live output to support closed-loop automation, which requires the ability to check how the network responds to changes. Domain intelligence services interact with the domain data collection services, domain control services, domain analytics, and domain orchestration services [245].

6.4.2.2 Domain Analytics

Domain analytics services produce domain-specific recommendations based on the data obtained by several sources, including domain data collection services [246].

6.4.2.3 Domain Intelligence

These are responsible for driving a domain's intelligent closed-loop automation by supporting automated decision-making and variable levels of human oversight with the help of autonomous management [247]. The following are the different types of intelligence services: (i) Decision assistance (ii) Decision-making (iii) Assistance in the plan of action

6.4.2.4 Domain Orchestration

Domain orchestration is a collection of management services that enable automate workflows and processes within a MD to control the life-cycle management of managed customer and resource-facing services. It also monitors the network services and virtual resources handled by the MD, further governed by policies and several other sources of information [248].

6.4.2.5 Domain Control

Each entity is controlled individually by the domain controller. The services are provided in the domain orchestration group by the management functions to change the configuration or the state of a consumed service and the controlled entity. The domain control category also provides services for managing virtualized resources [234].

6.4.2.6 E2E Data Collection

The availability and quality of customer-facing services are tracked by the E2E service data collection services that help monitor the quality of the actual E2E service and check the experience of the end-user based on updated data. The MDs' data collection services provide these data that control the services constituting the E2E service [249].

6.4.2.7 E2E Analytics

The E2E service analytics services provide the root cause analysis and E2E service impact and the generation of service-specific predictions. In addition, E2E service analytics includes testing service-level specifications and monitoring key performance indicators [250].

6.4.2.8 E2E Intelligence

The E2E intelligence services provide intelligent closed-loop automation in the E2E service MD that allows human oversight and variable levels of automated decision-making [251]. The following are the different types of intelligence services: 1) Decision-making assistance 2) Making the decisions 3) Action planning

6.4.2.9 E2E Orchestration

These are responsible for catalog-driven E2E orchestration of several MDs to modify/create/delete the customer-facing services across the domains. A service model shows how the different service components are connected and how they are related to the MDs [214].

6.5 Importance of ZSM for 5G and Beyond

The recent advancements in IoT technology increase the number of connected devices [252]. As the number of devices increases, there is a need to improve network infrastructure to ensure good communication or connectivity among geographically spread devices [253]. These advancements should enable real-time operations to be performed with minimal latency and improved performance. To be successful in achieving these goals, a suitable communication medium is required. 5G and beyond is the promising next-generation network that enables various enhanced capabilities such as ultralow latency, high reliability, seamless connection, and mobility support [254]. To meet enterprise requirements, 5G is built with service-aware globally managed infrastructures, highly programmable. SDN, NFV, MEC, and NS are critical foundations for 5G and beyond [255]. New business models such as multidomain, multiservice, and multitenancy will emerge in 5G and beyond due to new technologies, thus bolstering new industry dynamics. The existing infrastructure will result in a complex 5G architecture in terms of operations and services [20].

Traditional network management techniques do not fulfill the new paradigm; hence, the need arises for an efficient end-to-end automated network system capable of providing faster services to end-users [207]. The goal of automation is to drive services through an autonomous network governed by a set of high-level policies and rules. Enabled by the ZSM implementation, 5G and beyond networks can be operated independently, i.e. without human intervention [256]. Keeping the requirements in the account, ETSI developed the ZSM ISG in 2017. The ZSM objective is to create an underlying paradigm that enables fully autonomous solutions for network operation and service management of 5G and beyond networks. The ZSM comprises operational processes and tasks such as planning, delivery, deployment, provisioning, monitoring, and optimization that are executed automatically without human intervention [257].

7

Edge AI

Edge AI, also known as edge intelligence, combines the capabilities of edge computing with AI to facilitate 6G applications and services. After reading this chapter, you should be able to

- Understand the benefits of edge AI.
- Explore building blocks for edge AI.
- Learn future challenges in the area of edge AI.

7.1 Introduction

Edge computing promises to decentralize cloud applications while providing more bandwidth and reducing latency by moving computing resources and information technology (IT) functionalities from the centralized to the network edge [71]. These promises are delivered by moving application-specific computations between the cloud, the data-producing devices, and the network infrastructure components at the edges of wireless and fixed networks [258]. Meanwhile, owing to developments in computing infrastructure, big data, and data science, artificial intelligence (AI) and machine learning (ML) methods (especially deep learning) have been recognized as a crucial technology and found many practical applications, e.g. image and speech recognition, natural language processing, drug discovery, self-driving vehicles, and mobile communications and networking. However, the current deep learning methods assume that computations (e.g. deep model training and inference) are conducted in a powerful computational infrastructure [259], such as data centers with ample computing and data storage resources available.

As recent research and development go along, the term "edge" in edge computing itself remains a diffuse term. A commonly accepted definition of what the

6G Frontiers: Towards Future Wireless Systems, First Edition.
Chamitha de Alwis, Quoc-Viet Pham, and Madhusanka Liyanage.
© 2023 The Institute of Electrical and Electronics Engineers, Inc. Published 2023 by John Wiley & Sons, Inc.

edge is, where it resides, and who provides it is lacking across different research communities and researchers.[1] A common understanding is shared about its properties: as compared to the cloud, its features are closeness and on-premises (latency and topology), increased network capacity (effectively achievable data transmission rate), lower computational power, smaller scale, higher heterogeneity of devices, location awareness, and network contextual information [71]. Compared to the end devices (the final hop), it features increased computational and storage resources. It is an abstract entity to offload computation tasks and storage without the detour to the centralized cloud.

A rising area of tension arises from current AI and ML methods, which require powerful computational infrastructure [259] – a demand that is better satisfied in a data center and centralized cloud with ample available computing and data storage resources. However, sending the necessary raw data to the cloud puts pressure on the network in terms of network bandwidth and throughput and on the users in terms of cost, latency, and reliability. Meanwhile, organizations and end-users are usually less keen on sharing (potentially restricted) data with commercial cloud providers. It is reasonable since data may include private information, such as bank account, gender, and blood type. For example, the Self-Diagnosis App helps manage the COVID-19 self-quarantine situation in South Korea; however, this application requires users to report personal information, such as location, temperature, and symptoms, to the central server, which may cause privacy issues [260]. Since the edge is closer to end-users than the centralized cloud, these issues can be addressed by the fast-evolving domain of edge AI.

Put it simply, the amalgamation of AI and edge computing creates the concept of edge AI (i.e. edge intelligence). In edge AI, AI models can be learned either locally on end devices (e.g. internet of thing [IoT] devices, mobile phones, and sensors), edge nodes (e.g. edge cloud), or the centralized cloud. In other words, edge AI is not meant to be AI models at the network edge; the devices, edge, and cloud can work in close cooperation to achieve better performance by sharing the data and model updates throughout the network. Therefore, edge AI is expected to enable many potential use-cases, from AI applications, training, and inference at the network edge to AI techniques for optimizing the network edge. An example of this technology can be seen in the speakers of Google, Alexa, or the Apple Homepod, which has learned words and phrases through ML and then stored them locally on the device. When the user communicates something to applications such as Siri or Google, they send the voice recording to an edge network where it is passed to text via AI and a response is processed. Without an edge network, the response time would be seconds; with the edge, the times are reduced to less than 400 ms.

---

1 We deliberately renounce marketing-driven differentiation of edge vs. fog vs. mist computing in this work.

In a nutshell, edge AI refers to the use of AI for optimizing the edge and the training and deployment of AI services on the network edge. Edge AI takes and processes the data in close proximity to the edge and users.

7.2 Benefits of Edge AI

Edge AI is the processing of AI algorithms on edge, that is, on users' devices. It has several benefits, as follows:

- *Reduced Latency/Higher Speeds*: Inference is performed locally, eliminating delays in communicating with the cloud and waiting for the response.
- *Further Improvement to Security (Beyond "Edge Computing" Security Advantages)*: Edge AI processors can learn what normal network traffic patterns look like and be able to detect malware, attack signatures, and other types of malicious activity. They can quickly learn new attack patterns to adapt to next-generation threats that otherwise would not be possible [261].
- **Further Improvement to Reliability/Autonomous Technology**: The AI can continue to operate even if the network or cloud service goes down, which is critical for applications such as autonomous cars and industrial robots. Edge technology devices do not require specialized maintenance by data scientists or AI developers. The graphic data flows are automatically delivered for monitoring; therefore, it is an autonomous technology.
- *Further Improvement to Energy Consumption (Beyond "Edge Computing" Energy Consumption Advantages)*: The hardware design and "smartness" of edge AI chips can significantly reduce energy consumption. The data in edge AI chips are not required to be swapped between memory and processor (unlike traditional Von Neumann or stored-program chips) as edge AI chips usually rely on near-memory or in-memory data flow where logic and memory data are closer together. Additionally, companies that develop edge AI chips usually run ML algorithms as 8-bit or 16-bit computations, which can sometimes further reduce energy consumption by orders of magnitude. Qualcomm claims its edge AI-optimized chips can result in energy savings as much as 25× relative to conventional chips and standard computing approaches [262].
- **Further improvement to privacy (beyond "edge computing" privacy advantages) through distributed or decentralized AI techniques such as federated learning:** Federated learning mainly improves privacy through avoiding the need to send raw data over large distances and then store it at a remote centralized server. It makes the distribution of raw data on the clients (e.g. unmanned aerial vehicles [UAVs]) more feasible through training a shared model at a nearby server using a combination of locally computed updates.

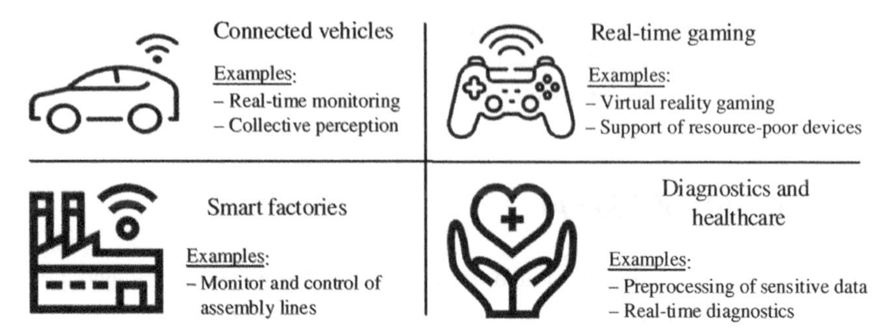

Figure 7.1 Use-cases of edge AI.

- **Further reduce communication overhead in certain scenarios by enabling aggregation frequency control:** In the case of training a deep learning model in an edge-computing environment where distributed models are trained locally first and then updates are aggregated centrally (e.g. federated learning), the control of updates aggregation frequency has a significant influence on the communication overhead. In such scenarios, the aggregation process can be carefully controlled and optimized through "aggregation frequency control" [263].

As highlighted in Figure 7.1, edge AI has gradually found its way to mainstream service domains, such as connected vehicles, real-time gaming, smart factories, and healthcare. From an infrastructure perspective, edge environments provide a unique layer for AI and also offer opportunities for existing technologies such as embedded AI or federated learning, which look at minimizing memory consumption on individual devices, increasing privacy by keeping data on the local device, as well as reducing communication needs between distributed entities. Those features serve as the foundation for the use-cases shown in Figure 7.1.

7.3 Why Edge AI Is Important?

The list of edge AI applications is long. Current examples include facial recognition and real-time traffic updates on smartphones, as well as semiautonomous vehicles or smart devices. Other edge AI-enabled devices include video games, smart speakers, robots, drones, security cameras, and wearable health-monitoring devices. Below are a few more areas where edge AI is expected to continue to be used.

- Edge AI will provide intelligence to the security camera detection process. Traditional surveillance cameras record images for hours and then store and use

them if necessary. However, with edge AI, the algorithmic processes will be carried out in real-time in the system itself, so the cameras will be able to detect and process suspicious activities in real time for a more efficient and less-expensive service.
- The autonomous vehicles will increase their capacity to process data and images in real-time for the detection of traffic signs, pedestrians, other vehicles, and roads, improving the levels of security in transportation.
- Edge AI will be possible to be used in image processing and video analysis to generate responses to audiovisual stimuli, or for real-time recognition of scenes and spaces, for example in smartphones.
- Edge AI will reduce costs and improve safety in terms of industrial internet of things (IIoT). The AI will monitor machinery for possible defects or errors in the production chain, while the ML will recompile data in real time of the whole process. Edge AI will be used for the analysis of medical images in emergency medical care.
- The deployment of 6G technology networks will mean greater speed and very low latency for mobile data transmission, making edge AI more useful. There are no application limits for edge AI technology. After the crisis of COVID-19, the ingenuity of the companies has led to deploy solutions based on AI to provide accurate information in real time. In healthcare, for example AI is helping with patient monitoring, testing, and treatment.

7.4 Building Blocks for Edge AI

Several building blocks have been identified to fully reap the potential of edge AI: edge computing, support for advanced edge analytics, edge inference, and edge training. These blocks are explained in more detail in the following sections.

7.4.1 Edge Computing

As a key building block of edge AI, edge computing offers computing resources and IT functionalities at the edge network to run AI models. Edge computing has the means to perform data processing and AI deployment close to the end-user. Since the edge-computing nodes are closer to the users, the traffic flow is also reduced. Edge computing also minimizes the bandwidth demands and latency in data storage and computation in IoT networks. In fact, IoT devices can offload their data to the edge (ES), located at the base stations and aerial components in the air (e.g. UAVs, high-altitude platform [HAP], and low earth orbit [LEO] satellites in aerial computing [165]), for further processing. Edge servers are typically equipped with rich computing and storage resources to handle IoT data tasks, train

AI models, and provide AI services for end-users, such as video and data analytics, data prediction and mining, autonomous internet of everything (IoV), edge hardware for AI, and distributed AI training and inference at the edge.

As information and communications technology (ICT) became affordable, there has been an enormous surge in the data generated by mobile phones, IoT devices, and industries. The volume of data generated resulted in the use of cloud and edge computing for storage and computational purposes. In this context, edge of things is a promising concept based on the integration of edge computing with IoT networks [264]. In the edge of things, the data acquired by several sensors are temporarily stored in the edge node for real-time analytics and predictions. The general architecture of the edge of things is depicted in Figure 7.2. Typically, the data generated from sensors in several IoT applications, such as smart homes/buildings, smart grids, smart healthcare, smart transportation, and industrial IoT, are stored in the edge nodes at regular time intervals. Once the data are processed in the edge nodes, they are dissipated to the cloud. Apart from improved latency, edge of things offers several other benefits, such as reduction of traffic to the cloud and improved reliability by the installation of applications in close proximity to edge devices.

7.4.2 Support for Advanced Edge Analytics

Edge AI is expected to enable many vertical applications that require advanced edge analytics and real-time decision-making. As a result, making use of AI

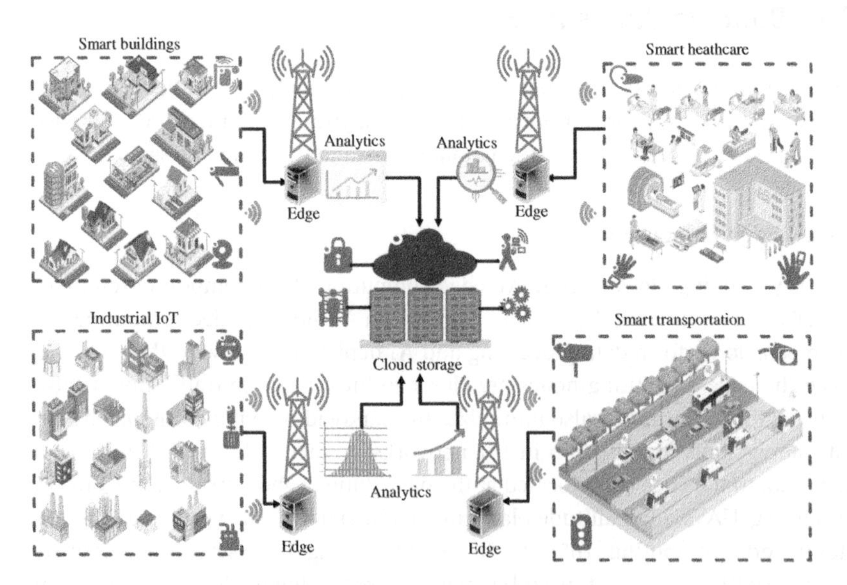

Figure 7.2 Architecture of edge of things.

algorithms for data processing and data analytics at the network edge is necessary. Edge computing is a means to provide dedicated hardware and learning frameworks for deploying AI services at the network edge. Moreover, network security and data privacy are major issues in edge AI. Several applications, such as connected vehicles, social media apps, and healthcare applications, use edge computing for less latency and higher inference accuracy through AI algorithms, especially deep learning with the ability to learn the representational features from raw data of IoT and mobile devices. These applications that generate sensitive data of the users, such as personal data, health, location, and utility services, may possess an elevated risk of compromised security. The distinct properties of blockchain include distributed nature, traceability, and immutability, making it an ideal solution to overcome security issues of edge AI applications.

The use of blockchain in edge computing and IoT (i.e. edge of things) has the potential to be the next revolution in ICT and edge AI, where application providers can provide safe, transparent, immutable, decentralized applications to users with reduced latency, real-time analytics, and accurate recommendations. For example, Figure 7.3 illustrates a single communication round of blockchain-enabled federated learning healthcare systems, where individual AI models are trained completely on local devices and blockchain helps coordinate the calculation of the global model via the block consensus among participants

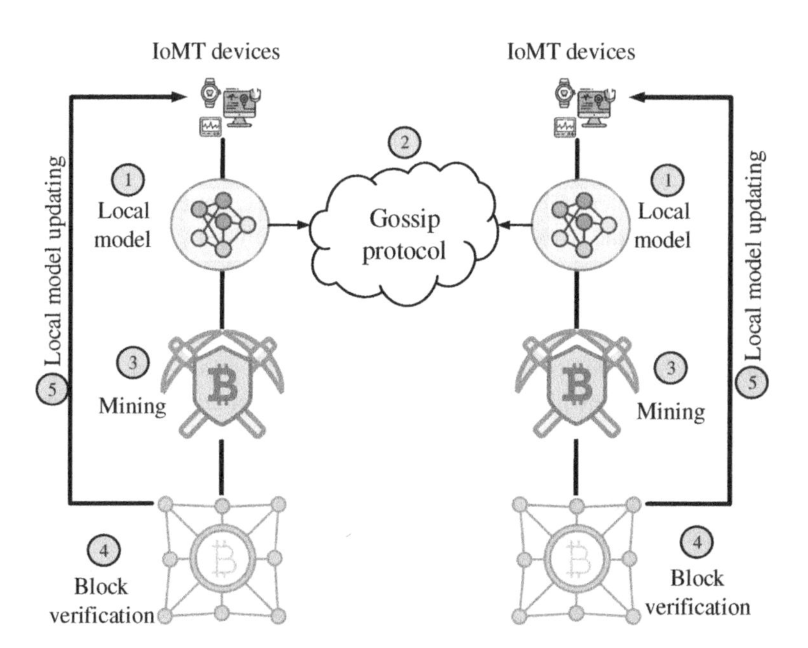

Figure 7.3 Illustration of a learning round of blockchain-enabled federated learning healthcare systems.

in a peer-to-peer manner [265]. The procedure in a global communication round of such a blockchain-enabled federated learning system consists of five main steps. In the first step of local training, each internet of medical things (IoMT) device trains the local learning model using its local data. The next step is model broadcasting and verification when each device adds its digital signature to the model and broadcasts the model to the other IoMT devices via some gossip protocols. The transaction of a device is then verified by all other IoMT devices in the network. Then, upon receiving the local models from the other devices, each IoMT device tries to mind the current block. After that in the step of block validation, the current block, if verified, is added to local ledgers of IoMT devices. Finally, upon receiving the verification, each device updates its local model and starts a new communication round.

7.4.3 Edge Inference and Edge Training

From the viewpoint of end devices (e.g. sensors, wearables, phones, and IoT devices), the network edge can provide a powerful platform to collect and process a large amount of data. Moreover, the issues of data quality and data heterogeneity can be well addressed through the deployment of advanced deep learning (DL) techniques at the network edge. For example, convolutional neural network is one of the most successful deep architectures with great capabilities to process high-dimensional unstructured data (e.g. images) as well as text, signals, and other continuous responses. As a result, convolutional neural network has been leveraged to design data-driven algorithms for improving and optimizing wireless networks, such as modulation classification of radio signals [131] and waveform recognition in integrated radar-communication systems [266]. As a result, edge AI is seen as a key enabler of intelligent and autonomous 6G networks in which many building blocks and network operations can be optimized via the deployment of AI-enabled and data-driven algorithms at the network edge. The proven and potential capabilities of edge inference and training at the network edge will bring opportunities to enable and improve emerging services and applications in 6G, such as holographic telepresence, extended reality, smart grid 2.0, and Industry 5.0. In the meanwhile, the efficient deployment of edge AI is reliant on many important factors, such as resource limitations of mobile devices, high communication burdens of edge cloud, and long latency of centralized cloud. For example, a three-step framework is proposed in [267] to reduce the computation cost and communication overhead in edge-device cotraining and inference, which is level 4 in the edge AI five-level architecture presented in Section 7.6. Experimental results in this work emphasize the importance of the design of deep neural networks in reducing the computation cost and communication overhead for edge AI. To further facilitate the deployment and applications of edge AI in 6G networks, more and more research efforts into edge inference and edge training are required to overcome its challenges [268]: hardware design, software platform, and edge AI architecture, just a few to list.

7.5 Architectures for Edge AI networks

This section will briefly discuss the architectures in edge AI. We will cover a generic end-to-end architecture, decentralized edge intelligence architecture, and device-level edge AI architecture.

7.5.1 End-to-End Architecture for Edge AI

For each new generation of networks, new services and capabilities have been introduced at the architecture level in order to meet more and typically more stringent demands. The mobile network was originally designed to deliver voice services. Since then, both the architecture and deployment of mobile networks have followed a centralized and hierarchical paradigm that reflects the nature of voice traffic and packet traffic of the mobile Internet. To realize the vision of "connected intelligence," 6G will break and shift these traditional paradigms toward a novel architecture and design that meet new requirements for the deep integration of communication, AI, computing, and sensing at the network edge with new integrated capabilities empowered by evolutionary, as well as, revolutionary enabling technologies. Under this new design philosophy, we introduce a holistic end-to-end (E2E) architecture for scalable and trustworthy 6G edge AI systems, as illustrated in Figure 7.4. By providing new wireless network infrastructures, enabling efficient data governance, integrating communication and computation at the network edge, as well as performing automated and scalable edge AI management and orchestration, the proposed E2E architecture will provide a scalable and flexible platform to support diversified edge AI applications with heterogeneous service requirements.

7.5.2 Decentralized Edge Intelligence

A 6G-based Internet of everything (and/or 6G-IoE) is an involving network of interconnected heterogeneous cyber-physical systems or smart objects. A report reveals that IoE will rise rocket high; for instance more than 125 billion devices will be connected to the Internet by 2030 or so. Such high-scale interconnected networks would require mega-corporations that can provide advanced serviceable AI independently. This has pushed AI intelligence to the extreme edge that will improve several adaptive parameters, such as efficiency, throughput, latency,

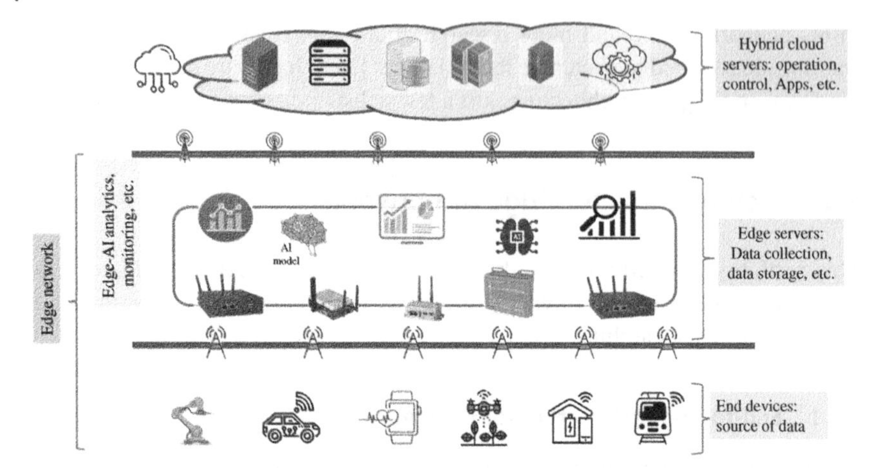

Figure 7.4 End-to-end Edge-AI architecture. Source: AAEON Technology Inc.

learning, accuracy, and processing, for 6G-based services and applications. Moreover, to achieve all these adaptive parameters, the real-time distribution of data and AI workload is required in the fast-changing and large-scale heterogeneous networks. Therefore, future 6G networks require a new decentralized architecture, as shown in Figure 7.5.

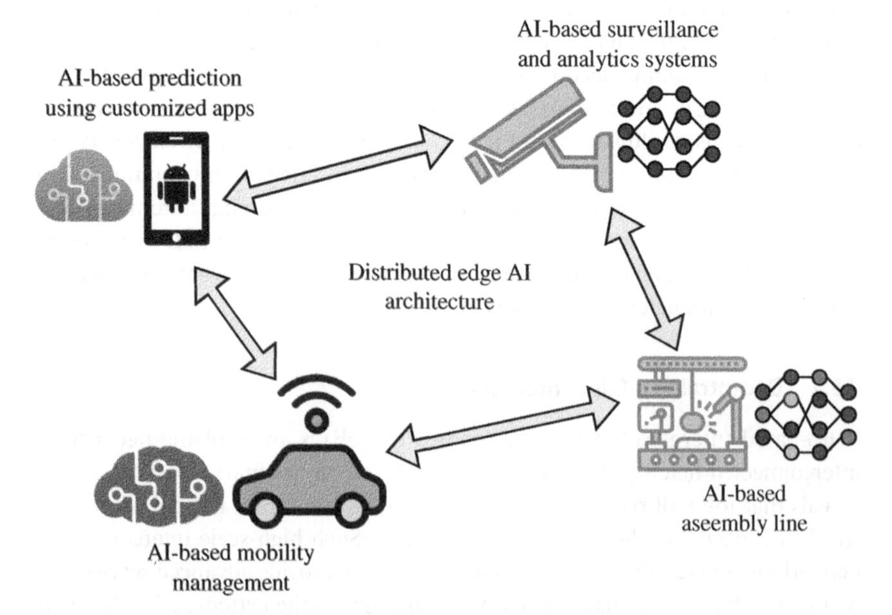

Figure 7.5 Decentralized Edge-AI architecture.

7.6 Level of Edge AI

With a hierarchy of end devices, edge nodes, and centralized clouds, AI models can be trained and inferred in various locations. Recent studies showed that the cooperation between these three layers can significantly improve energy consumption, latency, and data privacy. For example, running AI models on the centralized cloud can get numerous benefits from powerful computing resources and storage but raise the issue of high communication overhead and data privacy. It is evident because the data need to be traversed from end devices and edge nodes to the cloud over mobile access networks with limited bandwidth and communication resources. Instead, a part of the entire data can be shared with the cloud for model training, while model inference can be executed on the edge and end devices. Therefore, edge AI can be classified as multiple levels depending on how the data is shared, the AI model is trained, and inference is made. As a promising definition, there are several levels of edge AI [56, 263], as illustrated in Figure 7.6 and explained in the following.

- **Level 0 – Cloud AI**: The entire data are uploaded to the centralized cloud for model training and inference. Cloud AI enjoys considerable advantages of powerful resources (i.e. computing, caching, and storage) of the cloud. Cloud AI has a long history of development and has found many applications in the fields of engineering. Notable, cloud deep learning has been recognized as a crucial technology for many practical applications, e.g. image and speech recognition, natural language processing, drug discovery, self-driving vehicles, and 6G communications and networking [269].
- **Level 1 – Cloud-Edge Co-Inference and Cloud Training**: The AI model training takes place in the cloud and inferencing takes place in both the network edge and in the cloud (in a cooperative manner). The data generated by end devices and at the edge are shared with the cloud for centralized learning, and the trained model is then distributed to the edge and devices for inference.
- **Level 2 – In-Edge/On-Device (Co-)Inference and Cloud Training**: The AI model training takes place in the cloud and inferencing takes place within the network edge or end devices. Similar to level 2, the cloud collects the data generated by distributed IoT and mobile devices at the edge for centralized learning. Edge inference is typically preferred to on-device inference as edge nodes usually have higher computing capabilities, energy, and storage. In fact, edge inference can be enhanced by optimizing edge processing and transmissions from the edge to devices. Moreover, edge-device co-inference can achieve better performance than individual ones since massive data and computing capabilities of both the edge and devices can be exploited.

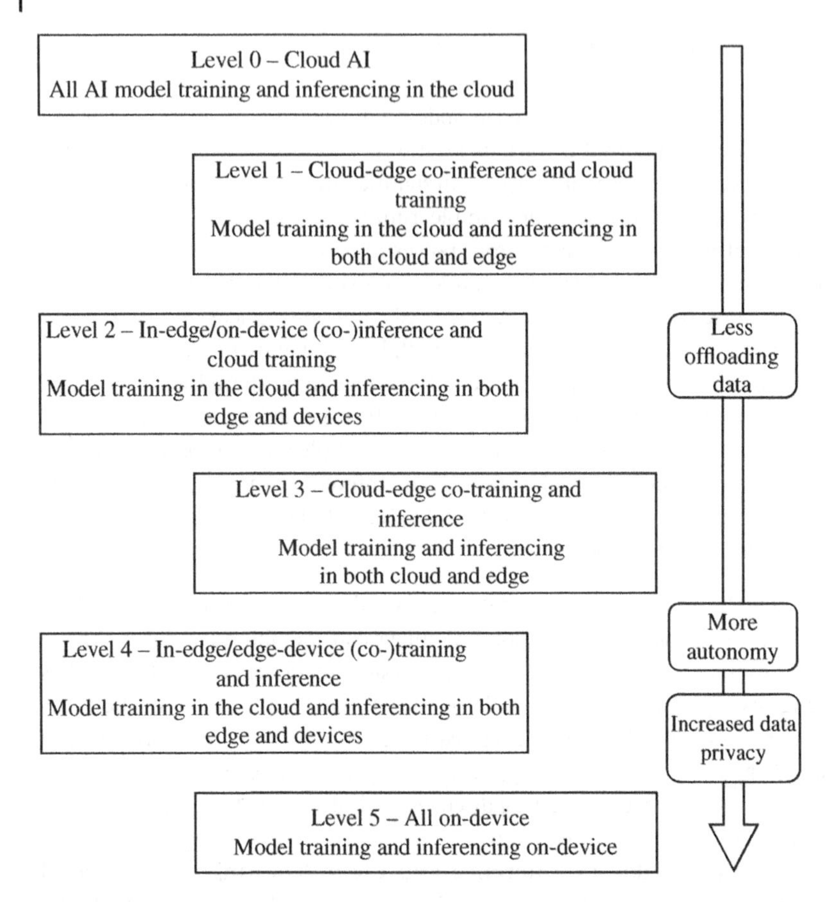

Figure 7.6 Edge AI five-level categorization.

- **Level 3 – Cloud-Edge Co-Training and Inference**: The AI model training and inferencing take place in both the network edge and in the cloud (in a cooperative manner). This level belongs to hybrid centralized-distributed model training as only a part of the data must be shared with the centralized cloud. In this regard, AI models can be either centralized learning with the cloud or distributed learning with edge computing.
- **Level 4 – In-Edge/Edge-Device (Co-)Training and Inference**: The AI model training and inferencing take place on the network edge and on-device (in a cooperative manner). The distributedness of this edge AI level is, respectively, higher and lower than all-in-edge learning and all-on-device learning (level 5). Edge-device cooperation helps to diversify AI services and reduce on-device training costs in all-on-device learning. Meanwhile, data privacy can be enhanced as most data remain local at end devices.

- **Level 5 – All On-Device**: The AI model training and inferencing take place completely on-device. This level corresponds to distributed AI, where the data need not be shared with the edge and cloud for further processing and learning. One example of on-device learning is federated learning, a promising AI concept for collaborative learning across distributed devices without requiring all data to be stored and processed in the cloud.

7.7 Future Cloud Computing Perspective

The perspective of future cloud computing is shaped by the collaboration between future clouds and future edge nodes. The main reasons for shifting tasks from the cloud to the edge are latency, bandwidth, locality of data, scalability, accessibility, security, and fault tolerance. This collaboration is not only a technical aspect but also involves business and stakeholder challenges. An important nontechnical challenge is the potential competition between cloud and edge, which need to work together closely to provide the best possible services to customers. The rest of the chapter goes into detail about the technological challenges involved from the cloud perspective. These challenges are the following: (i) resource management; (ii) energy constraints and efficiency; (iii) security, trust, and privacy, and (iv) intermittent connectivity.

7.7.1 Resource Management

Although the total volume of edge servers may provide a large amount of resources, the necessary locality of edge servers limits the amount of available resources compared to a cloud environment quite drastically. That limits the number of different ML tasks that may be executed simultaneously due to hardware limitations and the execution latency of the tasks. While the number of ML tasks supported by the edge can be increased through on-demand loading and execution of the corresponding models, this may increase the latency of tasks that need to be loaded. Thus, for latency-critical tasks, proactively reserving resources might be a necessity. As the reservation of resources could drastically limit the number of supported ML tasks, the necessity of reserving resources should be determined based on the risk of failure together with the consequences of that failure. Especially when resources are sparse, the management of these resources in critical situations might be challenging.

Importantly, AI tasks at the edge are often embedded in larger settings, e.g. as part of a control workflow. Accordingly, different services hosted on edge resources have to interact with each other. Not taking into account that data items may have to be forwarded to services may lead to the fact that these services are not ready when receiving data and/or may not possess the computational resources for handling data items. Accordingly, a backpressure of data items might occur. This is especially the case in stream-processing scenarios, e.g. when new data items need to be classified and – based on the classification – forwarded to different destination services. Thus, it might be interesting to observe and analyze the execution of tasks depending on their priority level given different loads on the edge.

7.7.2 Energy and Operational Constraints

In general, edge nodes are expected to be less efficient (in terms of energy and cost) than cloud data centers. That is, as the economies of scale might work against edge data centers, e.g. by allowing for better cooling. Both, edge nodes and cloud data centers, have the possibility to deploy renewable energy, but it is unclear where these units are producing energy more efficiently. On the one hand, large data centers provide more opportunities to not only deploy renewable energy, increasing their sustainability and environment compatibility but also consume a significantly higher amount of energy compared to edge nodes. On the other hand, edge nodes might harvest their own energy easier than cloud data centers due to their small scale and geographical dispersion, even though their capacity for renewable energy units is quite limited.

In addition, the data need to be transferred via the Internet, which also consumes energy, increasing the possibility of edge nodes to surpass energy-efficient cloud data centers: Edge nodes can preprocess data, e.g. can detect anomaly cases and transmit only relevant data or adapt sensing rates dynamically. However, energy-saving largely depends on the amount of processing that can be done at the edge. Another possibility is to move the training process partially to the edge, utilizing methods such as federated learning (see Chapter 4). In that case, only learned parameters (e.g. gradients) need to be transmitted rather than raw data, which are multiple orders of magnitude larger.

Another potential advantage of the edge nodes is the more specially designed hardware compared to generic cloud server racks, as this specially designed hardware is becoming more popular in the market (e.g. hardware accelerators) and utilizes limited processing and graphical capacities more efficiently.

7.7.3 Security, Trust, and Privacy

As shown in Figure 7.7, edge AI offers both opportunities for higher trust, security, and privacy, and also adds additional challenges. The security might be increased as some attacks might be detected early and countermeasures can be taken. However, the number of possible targets and attack vectors is much higher, which leads to a higher potential for attacks.

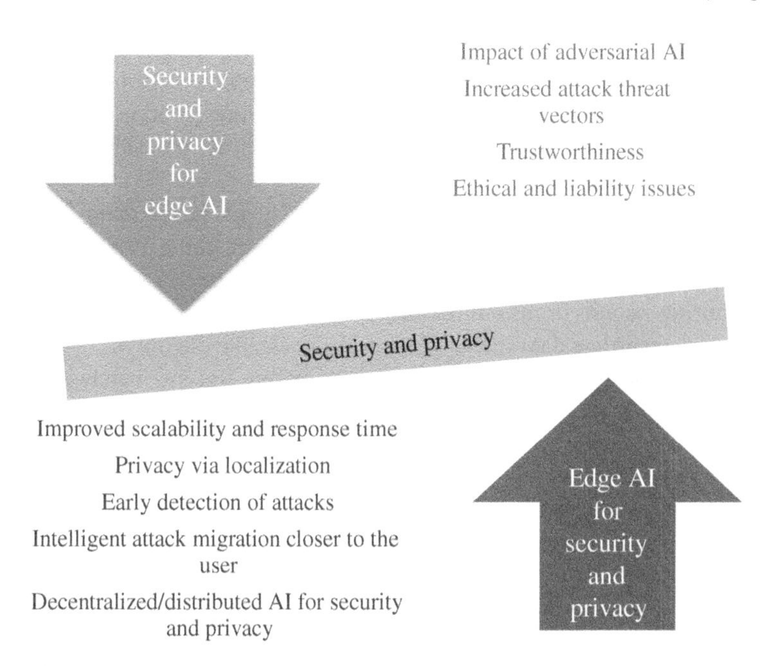

Figure 7.7 Trust, security, and privacy of edge AI.

Edge AI might have an issue with trust in the system, which is caused by its distributed nature and the potential liability issues. However, the trust in edge servers might be increased through open implementation and specifications. Additionally, it might be necessary to differentiate between devices under the control of the user and resources provided by providers (e.g. Amazon Web Services [AWS], Google Cloud Platform [GCP], Azure). One possible solution for increasing trust is the introduction of a reputation system known from other areas that assess and manage the trust. However, it will take new players a while to build that reputation.

Another property of edge AI is that the data required to perform the ML tasks are kept local at the edge servers. This led to the common assumption that the privacy and trust in edge servers are higher compared to the cloud. In contrast to the common assumption, the edge does not guarantee privacy but raises new challenges to ensure privacy. That is, as the trust of edge nodes is harder to keep and manage compared to a cloud provider, thus also aggravating data security and privacy, even though the locality of the data will prevent some attacks on the data.

7.7.4 Intermittent Connectivity

In general, we also consider the possibility of poorly connected edge servers, i.e. edge servers in regions where the connection to the Internet is poor. Also, in those areas, edge intelligence has a huge potential and can be a driver toward digitalized nonurban areas. Some examples of such systems include water and air pollution monitoring and natural disaster (e.g. wildfire, flood, and volcanic eruption) prediction. For instance, in water pollution monitoring, the identification of substances in the water could be made using edge resources.

The lack of reliable network connectivity in those areas could be the main driving factor in environmental monitoring scenarios in which streaming data have to be processed in near real time. Intermittent connectivity and energy constraints prevent monitoring systems from continuously transmitting raw data to the cloud for processing, whereas less data-intensive control signals like the output of ML algorithms can still be communicated under interruptions and low bandwidth availability. In addition, the actors which rely on the outcome of ML algorithms are often close to the data sources. By not sending the data to the cloud, a complete processing step can be saved, e.g. starting countermeasures automatically. In the previous example of water pollution monitoring, the services running at the edge can then inform pumps or valves to open or close, depending on the scenario.

7.8 Role of Edge AI in 6G

The evolution of the fifth-generation networks toward the 6G era shapes the perspective of future networking. The developments along this path not only encompass network communication (e.g. speed, coverage, and resilience) but also quality and delay in computations. To unlock the true potential of future networks with such a complicated structure, various technologies, at both hardware and software levels, need to coexist and cooperate. These include, for example the creation of edge computing and communication fabrics or using self-learning technologies for dynamic network orchestration.

7.8.1 Communication and Computation with Human-in-the-Loop

With the wide dissemination of smartphones and other personal carry-on devices, their significant computation, communication, and sensing capabilities become valuable to solve challenges in networking. The examples include local data acquisition as in federated learning, reducing the communication overhead as in device-to-device caching, and cooperating in executing computationally intensive tasks. In future networks, mobile devices can decide if, when, where, and which fraction of a specific task to offload to a server at cloud or edge. Besides, they can take the role of computational worker, form pools of resources, and divide the tasks based on their preferences to optimize their utility and performance.

While humans or human-driven devices may significantly contribute to edge AI, such involvement raises several challenges. To model such challenges, one can use multiagent systems. Subsequently, to address them, one can use various mathematical tools such as control and game theory. Furthermore, ML and AI play significant roles if there is some uncertainty and lack of information. In all of the abovementioned steps, the specific characteristics of humans should be taken into account [270]. In particular, humans often act based on heuristics and irrational influences, taking into account the factors such as social norms and peer pressure. When dealing with self-interested entities, it is essential to consider mutual trust and to respect the welfare of each entity [271]. Finally, using humans as a data source, e.g. using body sensors or GPS, proliferates privacy concerns that strongly couple with legal and ethical challenges.

7.8.2 Critical but Conflicting Actors and Applications

In the transition from the current systems and networks to 5G beyond, the future needs of the societies become the driving force that creates use-cases. As such, building innovative technology to address the society-driven use-cases becomes imperative [272]. To some extent, it stands opposite to the current use-cases such as low-latency and reliability that are generated by technological advances rather than taming directly from the society. Examples include the current vertical trends, including resource-efficient manufacturing, green energy generation and distribution, organic agriculture, and optimization of retail logistics.

Heterogeneous actors are expected to build and consequently share the massive edge-computing infrastructure to serve their wide range of demands. Despite having some common goals to achieve, such actors often exhibit conflicting interests; i.e. more benefit for one attribute may reduce that of the other (here, the utility can correspond to higher monetary return, sustainability, improved environmental factors, and the like). Finding a Pareto optimal and stable solution to this problem is significantly challenging as different utility measures are often conflicting. The problem becomes aggravated in practice as it involves several decision-makers instead of a single central authority. That is not only because of self-interest but also because of information asymmetry and different types/preferences [273]. AI can be a solution to this problem, as it enables distributed systems to interact, learn, and make decisions – rendering smart systems inseparable blocks of edge intelligence.

7.8.3 Edge AI and Emerging Technologies

Next-generation networks beyond 5G encompass several technologies whose efficient deployment depends strongly on reliable cloud infrastructure as well as edge intelligence. These include, among others, joint communication and sensing, campus networks, Open Radio Access Network (O-RAN), intelligent reflecting surface (IRS), to name just a few. For example, the technology of joint communication and radar sensing is implementable in two networking architectures, namely small cell networks, and cloud RAN [274]. While the latter is amenable to the cloud infrastructure, both implementations greatly benefit from edge intelligence. That is because joint communication and sensing networks necessitate swift signal processing and precise pattern recognition, both of which are computationally complex.

Another example is campus networks which cover a geographically limited region to cover the communication requirements specific to that area. For example, a manufacturing company can integrate a campus network in response to the need for secure, reliable, and persistent industrial communications with ultralow latency. Other applications of campus networks include agriculture fields, construction sites, hospitals, and the like. The 5G technology, together with the edge computing capacity and AI, is the driver of campus networks. They enable not only secure and stable communication, fast and no-failure computation, but also reliable and efficient performance, even in the absence of precise information. The reliable performance of several other technologies (e.g. IRS) depends on edge intelligence as well. IRS technology relies on the optimal beam configuration, which might happen repeatedly. As a result, the required low-delay computation can be handled by the edge.

7.8.4 Technology Meets Business

Recently, the discussion around 5G and beyond has been largely driven by the potential use-cases and the over-arching goal to build real-time integrated edge computing, AI, and communication services that respond to the dynamic needs of the applications. These anticipated solutions, from everyday life to smart traffic and medical advantages, are significant but need evaluations within the context of technology and novel business models.

We identified three examples where AI technologies have a role in entirely new functions; however, hardly any business cases and models are yet defined for them:

1. *Interpreting the results of joint sensing and communication capabilities in future networks*. Future higher-frequency communications allow some level of "radar-like" recognition of the environment. This represents a significant change to what current networks do or how they operate, and the security issues involved are substantial:
2. *Optimal link-level communication details discovery through ML*. This, in theory, is possible. However, the usefulness of these technologies is still questionable,

along with the necessary learning costs offset by the optimization benefits accrued.

3. *Interoperability and collaborative use of data and AI technologies.* Current systems are largely run within single organizations, but we raise the question: “what can we do to enable more sharing technologies and interoperable interfaces.” There is a need for various stakeholders to identify and discuss relevant security, privacy, and ethical issues and tools to respond to them in a trustworthy manner.

Acknowledgment

Dr. Pardeep Kumar with the Department of Computer Science, Swansea University, UK, has partly contributed for this chapter.

8

Intelligent Network Softwarization

Network softwarization emerged with the advent of 5G. These functionalities are expected to evolve toward intelligent network softwarization with the dawn of 6G networks. After reading this chapter, you should be able to

- Understand the technological development toward intelligent network softwarization.
- Understand concepts such as Service Function Chaining (SFC), Programmable Data Planes (PDP), and in-network computing.

8.1 Network Softwarization

The evolution to 5G and future mobile telecommunication networks is characterized by a significant surge in demands in terms of performance, flexibility, portability, and energy efficiency across all network functions. Softwarized network architecture integrates the principles of Software-Defined Networking (SDN), Network Function Virtualization (NFV), and cloud computing to mobile communication networks [275]. The softwarized network architecture is designed to provide a suitable platform for novel network concepts that can meet the requirements of both evolving and future mobile networks.

The underlying principle of the SDN architecture is the decoupling of the network control and data planes. Using this principle, network control functions are logically centralized, and the underlying network infrastructure is abstracted from the control functions. The introduction of NFV offers a new paradigm to design, deploy, and manage networking services based on the decoupling of the network functions from proprietary hardware appliances and providing such services on a software platform. However, the separation of control and data planes as well as the virtualization of network functions and programmability introduce

6G Frontiers: Towards Future Wireless Systems, First Edition.
Chamitha de Alwis, Quoc-Viet Pham, and Madhusanka Liyanage.
© 2023 The Institute of Electrical and Electronics Engineers, Inc. Published 2023 by John Wiley & Sons, Inc.

a number of novel use cases and functions on the network. This will further usher in new stakeholders into the networking arena and hence, will obviously alter the approach to security management in 5G and B5G mobile communication networks.

Softwarized network architecture integrates the core principles of SDN, NFV, and cloud computing (CC) into a design of programmable flow-centric mobile networks providing high flexibility. This is a significant improvement over 4G LTE networks. Softwarized networks offer many advantages, such as a uniform approach to Best Effort and Carrier Grade services, centralized control for functions that benefit from a network wide view, improvement in flexibility, and more efficient segmentation. It also provides an enabling platform for automatic network management, granular network control, elastic resource scaling, and cost savings for backhaul devices. With softwarized networks, resource provisioning is done on-demand, hence, allowing elastic resource scaling across the network.

The developments of SDN, NFV, and CC will pave the path for various other services to be migrated to virtualized platforms. For instance, cloud factories will combine the resources in multiple physical locations to a virtualized platform and coordinate their operation to facilitate manufacturing through cloud-based technologies. It is evident that this type of a virtualized network requires end-to-end orchestration of network and services. Furthermore, service network automation, together with network slicing and multiaccess edge computing, leads mobile networks to be fully softwarized 5G and B5G networks. Network softwarization facilitates self-organizing, self-configurable, and self-programmable flexible network infrastructure to facilitate emerging heterogeneous applications and use-cases [276]. In line with these developments, 6G is expected to facilitate intelligent network softwarization by harnessing the capabilities of AI techniques such as machine learning and deep learning. The path from traditional nonsoftwarized networks toward intelligent softwarized networks is illustrated in Figure 8.1.

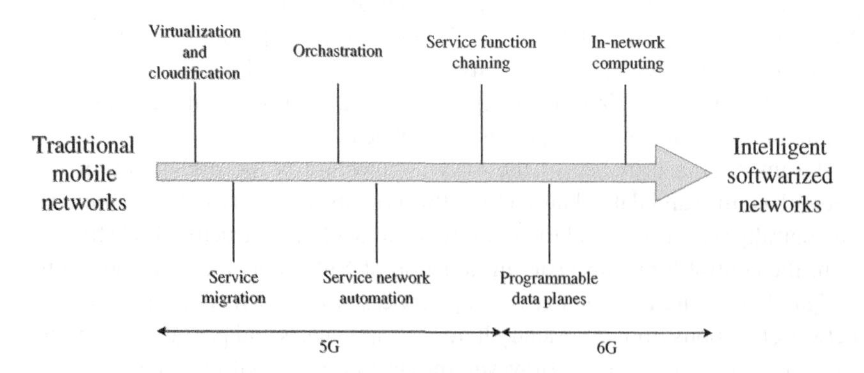

Figure 8.1 Toward intelligent softwarized networks.

8.2 Intelligent Network Softwarization

6G networks are envisioned to be more complex as it will support more network nodes, heterogeneous connected devices, and advanced technologies and applications compared to 5G networks. These include content caching, edge intelligence, traffic forecasting, wireless power sharing and transfer, THz communication, large intelligent surfaces, advanced data analytics, and zero-touch network and service management. To facilitate such developments, 6G is expected to have a dynamic and intelligent network architecture, which will bank on the recent developments of AI and move network softwarization toward intelligent network softwarization [16].

Several technologies are being developed in line of making network softwarization more intelligent. These include Service Function Chaining (SFC), Programmable Data Planes (PDP), and in-network computing.

8.2.1 Service Function Chaining

Network services, which are end-to-end functions used by operators of a softwarized network, require virtual network functions (VNFs) to be executed in a meaningful order. This order is highly dynamic as it depends on factors, such as user requests, user/traffic classifications, and network policies. Hence, the functionalities of network services are designed as a set of chained VNFs. The process of automating the order of the VNF executions, allocating the required physical network resources (e.g. computing and storage to facilitate the VNF execution), and executing these VNFs by sending network packets through the ordered VNFs to provide end-to-end network services is known as SFC [277, 278]. Therefore, each network function or a service function, which is arranged in a meaningful order, is required to perform a prescribed set of actions to their received packets. For instance, providing network security and protection requires service functions, such as firewalls, virus, and malware scanning functionality, and deep packet inspection functionality, which should be ordered in the meaningful order.

The Internet Engineering Task Force (IETF) that provided SFC architecture is illustrated in Figure 8.2 [279]. Accordingly, the architectural principals of SFC are also listed below.

- Topology independence: independent of the underlying network topology.
- Plane separation: SFC is independent from packet handling operations.
- Classification: SFC can treat traffic based on the specified traffic classification rules.
- Shared metadata: Metadata are shared with service functions.
- Service definition independence: SFC architecture is independent of the service functions.

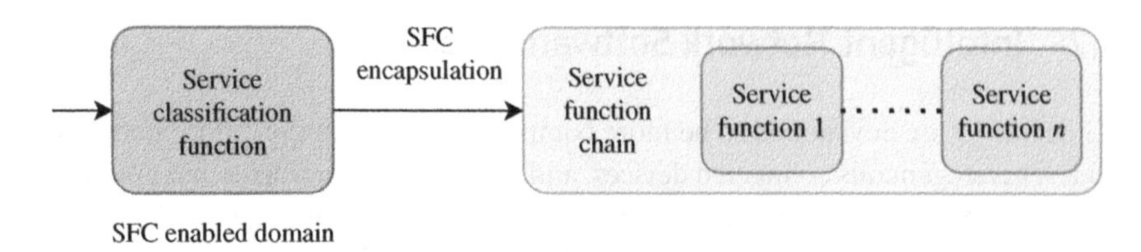

Figure 8.2 Service Function Chain (SFC) architecture.

- Service function chain independence: The creation, modification, execution, or deletion of one SFC does not interfere with the operation of another SFC.
- Heterogeneous control: SFC allows service functions to adapt local policies and classifications.

8.2.2 Programmable Data Planes

SDN, which enables control-plane programmability can also enable data-plane programmability [280]. Control-plane programmability offers flexibility to 5G and beyond networks, while data plane programmability facilitates the customization of network protocols and functionalities, dynamic packet filtering and processing, and dynamic reconfiguration of run-time processing logic through network description languages.

The 5G and beyond networks will be highly dynamic, ubiquitous, self-configurable, and self-managed to provide seamless coverage for future communication applications. Conventional methods such as sample and probe-based methods fetch data in a poll-based fashion to report the state of the network. This can also utilize network resources and adversely affect the performance of the data transmission network. However, PDPs can use the P4 programming language [281] and enable In-band Network Telemetry (INT) to collect and report the state of the network data plane [280].INT can be performed in three methods.

- INT-XD: Metadata is directly exported from the data without performing any modifications to the packet.
- INT-MX: INT instructions are integrated to the packet header. Network nodes can comply with the embedded instructions to send metadata for monitoring.
- INT-MD: INT instructions together with metadata are incorporated in the packet. These are forwarded through each hop until it reaches the network node located prior to the destination. This node removes the INT instructions and the metadata from the packet and forward these information to the network monitoring system.

Many types of data reports can be generated through PDP using the P4 language. These reports can provide real-time telemetry, which is instrumental toward the

Zero-touch Network and Service Management (ZSM) functionality of emerging 6G networks. Furthermore, INT is instrumental toward the detection of various attacks, such as distributed denial-of-service (DDoS) attacks, to the network.

8.2.3 In-Network Computing

In-network computing is able to distribute computing functions to network nodes and devices [282]. This concept was initially developed as active networks that can perform the necessary processing of packets while forwarding them. However, this concept did not see light due to processing limitations of traditional networks. However, with the emergence of powerful processors in 5G and B5G network devices, such as routers and firewalls, SDN, the emergence of multiaccess edge computing, and edge intelligence, the concept of active networking is being realized through in-network computing. Through this, redundant computational resources of network equipment are utilized to satisfy the computational requirements of network functionalities and applications. The concept of in-network computing is illustrated in Figure 8.3.

In-network computing is primarily expected to be realized through devices such as Field Programmable Gate Arrays (FPGAs), Smart Network Interface Cards (SmartNICs), and Switch Application-Specific Integrated Circuit (Switch-ASICs) [283]. Newly proposed methods, such as NetCache [283], extend the data transmission functionality of networks and shift toward implementing application-level functionalities in the network. This can significantly reduce the latency and improve the throughput of the network. Another method known as DAIET [284] aggregates application data in network data plane. This is based on SDN and the P4 programming model [281]. The application data are encapsulated in UDP packets, whereas network switches that receive these packets parses the message and aggregates data to the packets. Through the utilization of in-network computing, data processing delays can be reduced while achieving a high-data

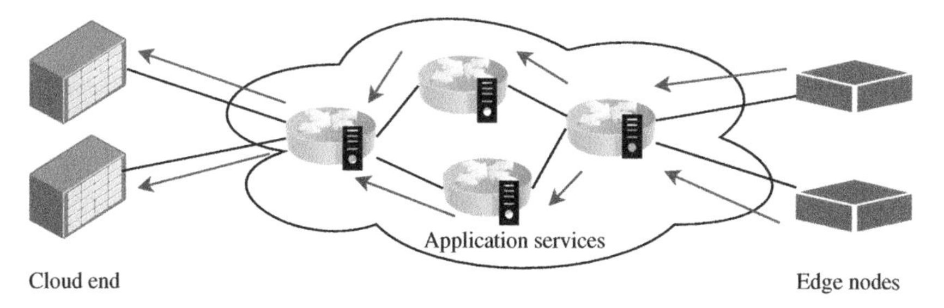

Figure 8.3 In network computing – Application Services are executed in the network unlike with traditional cloud computing where application services are performed at the cloud end.

reduction rate. SwitchAgg [285] is another in-network computing method that integrates packet load analysis and data aggregation capabilities to network switches to reduce network traffic and computational overload of servers. Similarly, in-network computing can also efficiently handle the processing of artificial Neural Networks (NNs) by offloading the computational overload to the processors of network nodes and devices [286]. This can also facilitate many of the AI-based 6G applications and use-cases. However, methods to efficiently deploy in-network computing without interfering with the processors of network devices and their operations remains to be an area for future research and development.

9

6G Radio Access Networks

In this chapter, we discuss the three key concepts of 6G Radio Access Networks (RANs), including Aerial Radio Access Network (ARAN), AI-enabled Radio Access Network (AI-RAN), and Open Radio Access Network (O-RAN). After reading this chapter, you should be able to:

- Understand the key requirements for the design of RANs in future 6G wireless systems.
- Understand three 6G RAN concepts: ARAN, AI-RAN, and O-RAN.
- Explore key features and applications of three 6G RAN concepts.

9.1 Key Aspects and Requirements

An RAN is an important part of any wireless communication system that connects end users with the other components of a network through radio communication links. RANs have evolved through several network generations: Advanced Mobile Phone System (AMPS) in 1G, Global System for Mobile Communications (GSM) in 2G, Universal Mobile Telecommunications Service (UMTS)/International Mobile Telecommunications-2000 (IMT-2000) in 3G, LTE/Worldwide Interoperability for Microwave Access (WiMAX) in 4G, and New Radio (NR) in 5G [71]. Now, due to massive Internet of Things (IoT) connection, diverse quality of service (QoS) requirements, and numerous use cases, 6G RANs are expected to support flexibility, massive interconnectivity, and energy efficiency:

- *Flexibility* allows 6G RANs to reconfigure multiple network functions, services, and resources in order to meet use cases in 6G with diverse QoS requirements, latency, reliability, and network density.

6G Frontiers: Towards Future Wireless Systems, First Edition.
Chamitha de Alwis, Quoc-Viet Pham, and Madhusanka Liyanage.
© 2023 The Institute of Electrical and Electronics Engineers, Inc. Published 2023 by John Wiley & Sons, Inc.

- *Massive interconnectivity* enables 6G RANs to support data transmission and sharing among a very large number of IoT devices while meeting key requirements, such as spectral efficiency, network density, and energy efficiency.
- *Energy efficiency* promotes more efficient energy use in 6G and increases the operation time of battery-limited devices (e.g. tiny IoT devices, sensors, aerial components, and autonomous vehicles), thus achieving cost-effective and sustainable operation.

9.1.1 Flexibility

Several aspects need to be considered to enable flexibility in 6G RANs, including frame structure, cell size, bandwidth, network virtualization and softwarization, and resource management.

- *Frame Structure*: Cyclic Prefix-based Orthogonal Frequency-Division Multiplexing (CP-OFDM) has been investigated in 5G NR to support diverse scenarios and requirements [287]. In particular, LTE uses CP-OFDM for downlink transmissions only and has only one numerology (i.e. subcarrier spacing of 15 kHz). In 5G NR, subcarrier spacing is given by 15×2^n, where $n = \{0, \dots, 5\}$ is a numerology parameter. The adoption of CP-OFDM waveform and numerology, along with other physical layer technologies, allows flexible RANs for 6G to meet more stringent requirements. Therefore, it is expected that scalable OFDM will lay the foundations for the design of the 6G physical layer and the flexibility of 6G RANs.
- *Cell Size*: 6G RAN flexibility can be achieved by cell sizing via the deployment of macrocells, small cells, tiny cells in terahertz (THz) communications and aerial base stations (e.g. UAVs, HAPs, and satellites). For example, the dynamicity of low-altitude platform (LAP)/high-altitude platform (HAP) topology implementation and the networking overlay among the LAP, HAP, and Low-Earth Orbit (LEO) communications systems help 6G ARANs to adapt flexibly to diverse requirements of terrestrial and aerial users. Moreover, as a key enabler of 6G where cell boundaries do not exist, cell-free networks help further increase RAN flexibility in 6G.
- *Bandwidth*: Bandwidth allocation plays an important role in achieving RAN flexibility. For example, an overloaded cell can borrow bandwidth resources from neighbor underloaded cells to increase its capacity to serve a large number of users. The need for efficient bandwidth allocation is further emphasized when multiple frequency bands are utilized in 6G, such as sub-6G GHz, millimeter wave (mmWave), THz, and visible spectrum. For instance, visible light communications (VLC), also known as Li-Fi, can be used to support both indoor communication scenarios and user positioning [288]. Moreover, the integration of UAVs and VLC promises to provide communication and

illumination simultaneously and further improve the network performance and flexibility [289].

- *Network Virtualization and Softwarization*: It is envisioned that network virtualization and softwarization are integral components to improve 6G RAN flexibility. Network virtualization has been adopted in 5G to enable different tenant networks to coexist in a unified network infrastructure. The authors in [290] propose a digital twin-based virtualization architecture for flexible network management. In particular, the framework consists of six layers, including data collection, abstraction-1, local processing, abstraction-2, slice-level processing, and digital twin model control. Layer 1 is responsible for collecting data from physical entities, and layer 2 fuses data from different sources and creates digital twins for individual users. Next, the output data is processed and aggregated at the edge network in layer 3. After that, the aggregated data are forwarded to layer 4 to update the digital twins of slices, which are then processed in layer 5 to create service-specific predictions and slice-level decision controls. Finally, the digital twin models are updated based on resource availability, network management, and service provisions and demands. Through this architecture, 6G RANs can combine advantages of network slicing and digital twins to enable service provisioning and achieve user-centric networking. Moreover, artificial intelligence (AI) is a promising tool for resource management optimization and performance improvement in virtual and softwarized 6G RANs.
- *Resource Management*: 6G RAN flexibility can be realized via efficient management of network resources. Indeed, resource management has played a vital role in any new network generations, from 4G to 5G and 6G, so that the network be adaptive to traffic loads, QoS requirements, and channel conditions. Furthermore, resource management becomes more important for 6G RANs due to the coexistence of multiple resources in future network systems, including radio, computation, caching, control, sensing. For example, different AI techniques (e.g. machine learning, deep learning, federated learning, and reinforcement learning) can be used to solve optimization problems related to computing and resource management in edge computing systems, mobility and handover management in ultradense 6G networks, and spectrum management in heterogeneous networks [90].

9.1.2 Massive Interconnectivity

The need to provide interconnectivity among end devices in 6G is due to the emergence of many new services, applications, and a number of exponentially increasing connected devices. It is expected that the number of connected devices will be 27 billion by 2024 and the connectivity density will be 10 million devices per square kilometer [291]. The diversity of IoT applications will also expand from simple smart home solutions to more advanced applications such as mission-critical, smart healthcare systems, fully autonomous driving, and metaverse. These application scenarios demand various performance requirements such as low latency, ultrareliability, high security, and high data rates [292]. In order to design 6G RANs for future IoT, several promising technologies can be identified, such as edge intelligence, reconfigurable intelligence surfaces, Aerial Access Networks (AANs), THz communications, massive ultra reliable low latency communication (uRLLC), and blockchain. For instance, for intelligent IoT applications, federated learning can be used to improve the data privacy of IoT devices. An example of federated learning for smart healthcare is illustrated in Figure 9.1, where homes integrate the trained global model with their local health data to have different personalized models.

9.1.3 Energy Efficiency

Due to the massive number of IoT devices and the proliferation of mobile data, energy efficiency has been an important metric for the design and development of

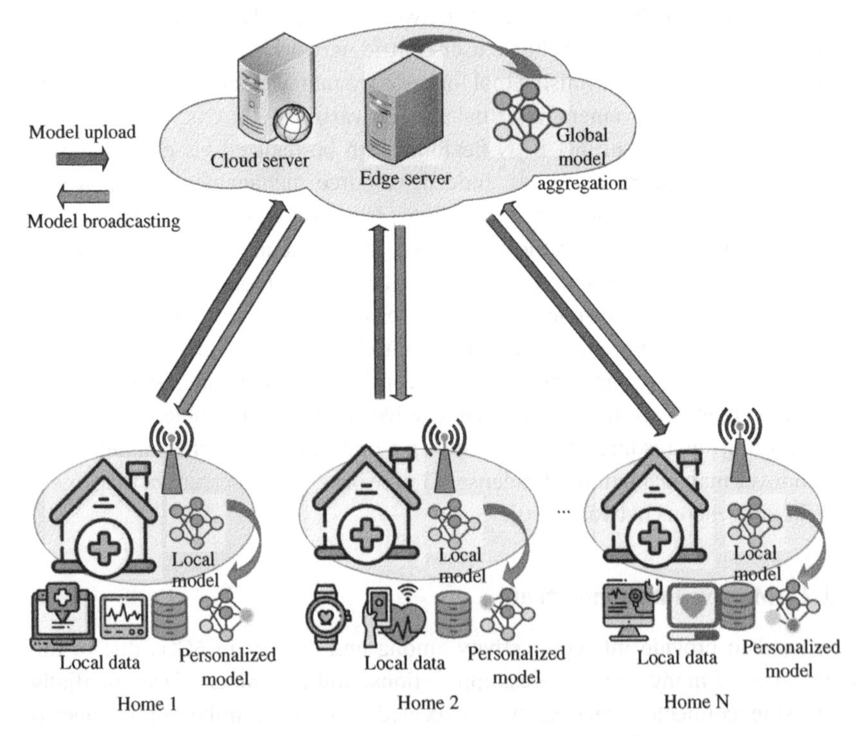

Figure 9.1 Edge intelligence with federated learning for personalized healthcare services in IoT.

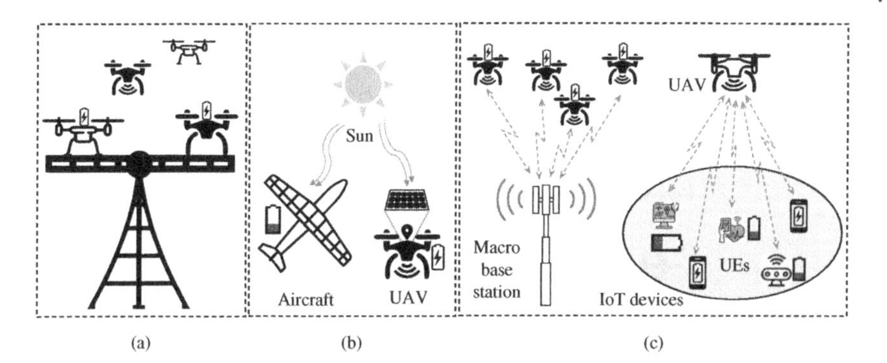

Figure 9.2 Three scenarios of energy refills in aerial access networks. (a) Charging station. (b) Charging station. (c) Wireless charging.

5G networks as well as future 6G wireless systems. Compared to 5G networks, 6G is expected to increase energy efficiency at least 10 times [76], and at the same time, increase multiple times over 5G in terms of peak data rate, user-experienced data rate, area traffic capacity, connection density, latency, spectrum efficiency, and mobility. Energy efficiency is further emphasized when 6G is expected to provide various services of communication, computing, caching, control, and sensing. For instance, mobile and IoT users in 4G typically have communication services only, while those in beyond 5G and 6G have many emerging applications, for example virtual reality (VR), augmented reality (AR), autonomous vehicles, and mobile blockchain, thus significantly increasing the amount of power and energy consumption. Moreover, energy efficiency is a key metric for new network scenarios, such as AANs, Internet of Bio-Nano Things (IoBNT), and Internet of Nano-Thing (IoNT) systems. In particular, aerial components, e.g. UAVs and aircrafts, consume a large amount of energy for hovering in the air and this amount can be many times larger than that for communication and computing purposes. In this regard, energy refilling is a primary issue in order to achieve sustainable solutions for AANs. As illustrated in Figure 9.2, aerial components can harvest energy at charging stations and self-recharge through the use of energy harvesting and wireless charging technologies. Although many studies have been conducted to improve the energy efficiency of 4G RANs and 5G RANs, there are still many challenges for the design of energy-efficient 6G RANs due to new services, network deployments, and massive connectivity and data availability.

9.2 Aerial Radio Access Networks

In the context of comprehensive 6G access infrastructure, ARANs are positioned in the aerial communication layer to serve high-altitude and terrestrial users.

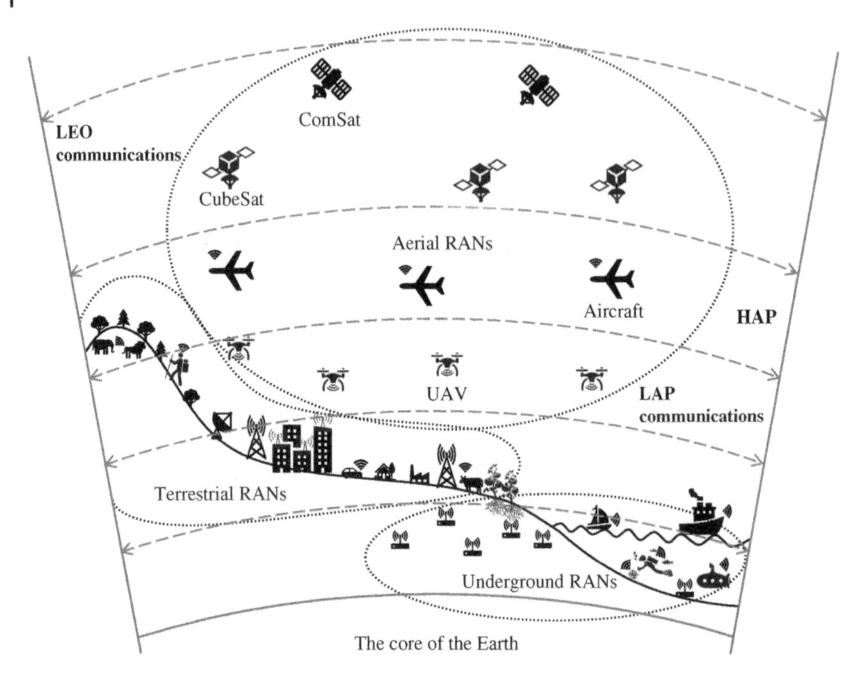

Figure 9.3 ARANs in a comprehensive 6G access infrastructure.

As illustrated in Figure 9.3, ARANs encompass three systems, including LAP communications at an altitude of 0–10 km, HAP communications at an altitude of 20–50 km, and LEO communications at an altitude of 500–1500 km above relative sea-level [122]. Compared to terrestrial RANs, underground RANs, and water RANs, ARANs serve a large coverage space with highly dynamic and mobile users for in-flight infotainment, aerial surveillance, flying vehicular control, and isolated populations.

The literature review reveals several aerial communication classes that are closely related to ARANs. Some similar terms to aerial communications that support mobile networks are summarized as follows:

- *Drone/UAV RANs (DA-RANs)*: DA-RAN stands for a drone/UAV-assisted RAN, where drones help to extend the coverage area and capacity of terrestrial access infrastructure. In this context, drones can connect either directly to terrestrial base stations or via a head node in (multihop) mesh, star, tree, and chain topologies. Unlicensed/licensed spectrum can be exploited by drones in DA-RANs [293].
- *Flying Radio Access Networks (FlyRANs)*: FlyRAN stands for FlyRAN, where aircraft and airships equipped with radio transceivers are utilized as aerial base stations to provide mobile services in underserved areas. In FlyRAN, the LAP

tier exploits the unlicensed/licensed spectrum, while the HAP tier typically exploits the licensed spectrum to avoid unmanaged conflicts with terrestrial systems [294]. The term "flying ad hoc network (FANET)" is covered by this definition [295].

- *Satellite–Terrestrial Integrated Networks (STINs)*: STIN represents a STIN by which satellite communications supplement the terrestrial systems to offer global seamless and ubiquitous Internet services to users in isolated areas. STIN utilizes the licensed spectrum for ground–air connections. The airborne infrastructure includes multiple tiers such as LEO, Medium Earth Orbit (MEO), and Geostationary Earth Orbit (GEO) satellite constellations [296].
- *ARAN*: ARAN is a multitier and hierarchical aerial access infrastructure combining the FlyRAN and LEO communication systems of STIN to provide radio access medium from the sky to the end users for the Internet services through aerial base stations equipped with heterogeneous wireless transceivers. The aerial base stations can be UAVs, drones, balloons, and airplanes. The definition of ARAN was first introduced in [297] with the original name AAN at the IEEE International Conference on Communications (ICC) in June 2020.

Briefly, these are DA-RANs focusing on LAP communications [293], FlyRANs aiming at LAP/HAP communications [294], and STINs involving satellite communications [296]. Because each of the similar terms represents a partial ARAN tier that is used for special applications, the interconnection among these systems is weak and asynchronous. Hence, the introduction of an ARAN definition is a significant contribution toward unifying the aforementioned systems into a common model.

Figure 9.4 shows a wide perspective to demonstrate a detailed ARAN architecture in the context of a complete user-core path. Typically, an ARAN architecture comprises three segments as follows:

- The main segment is a cross-tier networking infrastructure shared among aerial base stations at the three platforms, including LAP, HAP, and LEO.
- A front-end interface provides terrestrial and aerial access points that gather user connections.
- A back-end interface bridges the ARAN infrastructure to the terrestrial core networks.

Note that LEO communication systems use satellite links to contact the terrestrial network through ground stations, while LAP and HAP systems use mobile (wireless) links to contact aerial base stations and terrestrial base stations (e.g. gNBs in 5G and eNBs in 4G) directly.

In the main segment of an ARAN, LAP systems are at the lowest tier. Drones and UAVs in LAP systems act as aerial base stations, providing connectivity directly to

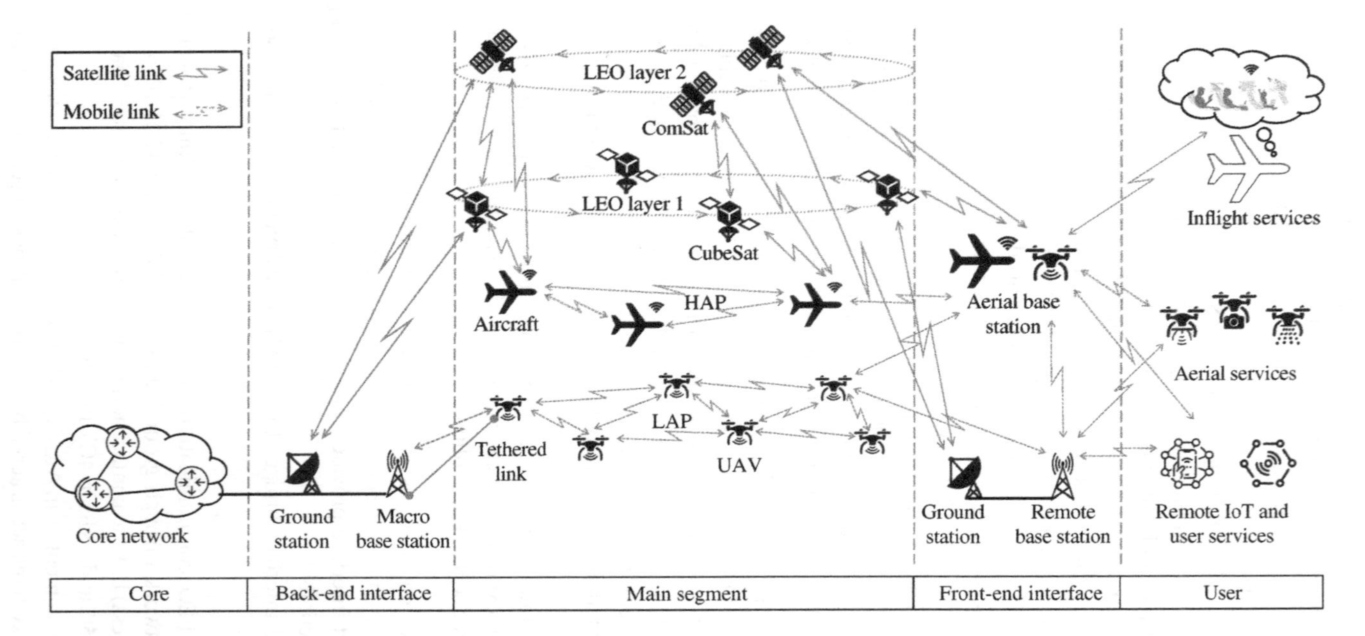

Figure 9.4 System architecture of ARANs.

aerial and terrestrial end users or remote base stations using wireless technologies such as 5G NR and Wi-Fi. Conversely, these aerial base stations may utilize either satellite technology to connect with LEO communication systems or wireless technologies to the terrestrial macro base stations for backhaul transmission to the core networks. LAPs can be classified by size, range, speed, and endurance. According to the US Department of the Army [298], LAPs are classified into five categories, i.e. small < medium < large < larger < largest groups. Most LAPs are relatively lightweight and cost-effective devices that can be deployed quickly and flexibly. However, they have relatively low endurance with limited energy and networking resources. To mitigate these issues, tethered technology can be considered a feasible approach to either establish a reliable broadband backhaul through a fiber cable or energize LAPs by a powerline connection between the LAPs and terrestrial stations [299]. Nevertheless, LAPs effectively support time-sensitive and event-based scenarios such as emergency and rescue, aerial surveillance, traffic offloading, and mobile hotspots at public gatherings.

In the middle tier of the ARAN main segment, HAPs are defined by the ITU Radio Regulations (RR) as radio stations located at a specified, nominal, fixed point relative to the Earth.[1] The 2, 6, 27/31, and 47/48 GHz frequency bands were assigned for HAP communications in bidirectional HAP-terrestrial links [122] at three world radio communication conferences (WRC-97, WRC-2000, and WRC-12). HAP systems serve aerial and remote terrestrial end users with wireless technologies in a wide coverage area from high altitude. For backhaul transmission to terrestrial core networks, HAP systems mostly utilize satellite technology via the LEO communication systems. In practical situations, some industries have implemented trial projects using lightweight solar-powered aircraft and airships to provide stable broadband services in rural and remote areas [300].

At the top of the ARAN main segment, two-layer LEO communication systems consist of miniaturized satellites below (i.e. CubeSats) and communication satellites above (i.e. ComSats), orbiting at an altitude of 500–1500 km. CubeSats aim to provide low latency and high throughput Internet services, while ComSats are designed for high coverage and service availability. These two LEO layers interact with each other through interlinks, which provide redundancy, backup, and collaboration interfaces for dynamic network organization. Compared with other satellite classes, LEO satellites are characterized by lower latency, cost-efficiency, and quick production and deployment. Most LEO satellites operate on Ku, Ka, and V bands to provide Internet connection to aerial and ground stations within tens of ms latency [301]. Unlike the LAP/HAP systems, LEO communication systems typically do not provide services to the end users directly. Both fronthaul

1 https://www.itu.int/en/mediacentre/backgrounders/Pages/High-altitude-platform-systems.aspx

to the end users and backhaul to the core networks are satellite transmissions through ground stations and very small aperture terminals (VSATs). To orchestrate intertier networking operations in an ARAN, LEO communication systems additionally support backhaul tunnels, allowing LAP/HAP systems to connect to the core networks. Recent years have witnessed several emerging commercial satellite projects on LEO communication systems such as OneWeb, Telesat, and Starlink, with hundreds of satellites successfully launched into orbit and thousands of satellite launches planned for the near future [302].

In ARANs, the front-end interface includes aerial base stations at the LAP/HAP tiers, remote base stations, and ground stations, enabling the end users to access the networks. The aerial base stations and remote base stations are ARAN access points that provide wireless links directly to aerial and terrestrial users. Meanwhile, the ground stations are end-points of satellite links from LEO communication systems that deliver traffic to and from the remote base stations. Depending on prevailing circumstances, the aerial base stations and remote base stations may cross-serve both aerial and terrestrial services, as illustrated on the left side of Figure 9.4. Conversely, the back-end interface includes macro base stations and ground stations to accommodate backhaul traffic toward the core networks. Typically, the macro base stations help transfer traffic flows from LAP systems, while the ground stations support traffic forwarding from (HAP systems to) LEO communication systems; see the right side of Figure 9.4.

9.3 AI-enabled RAN

It is quite challenging for conventional RANs to achieve stringent requirements of future 6G wireless systems, such as peak data rate of more than 1 Tbps, end-to-end delay of less than 0.1 ms, reliability of 99.99999%, connection density of more than 10^7 devices per square kilometer, and mobility of more than 1000 kmph. To tackle the existing challenges and open more capabilities for future RANs, AI has been emerged in RANs to create a new concept, namely AI-enabled RANs. At the latest 3GPP RAN plenary meeting, 3GPP reached an agreement on including AI for 5G NR, which will have a great impact on the development and standardization of AI-native 6G RANs in the near future (around 2025). In the following, we first present a three-tier computing architecture that allows AI to be deployed in RANs in a hierarchical manner. Then, we analyze some promising applications of AI for future RANs, including data processing and resource allocation.

With the emergence of cloud computing, edge computing, and on-device intelligence, storage resources and computing capabilities can be provided anywhere in the network. More specifically, the network is typically composed of three layers,

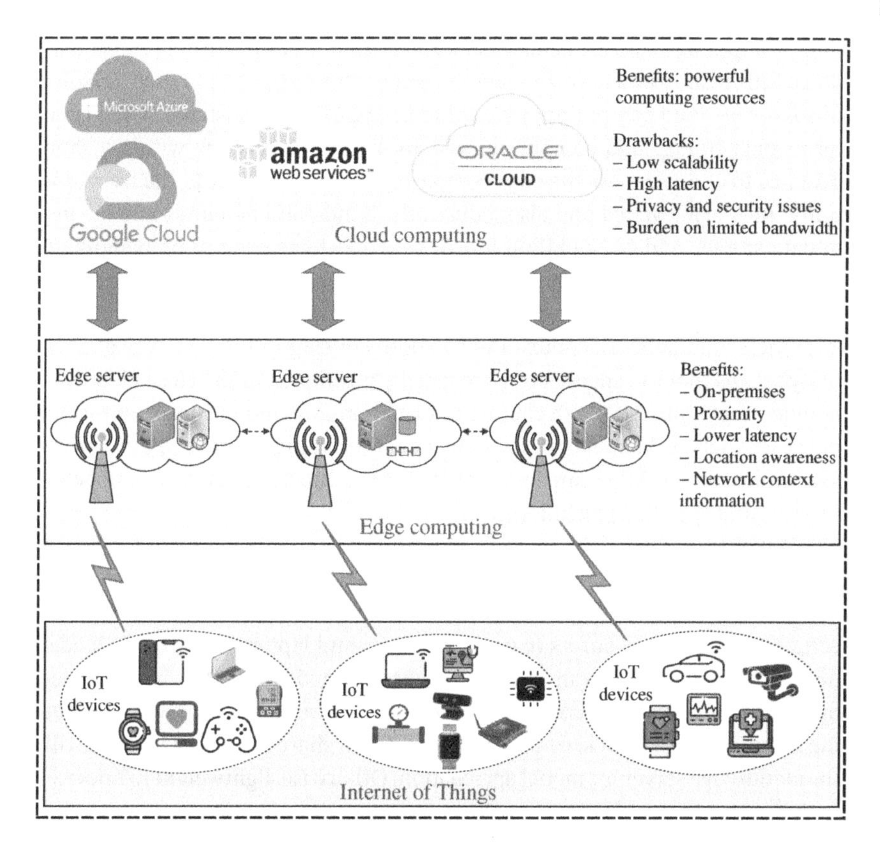

Figure 9.5 Illustration of a three-tier computing infrastructure, including cloud computing, edge computing, and IoT device computing. Source: Apple Inc.

including cloud computing layer, edge computing layer, and IoT device layer, as illustrated in Figure 9.5:

- *Cloud Computing Layer*: This layer includes cloud computing servers, such as Amazon Web Services, Google Cloud Platform, and Microsoft Azure, all of which have powerful computing and memory resources. Thus, this layer can provide an abundant amount of resources for centralized learning algorithms with massive data. However, cloud computing suffers from major disadvantages, such as low scalability, high latency, privacy and security issues, and extreme burden on limited bandwidth. For example, the privacy and security issues caused by the high concentration of data information leave the cloud highly susceptible to violent attacks, and data/application offloaded to the cloud through wireless environments can be overheard by eavesdroppers [71].

- *Edge Computing Layer*: It is necessary for computation nodes to perform data processing near to end users in order to decrease the transmission time. To solve this issue of cloud computing, edge computing came into existence, which performs data storage and computing tasks in their edge network within a short distance to end users. As the edge computing nodes are closer to the users, the traffic flow is alleviated and also reduces the bandwidth demands and latency in data storage and computation in IoT network. Edge computing paradigms, e.g. fog computing and multi-access edge computing (MEC), can be characterized by some features, namely, on-premises, proximity, lower latency, location awareness, and network context information. For example, the feature of proximity means that the edge servers are usually positioned in the close vicinity of mobile users; thus, MEC can capture information from mobile users for further purposes such as data analytics and big data processing. Moreover, this feature also indicates that MEC can access mobile users directly, thus providing better services and specific applications.
- *IoT Device Layer*: IoT devices, such as sensors and mobile phones, can generate or gather data from the physical environments and then transmit it to the nearby edge servers via access points or base stations [264]. IoT devices with certain computing resources (e.g. smartphones and laptops) can act as an edge node to process local data and train local learning models. This can be illustrated by the federated learning concept in Figure 9.1, where IoT devices with computing resources train local learning models and then share the model updates with the cloud/edge server for model aggregation. Otherwise, lightweight IoT devices with little/no computing resources (e.g. tiny sensors and backscatter nodes) can participate in the learning process by offloading their data and computing workload to the edge and cloud computing servers via their representative gateways.

Given the availability of computing resources, ranging from the IoT devices to the edge network and the centralized cloud, AI mechanisms can be deployed to optimize the operation, management, and performance of future RANs. Different AI techniques (e.g. machine learning, deep learning, reinforcement learning, and federated learning) may have numerous applications, such as network control, network security, network applications, signal processing, cyber-attack detection, caching and computation offloading, resource allocation, and network management. Due to its learning capabilities, computing resource availability, and massive data, AI-based solutions have distinct advantages over conventional model-based schemes in RANs. It is believed that AI has three main use cases in RANs as discussed below [269].

Algorithmic Approximation: A common limitation preventing algorithms from finding the optimal solution is the difficulty of real-time executions;

therefore, they are impractical for real-time implementation. Several approaches, e.g. heuristics, metaheuristics, and problem decomposition, have been proposed to optimize the trade-off between computational complexity and performance. However, the real-time implementation of the underlying algorithms is quite challenging. For this case, the use of AI techniques appears to be a promising solution. In particular, the data generation and training phases can be executed offline while the system operates in real time by using the trained model. For instance, the work in [131] proposes a DL architecture for automatic modulation classification (AMC), namely, MCNet, which was 93.59% accurate at a signal-to-noise ratio (SNR) of 20 dB with an inference time of only 0.095 ms.

Unknown Model and Nonlinearities: Many physical phenomena cannot be accurately modeled. Therefore, conventional model-based algorithms usually fail to obtain efficient solutions. For instance, fiber nonlinearities (e.g. signal distortion and self-phase modulation) in optical systems together with the adoption of coherent communication render model-based methods ineffective for network optimization. To mitigate the nonlinearities and perform signal detection, AI techniques (e.g. an end-to-end learning approach) can be utilized with very low bit error rates (BERs). The end-to-end learning approach [303] has found many applications in scenarios in which the channel model is unknown or well-established mathematical models are unavailable. Another application is the use of DL to address hardware nonlinearities in multiple-input and multiple-output (MIMO) systems, such as hardware impairments. Nonlinearity was also observed in MIMO systems with low-bit analog-to-digital converters. In an attempt to mitigate this nonlinear effect, the work in [304] proposes a deep neural network model to jointly optimize the channel estimation and training signal. The model outperformed the linear channel estimator in various practical settings.

Algorithm Acceleration: Another direction intelligent signal processing has been taking is to use AI to facilitate and accelerate existing algorithms. This approach differs markedly from the two scenarios discussed above in that an existing model-based algorithm is completely replaced by an AI-based algorithm, i.e. an end-to-end learning paradigm. For instance, many DL-based algorithms have been proposed to improve and accelerate near-optimal detection schemes. The work in [305] proposes a DL model, namely, FS-Net, to initialize the highly reliable solution for the tabu search (TS) detection scheme, and also proposed an early termination scheme to further accelerate the optimization process. Compared with the original TS scheme, the DL-aided TS detector can reduce the computational complexity by approximately 90% at an SNR of 20 dB with similar performance.

9.4 Open RAN

Open RANs, also known as O-RANs, have been considered a promising RAN concept for future 6G wireless systems. O-RAN is centered on the concept of openness and intelligence of the network elements. Therefore, O-RAN is considered a promising RAN technology to overcome the limitations of existing RANs, such as virtual RAN and cloud RAN, as illustrated in Figure 9.6. Many believe that O-RAN has numerous benefits compared with the existing RAN technologies. In particular, traditional RANs largely rely on proprietary hardware, software, and radio interfaces. The feature of openness is to allow the involvement of smaller and new players in the RAN market to deploy their customized services, while the feature of intelligence is to increase the automation and performance through optimizing the RAN elements and network resources. Moreover, O-RAN offers many RAN solutions and elements to the network operators so that the network will be more open and flexible. Furthermore, because of the virtualization feature, the network operators can shorten the time-to-market of new applications and services so as to maximize the overall revenue.

There are two major O-RAN organizations, including the Telecom Infra Project (TIP) and the O-RAN alliance. The TIP's O-RAN program is an initiative that focuses on developing solutions for future RANs based on disaggregation of multivendor hardware, open interfaces, and software.[2] Besides O-RAN, the TIP also runs several projects with the main aim of providing Internet and wireless access over the globe. In particular, the TIP's product project group divides the

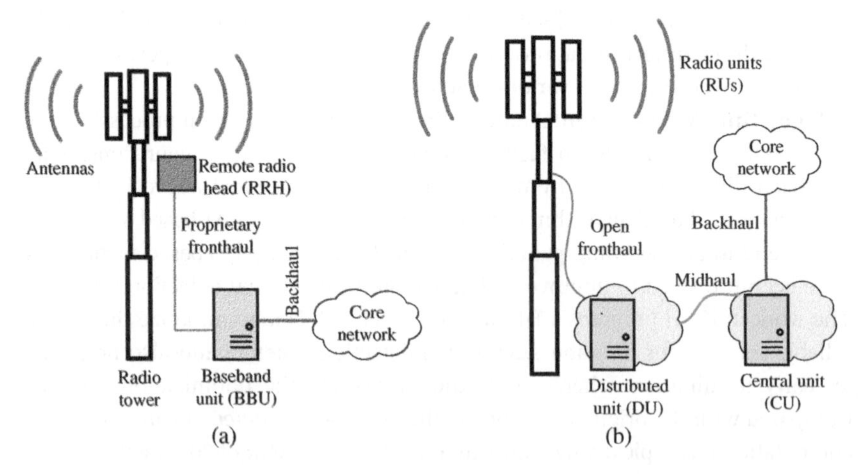

Figure 9.6 A high-level illustration of traditional RAN (a) and O-RAN (b).

2 https://telecominfraproject.com/openran/

end-to-end network into three segments, including access (e.g. fixed broadband and O-RAN), transport (e.g. nonterrestrial network, optical communication, and wireless backhaul), and core and services (e.g. end-to-end network slicing and open core network). The TIP also establishes a solution project group to codify network elements and technologies so that the network solutions can be used by different operators, service providers, and stakeholders. Further, the TIP has a software project group to develop open-source solutions for open and interoperable 6G RANs. The other O-RAN organization is the TIP, whose purpose is to develop more competitive RAN solutions, which are characterized by several principles of openness, intelligence, virtualization, and interoperability.[3]

The high-level architecture of O-RAN is represented in Figure 9.7. In terms of the O-RAN interface, there are four main interfaces to connect these elements, including A1, O1, open fronthaul M-plane, and O2. More specifically, the open fronthaul M-plane interface is to connect Service Management and Orchestration Framework (SMO) ad O-RAN radio unit (O-RU). A1 is to connect nonreal-time RAN intelligent controller (RIC) located in the SMO framework and near real-time RIC for RAN optimization. O1 is to support all O-RAN network functions when they are connected with SMO, and O2 is to connect SMO and O-Cloud for providing cloud computing resource and workflow management. According to [306], there are different deployment scenarios of the O1 interface, such as flat, hierarchical, and hybrid models, by which the SMO framework can provide numerous

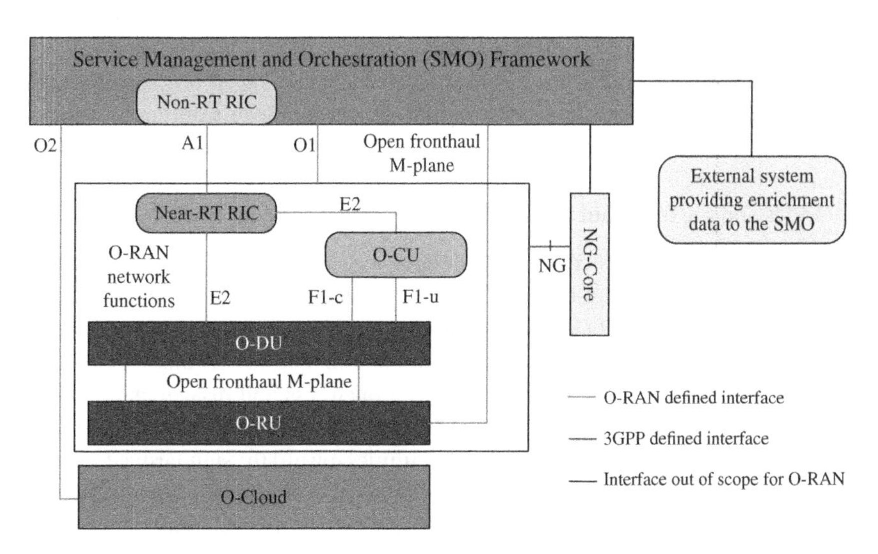

Figure 9.7 The high-level architecture of O-RAN proposed by the O-RAN alliance.

3 https://www.o-ran.org/

management services, for example provisioning management services, trace management services, and performance management services.

The main elements of the O-RAN architecture include SMO, RIC, O-Cloud, O-RAN central unit (O-CU), O-RAN distributed unit (O-DU), and O-RU.

- *SMO*: The SMO framework is a core component of the O-RAN architecture, whose main responsibility is to manage the RAN domain, such as the provision of interfaces with network functions, near-real-time RIC for RAN optimization, and O-Cloud computing resource and workload management. As mentioned earlier, these SMO services can be performed through four interfaces, including A1, O1, O2, and open fronthaul M-plane.
- *RIC*: This logical function enables O-RAN to perform real-time optimization of functions and resources through data collected from the network and end users in real time (between 10 ms and 1 s).
- *O-Cloud*: This is a physical computing platform that can be used for O-RAN functions, such as near-real-time RIC, O-CU control plane, O-CU user plane, and O-DU.
- *O-DU*: This logical node has functionalities of the physical and MAC layers. This element terminates the E2 with F1 interfaces.
- *O-CU*: This is a logical node in the O-RAN architecture and hosts all the functions of both the control plane and data plane. These two O-CU planes connect with the O-DU logical node via the F1-c interface and F1-u interface, respectively. For low latency services, O-CUs can be collocated with the O-DU at the same edge node.
- *O-RU*: This logical node has physical layer and radio signal processing capabilities to connect with the SMO framework via the open fronthaul M-plane interface and connects with end users via radio interfaces. Therefore, users can offload their computation tasks/data to the edge node via wireless links to O-RUs and the fronthaul that connects O-RUs and the edge node.

The benefits of O-RAN come at challenges, e.g. security, deterministic latency, and physical layer and real-time control. For example, the work in [307] discusses six security challenges of O-RAN, including disabled over-the-air ciphers, unauthorized access, unprotected synchronization, vulnerabilities among O-RAN elements, lack of security measures for O-RAN openness and disaggregation, and open-source software security threats. All these challenges demand significant effort from the research and industry communities toward the standardization and implementation of O-RAN in future 6G wireless systems. In this regard, O-RAN has distinct features that bring intelligence to future networks, while AI, particularly deep learning and machine learning, helps overcome various challenges of 6G O-RANs via intelligent and data-driven solutions [308], such as resource management, mobility management, and spectrum management.

Part III

Technical Aspects

10

Security and Privacy of 6G

6G is envisaged to rely on the advancements of AI and data analytics to provide personalized and fully automated seamless communication services. However, this may lead to several security and privacy issues and concerns. After reading this chapter, you should be able to

- Understand the security threat landscape of future 6G networks.
- Understand key 6G security and privacy requirements.
- Identify the possible security solutions which can be implemented in 6G network.
- Understand key 6G privacy challenges and possible solutions.

10.1 Introduction

Irrespective of the advancements of networking and communication technologies, security is always an important feature to consider to ensure the resilience and reliability of networks. Therefore, it will be helpful for the research community to identify the security-related research directions in the envisioned 6G networks. Since the standard functions and specifications of 6G are yet to be defined, there is still limited literature that provides security and privacy insights beyond 5G networks. Furthermore, it is necessary to systematically build on 5G research and consolidate existing emerging research toward 6G security realization.

The security and privacy considerations in the envisioned 6G networks need to be addressed in many areas. There are specific security issues that may arise with the novel 6G architectural framework, as stated above. In addition to that, there are many hypes on blending novel technologies such as blockchain, visible light communication (VLC), TeraHertz (THz), and quantum computing features

6G Frontiers: Towards Future Wireless Systems, First Edition.
Chamitha de Alwis, Quoc-Viet Pham, and Madhusanka Liyanage.
© 2023 The Institute of Electrical and Electronics Engineers, Inc. Published 2023 by John Wiley & Sons, Inc.

in 6G intelligent networking paradigms in such a way to tackle the security and privacy issues. Therefore, 6G security considerations need to be also discussed concerning the physical-layer security (PLS), network information security, application security, and deep learning-related security [309, 310].

10.2 Evolution of Mobile Security

The early generations of mobile networks (i.e. 1G, 2G, and 3G) encountered significant security and privacy challenges, including cloning, illegal physical attacks, eavesdropping, encryption issues, authentication and authorization problems, and privacy issues [39]. Then, the security threat landscape evolved with more-advanced attack scenarios and powerful attackers. The evolution of the security landscape of telecommunication networks, from 4G toward the envisioned 6G era, is illustrated in Figure 10.1. 4G networks faced security and privacy threats mainly due to the execution of wireless applications. The typical examples include media access control (MAC)-layer security threats (e.g. denial of service (DoS) attacks, eavesdropping, replay attacks), and malware applications (e.g. viruses, tampering into hardware).

In the 5G architecture, security and privacy threats are caused at access, backhaul, and core networks [311]. Cyberware and critical infrastructure

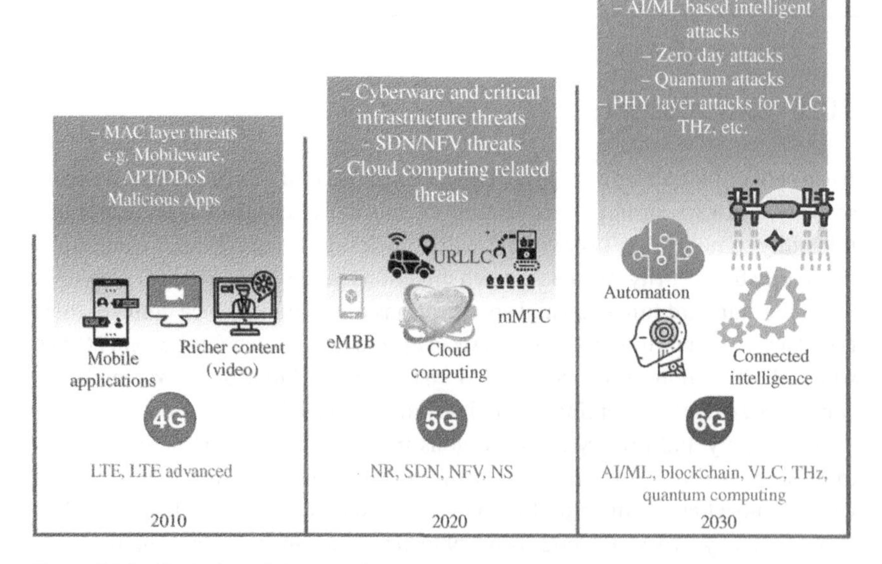

Figure 10.1 Evolution of the security landscape in telecommunication networks.

threats, network functions virtualization (NFV) and software-defined networking (SDN)-related threats, and cloud computing-related threats are the most common security issues in 5G [312]. There are numerous occasions that SDN may create security threats, such as by exposing critical application programming interfaces (APIs) to unintended software, the inception of OpenFlow, and centralizing the network control (i.e. subject to DoS attacks) [201]. Above all, the most significant driving force in the 6G vision is the added connected intelligence in the telecommunication networks accompanied by advanced networking and AI/ML technologies. However, the alliance between AI and 6G may also be a double-edge sword in many cases while applying for protecting or infringing security and privacy [313].

10.3 6G Security Requirements

In this section, we discuss the security considerations, 6G security vision, and the potential security Key Performance Indicators (KPIs). The last subsections describe the security landscape for the envisioned 6G architecture which is classified into four key areas such as functional architecture (i.e. intelligent radio (IR) and radio-core convergence), edge intelligence (EI) and cloudification, specialized subnetworks, and network management and orchestration.

10.3.1 6G Security Vision and KPIs

The vision of 6G networks is formed with many novelties and advancements in terms of architecture, applications, technologies, policies, and standardization. Similar to the generic 6G vision, which has the added intelligence on top of the cloudified and softwarized 5G networks, 6G security vision also has a close fusion with AI, which leads to security automation (Figure 10.2). At the same time, the adversaries also become more powerful and intelligent and capable of creating new forms of security threats. For instance, detecting zero-day attacks is always challenging, whereas prevention from their propagation is the most achievable mechanism. Therefore, the necessity will become more important than ever to incorporate intelligent and flexible security mechanisms for predicting, detecting, mitigating, and preventing security attacks and limiting the propagation of such vulnerabilities in the 6G networks. It is also equally significant to ensure privacy and trust in the respective domains and stakeholders. Especially, security and privacy are two closely coupled topics, where security relates the safeguarding of the actual data, and privacy ensures the covering up of the identities associated with those data. While security on its own is exclusive from privacy, vice versa is not

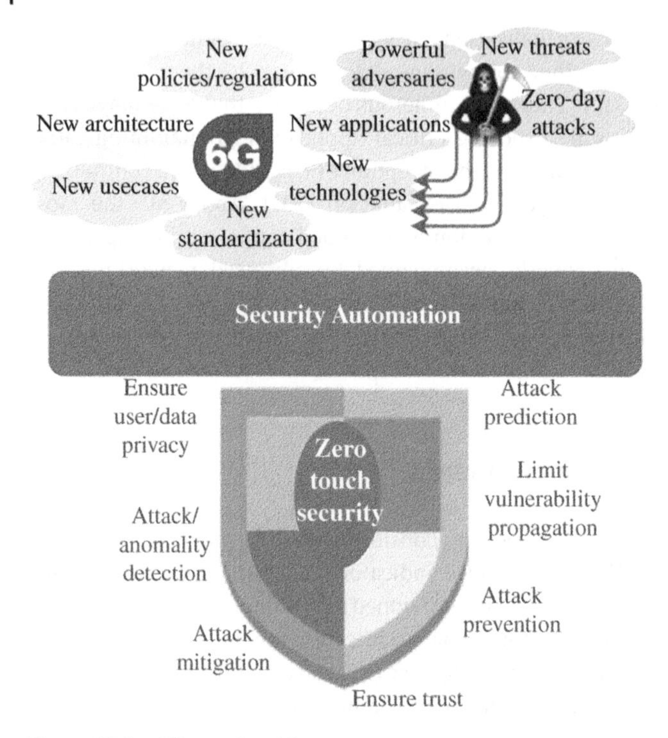

Figure 10.2 6G security vision.

valid: essentially, to assure privacy, there should always be security mechanisms that protect data. In the coming sections, we discuss how security and privacy complement each other for different aspects of 6G.

To set the scope of 6G, we also think that KPIs and Key Value Indicators (KVI) will help to take the dimensions of impact that go beyond the scope of deterministic performance measures into full account [19]. It is expected that 6G systems will incorporate novel aspects, such as integrated sensing, artificial intelligence, local compute-and-storage, and embedded devices [314]. These aspects will both lead to enhancements to existing KPIs, as well as require a whole new set of KPIs and KVIs which have not traditionally been associated with mobile networks, such as sensing accuracy, computational round-trip-time, and AI model convergence time. The KVIs will quantify the value of the new 6G-related technologies from the perspective of sustainability, security, inclusiveness, and trustworthiness stemming from the UN sustainable development goals [315, 316].

Therefore, we believe that the new aspects will have a significant impact on how security KPIs are designed and measured (as shown in Table 10.1). Various aspects should be considered for characterizing security, such as PLS, network information security, and AI/ML-related security [309].

Table 10.1 Security KPIs and 6G vision.

KPI	Description	6G impact
Protection level	The guaranteed level of protection against certain threats and attacks	More stringent due to the pervasive utility of 6G and burgeoning risk level
Time to respond (mean, max, etc.)	Time for security functions to counteract in case of malicious activity	Much smaller due to compressed timescale of 6G networks, e.g. an attack can cause havoc at an order or faster
Coverage	The coverage of security functions over the 6G service elements and functions	More challenging due to diverse 6G technologies and ultra-distributed functions
Autonomicity level	A measure of how autonomic security controls can act	Expected not only to be easier to implement with pervasive AI but also may be counter-beneficial due to AI security issues
AI robustness	The robustness of AI algorithms in the network hardened for security	More difficult to maintain consistently system-wide but more critical due to AI's role in 6G
Security AI model convergence time	Time for learning models working for security to converge	Although more advanced AI/ML models are emerging and hardware capabilities are improving, the data availability and complexity are challenging factors for this KPI.
Security function chain round-trip-time	Time for chained security functions to process for ingest, analyze, decide, and act (related to "Time to respond" KPI)	Security architecture in 6G supposed to be more distributed, leading to challenges. But at the same time, device-centric and edge-centric solutions will help.
Cost to deploy security functions (mean, max, etc.)	Various cost metrics for measuring the cost of deployment	Substantially increases due to complexity, thus harder to meet target KPI values

10.4 Security Threat Landscape for 6G Architecture

Undoubtedly, the massive emergence of connections in the future 6G networks will increase security and privacy vulnerabilities. Considering the foreseen technological, architectural, and application-specific aspects and their advancements in the future 6G networks, the threat landscape of 6G security is summarized in

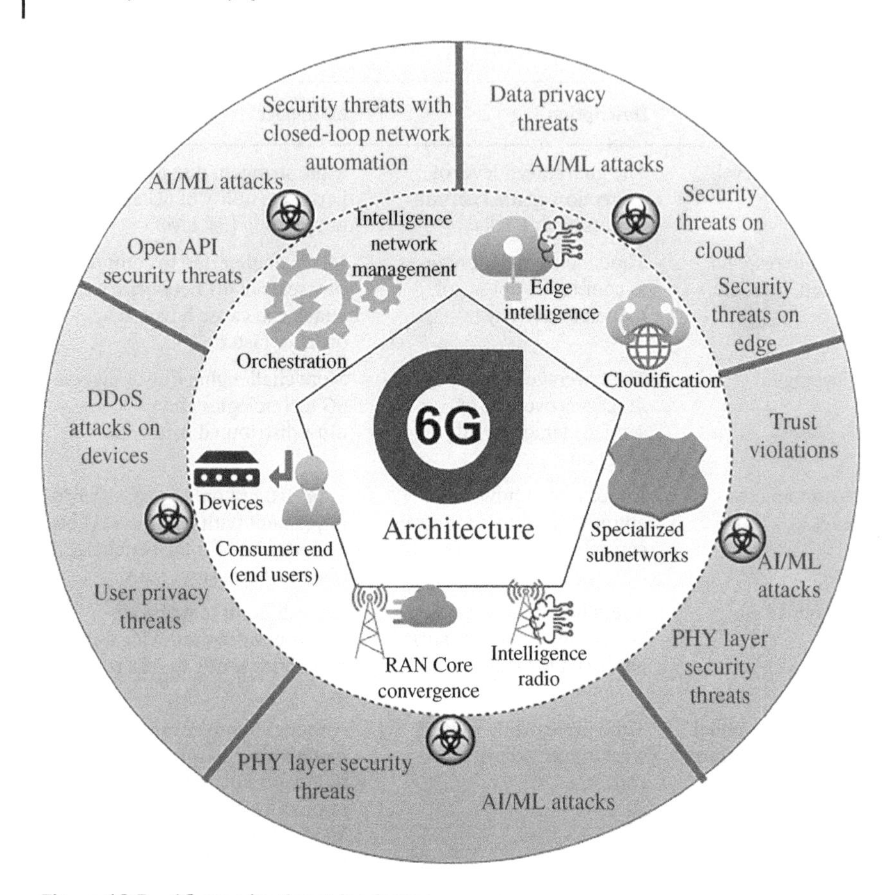

Figure 10.3 6G security threat landscape.

Figure 10.3. Since the attacks can be generalized based on the architecture rather than the technologies or the applications, we are taking this step forward to give the reader an insight into the security threat landscape on top of the envisioned 6G architecture.

Among various visionary 6G architectures proposed by the industrial and academic research community, We have identified the vision from Nokia Bell Labs as a realistic yet ambitious proposal to facilitate our security landscape analysis for 6G architecture [57]. As stated by Ziegler *et al.* in [57], after investigating the potential 6G architectural innovation, they decompose the data and information architecture into four segments, namely, *platform*, *functions*, *orchestration*, and *specialization*. In the infrastructure "platform" of 6G architecture, heterogeneous clouds need to create an agnostic, open, and scalable run-time environment to accelerate the hardware and improve data flow centrality. The "functional" architecture component includes the topics such as RAN core convergence and IR. The "specialized" part represents the architectural enablers of flexible off-load, subnetworks, and extreme slicing. The "orchestration" component includes the intelligent network management and the cognitive closed-loop and automation of 6G networks. In the rest of the section, we discuss the security considerations of these four 6G architectural components and how they are related at the consumer end.

However, in addition to the 6G architectural evolution, the advent and advancements of technologies may also pave the way to generate more powerful attackers who can create sophisticated attacks. For instance, while detecting AI-based malicious activities, distributed learning-based attack prediction methods offer promising solutions within the constantly changing environments [309].

10.4.1 Intelligence Radio and RAN-Core Convergence

The recent advances in state-of-the-art circuits, antennas, meta-material-based structures, and the dramatic evolution of AI techniques, including ML, data mining, and data analysis, have shed light on a novel path for the challenges expected in radio networks towards 6G. In this sense, providing intelligence beyond the already known intelligent spectrum access for cognitive radio networks is interesting for addressing novel radio network challenges. Thus, the envisioned IR will involve cutting-edge AI/ML techniques to address accurate channel modeling and estimation, modulation, beamforming, resource allocation, optimal spectrum access, automated network deployment, and management. Hence, the introduction of IR toward 6G will reduce implementation time and significantly reduce the cost of new algorithms and hardware [317]. With all these promising benefits of IR, security and privacy are becoming increasingly critical in wireless networks, especially for the increasing demands for mission-critical services. For example AI training can be manipulated in a spectrum access system by inserting fake signals so that a malicious party can take advantage of a large portion of the spectrum by denying the spectrum to other users. Also, attacks through the wireless channel, such as denial-of-service, spoofing, and malicious data injection, could affect the AI. Therefore, efficient detection of malicious training is critical for the proper performance of IR [318].

Besides, new network architecture paradigms are expected for 6G by harmonizing RAN and core functions. Given that different core functions are being distributed and virtualized to be implemented closer to RAN, which benefits low-latency services. At the same time, higher-layer RAN functions are being centralized, and RAN-Core functions can be combined (RAN-Core convergence) to simplify the network and facilitate the implementation of some services [319]. Thus, this convergence's security and privacy challenges and opportunities should be addressed toward 6G.

10.4.2 Edge Intelligence and Cloudification of 6G Era

The union between AI and edge computing is instinctive since there is a close interaction [320]. In specific 6G wireless applications, it is imperative to shift the computation toward the edge of the network. Whether AI/ML algorithms are used to acquire, store, or process data at the network edge, it is referred to as EI [321]. In EI, an edge server aggregates data generated by multiple devices that are associated with it. Data are shared among numerous edge servers for training models and later used for analysis and prediction. Thus, devices can benefit from faster feedback, reduced latency, and lower costs while enhancing their operation. However, as data is collected from multiple sources, and the outcome of AI/ML algorithms is highly data-dependent, EI is highly prone to several security attacks. Under such circumstances, trust is also required in EI services which are critical to ensure user authentication and access control, model and data integrity, and mutual platform verification [20]. In [322], it is demonstrated how Blockchain is used to secure distributed edge services to prevent resource transactions vulnerable to malicious nodes. Blockchain ensures the consistency of decomposed tasks and the chunks of learning data required in AI implementation.

Attackers can exploit the distributed nature and the respective dependencies on edge computing to launch different attacks such as data poisoning, data evasion, or a privacy attack, thus affecting the outputs of the AI/ML applications and undermining the benefits of EI [323]. Moreover, EI may require novel secure routing schemes and trust network topologies for EI service deliveries. Security in EI is closely coupled with privacy since the edge devices may collect privacy-sensitive data which contain user's location data, health, or activities records, or manufacturing information, among many others. Federated learning is one approach for privacy-friendly distributed data training in edge AI models, enabling local ML models. In addition to that, secure multiparty computation and homomorphic encryption (HE) for designing privacy-preserving AI model parameter-sharing schemes in EI services are also considered by researchers.

The fundamental architectural change in 5G, which has a cloud-native and microservice architecture, is expected to evolve with heterogeneous aspects in the cloud transformation toward 6G [57]. The heterogeneous clouds related to numerous service delivery platforms, including public, private, on-premises, and edge cloud may require proper coordination of communication resources and distributed computing through orchestration and network control. The security considerations may also differ based on the nature of each cloud environment and the stakeholders. Mainly the most common security issues include the violation of access control policies, data privacy breaches, information security issues, insecure interfaces and APIs, DoS attacks, and loss of data [324].

10.4.3 Specialized 6G Networks

As introduced in [57], the trend of having vertical industries in 5G for industrial automation will continue to 6G as subnetworks. These specialized 6G networks are expected to operate as stand-alone miniaturized networks for multiple application verticals (e.g. in-body, in-car, in-robot, subnetwork of drones). When the wireless interfaces enable subnetwork owners or infrastructure to use novel applications, those external communication interfaces may impose security vulnerabilities. To avoid unauthorized persons remotely taking control of the subnetwork functions, it will be essential to use strong and lightweight authentication and encryption algorithms together with methods for monitoring network security employing intrusion detection systems. Hierarchical and dynamic authorization mechanism will be more suitable to handle trust boundaries between the large networks and the miniaturized subnetworks. Use of trusted execution environments (TEE) may also guarantee the confidentiality and integrity of such closed subnetwork environments.

10.4.4 Intelligence Network Management and Orchestration

The extreme range of 6G requirements such as massive demand for increased capacity, extremely low latency, extremely high reliability, and support for massive machine-to-machine communication will demand a radical change in network service orchestration and management in 6G. With the support of AI, the new 6G architecture is expected to offer intelligent end-to-end automation of network and service management. The upcoming ETSI ZSM (Zero-touch network and Service Management) [325] architecture is paving the path toward such intelligence network management deployment beyond the 5G network. Below we discuss the key security challenges in such intelligence network management deployments under three aspects and summarize in Table 10.2.

Open API's security threats: 6G network is expected to support open APIs by continuing the trend developed in 5G networks [196, 326]. There are mainly three variants of open API attacks we identify in the current literature. (i) Parameter attacks lead to unauthorized exploitation of the data transferred through the API. The improper validation of API parameters may also lead to inject attacks on cross-domain data services. (ii) Identity attacks allow the attackers to exploit flaws in the authentication and authorization process. For instance, extraction of API keys and using them as credentials can result in identity-based attacks. Moreover, the unencrypted transmission of API messages may lead to (iii) man-in-the-middle attack. An attacker can intercept the unencrypted API messages and capture confidential information. In addition, these open APIs can also be vulnerable to DoS/distributed denial-of-service (DDoS) attacks. Here an

Table 10.2 Security challenges in intelligence network management and orchestration of 6G networks.

Aspect	Issue	Description	Solutions
Open API's security threats [196, 326, 327]	Parameter attacks	– Improperly validated parameters may lead to injection attacks on cross-domain data services. – Data injection, data manipulation, and logic corruption. – Manipulating network topology data to insert fake links, malicious nodes. – Continuous injection of false parameters may lead DoS attack to make the data services unresponsive.	– Input validation and user authentication. – Access control and rate limiting.
	Identity attacks	– Exploit flaws in authentication and authorization. – Extraction of API keys and using them as credentials. – Attack insecure E2E domain orchestration service to change configurations to fail SLAs, create new instances demanding more resources to exhaust the network.	– Authentication (signed JWT tokens, OpenID connect) – Authorization (Role-based access control, attribute-based access control, access control lists)
	Man-in-the-middle attack	– Obtain information from unencrypted transmission of API messages between the API consumer and provider. – Interception of API messages and revealing confidential information	– Use secure encrypted communication – Use of VPNs (e.g. IPsec, SSL/TLS, and HIP)
	DoS/DDoS attacks	– Make an API out of order by submerging it with a massive amount of requests	– Throttling/rate limiting the usage of APIs – Deployment of API gateways and microgateways – AI-based API security for proactive monitoring

Closed loop automation [196, 214, 226, 326, 327]	DoS attacks	– Fake heavy load on VNFs to increase the capacity of VM, which may lead to DoS	– Throttling/rate limiting on resources for VMs – AI-based resources level prediction
	Man-in-the-middle attacks	– Triggering a fake fault event and intercept the domain control messages to reroute traffic via a malicious switch	– Use secure encrypted communication – Use of VPNs (e.g. IPsec, SSL/TLS, and HIP)
	Deception attacks	– Intends to tamper transmitted data	– Use integrity validation mechanisms (e.g. Blockchain)
Intent-based interfaces [196, 328–330]	Information exposure	– Intercepting information of intents by an unauthorized entities to compromise system security objectives (e.g. privacy, confidentiality). This may lead to the launch of other attacks.	– Authenticating between intent producer and consumer (Signed JWT tokens, OpenID connect) – Controlled access via authorization controls (role-based access control, OAuth 2.0) – Secure communication via transport protocols (TLS 1.2)
	Undesirable configuration	– Changing the mapping from intent to action. Setting the security level from "High" to "Low"	– Input validation via user authentication.
	Abnormal behaviors	– Malformed intent could change the behavior, causing network outage	– AI-based proactive monitoring for abnormality detection
	Malinformed intent	– Changing the intent reduce the service quality.	– Intent format validation

attacker or a group of attackers can manipulate an API out of order by submerging it with many requests.

Closed-loop network automation: 6G networks may allow closed-loop network automation for the network's zero-touch management capabilities, such as monitoring the network to identify the fault and congestion occurrence. Then, it analyzes the data and acts accordingly to eliminate the identified issues. Thus, it creates a feedback loop of communication between monitoring, identifying, adjusting, and optimizing the network's performance to enable self-optimization. Closed-loop network automation in 6G will create security threats such as DoS, man-in-the-middle, and deception attacks [196].

Intent-based interfaces: Intent-based networking (IBN) is a novel concept which is proposed initially to introduce AI into the 6G mobile networks. The main idea of IBNs is to directly transform users' business intent into a network configuration, operation, and maintenance strategies using AI technologies. Using IBN concepts, 6G can effectively mitigate the typical limitations in the traditional networks in terms of efficiency, flexibility, and security. The critical security vulnerabilities with IBN may include information exposure, undesirable configuration, and abnormal behaviors.

10.4.5 Consumer End (Terminals and Users)

From the beginning of the advanced portable communication in early generations of wireless systems, they are dependent on a physical placing of symmetric keys in a subscriber identity module, also known as subscriber identity module (SIM) card. Although the encryption computations are moved from undisclosed to universal guidelines, the alternative cryptographic instruments are introduced for the shared verification process [331]. In accordance with the general standards, the 5G security model is still dependent on the SIM cards [332]. Although the SIM cards are getting smaller into the nanoscale, they still need to be inserted into devices/gadgets. This may limit the appropriateness of foreseen IoE paradigm in 6G. In a way, this challenge can be tackled by using embedded subscriber identity module (eSIMs) but introducing some physical measures issues. Another solution will be integrated subscriber identity module (iSIMs) which will be a part of system-on-chip in future gadgets. This will also face challenges due to the possible resistance coming from the telecom operators due to potential loss of control.

Typically, SIM cards rely on proven symmetric key encryption, which scaled well up to millions to billions of users. However, it has severe user privacy issues, Internet of Things (IoT), network authentication, and fake base stations. Therefore, 6G needs to consider a significant shift from symmetric crypto to asymmetric public/private keys and even to the postquantum keying mechanisms.

Already 5G plans to support authentication through a public-key infrastructure (PKI) and a set of microservices communicating over HTTPS. The authentication, confidentiality, and integrity for such communication are provided by transport layer security (TLS) using elliptic curve cryptography (ECC). Experiences that come from using these technologies in 5G, will shape the user and device authentication approaches in 6G.

10.5 Security Challenges with 6G Applications

6G is emerging as the network facilitator to a wide range of new applications which will drastically reshape the human society of the 2030s and beyond. However, these applications and services come with challenging performance requirements and extremely stringent security levels due to their critical nature and the need for a high-trust level. The interplay between the general performance expectations and security requirements becomes even more complicated with the emergence of skillful and ubiquitous attackers and nefarious activities. The envisaged capabilities of 6G could enable a myriad of possible novel applications and use cases. We extensively select the widely discussed ones and identify them as the most influential 6G applications (i.e. summarized in Figure 10.4) to elaborate on the security considerations. This set of applications is regarded as early deployment use cases, and applications of 6G within the current research literature [2, 12, 56, 333].

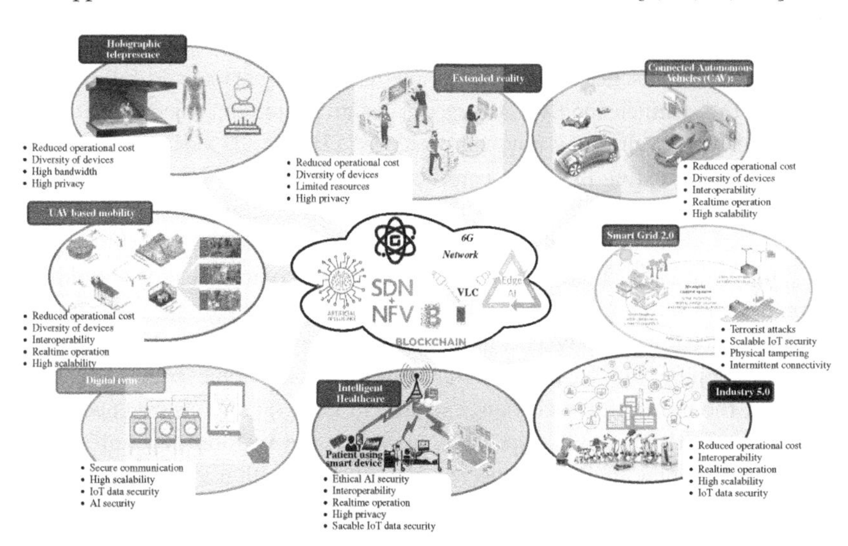

Figure 10.4 Key security requirements of prominent 6G applications. Sources: Arise Tech India; cutewallpaper.org; Srivatsan Sridhar/Fone Arena; Comau S.p.A.

10.5.1 UAV-based Mobility

Since 5G, Unmanned Arial Vehicles (UAVs) are popular in various application domains. With the support of 6G and AI-based services, UAV technologies will be used in new use cases such as passenger taxi, automated logistics, and military operations [334, 335]. Due to the limited available resources (i.e. processing and power) and latency-critical applications in UAVs, they should use lightweight security mechanisms to satisfy the low-latency requirements. Moreover, factors such as high scalability, diversity of devices, and high mobility have to be considered while developing the security mechanisms for UAVs. Since 6G will support AI and Edge-AI-based UAV functions such as collision avoidance, path planning, route optimization, and swarm control, it is essential to deploy mechanisms to mitigate AI-related attacks as well. Especially, protected integrity of control data is a vital requirement for proper operation. Due to the unmanned nature of UAVs, they are highly vulnerable to physical attacks. An adversary can physically capture the UAVs by jamming control signals or using physical equipment, then steal the important data contained within the UAVs.

Moreover, UAVs will have advanced computational and communication capabilities compared to other smart devices. Thus, a swarm of drones can be used to perform organized attacks. Such attacks can range from cyber-attacks to physical terrorist attacks [336, 337].

10.5.2 Holographic Telepresence

Holographic telepresence is a 6G application that can project realistic, full-motion, real-time, three-dimensional (3D) images of distant people and objects with a high level of realism rivaling the physical presence [17] (e.g. 3D video conferencing and news broadcasting [338]). A considerable bandwidth is required to enable holographic communication. When the number of holographic communication devices increases, the bandwidth requirements also increase proportionally. Thus, the security mechanisms used for holographic communication should not bring an extra burden on already overwhelmed bandwidths. Moreover, factors such as reduced operational cost and diversity of devices have to be considered while developing the security mechanisms for holographic communication. However, the most critical challenge related to holographic telepresence is the protection of privacy [339]. Especially, providing the required level of privacy when a holographic image is projected to a remote location is also an essential aspect to consider. Since the remote presenter cannot control the environmental settings of the projected location, additional privacy protection mechanisms should be implemented so that users can ensure privacy.

10.5.3 Extended Reality

XR is a term used to refer to all real and virtual combined environments which cover augmented reality (AR), virtual reality (VR), mixed reality (MR), and everything in between [59, 340]. 6G will support the advancements of XR by providing an opportunity to use it in various use cases, including virtual tourism, online gaming, entertainment, online teaching, healthcare, and robot control. Managing personal data is an important security aspect of XR, which will include people's credit card numbers or purchase histories and more personal information such as feelings, behaviors, judgments, and physical appearance. Thus, offering the required level of data responsibility is a critical requirement of 6G networks in terms of collection, storage, protection, and also sharing of personal data. Moreover, if fake or forged data are used in XR applications, the quality of user experience (QoE) in XR will fail. The factors such as high scalability, low overhead, and diversity of devices should be considered while developing the security mechanisms for XR. The security level or enforced security methods in XR applications can fluctuate significantly depending on the application. For instance, military applications may need the highest level of security (i.e. multifactor solid authentication, data encryption, user access control). In contrast, entertainment applications may require a lower level of security.

Another critical security issue explicitly related to XR is the fake experiences. If counterfeit or forged data is used in XR applications, the total XR experience will fail. Such incidents can even cause fatal results. For instance, fake experiences in critical XR environments such as surgery or military operation may lead to life-or-death consequences.

10.5.4 Connected Autonomous Vehicles (CAV)

Nearly 50 leading automotive and technological companies heavily invest in autonomous vehicle technology. The world moves forward to experience truly autonomous, reliable, safe, and commercially viable driver-less cars in the near future [341]. With the advent of connected autonomous vehicles (CAV) technologies, a new service ecosystem will emerge, such as driver-less taxi and driver-less public transport [56, 342]. The security issues in a complex CAV ecosystem can be categorized into three categories as vehicle level, CAV supply chain, and data collecting. Vehicle-level attacks can happen by hijacking vehicle sensors, vehicle-to-everything (V2X) communications, and taking over physical controls. Similar to UAVs, autonomous nature without human involvement will lead to the possibility of physical hijacking. However, autonomous vehicles have more advanced capabilities than UAVs. Therefore, emergency security measures can be integrated within a car. For instance, an automatic car stop during a terrorist attack is possible. 6G network can analyze the situation and deliver emergency signals to vehicles.

Moreover, new types of cyberattacks are possible due to V2X communications in the CAV ecosystem. Advance CAVs have a communication link with the car manufacturer. They can constantly monitor and instantly transmit software-related patches to mitigate any foreseen troubles over the air. However, vulnerabilities in the communication channels or forging the data downloaded from manufacturer cloud services can compromise the safety and security of the vehicles and their passengers.

The CAV ecosystem has a complex supply chain with different third-party service providers such as communication service providers (CSPs), road side equipment (RSE), and cloud service providers and regulators. Enabling a common standard of security requirements and enabling inter-operability is challenging. Privacy issues may arise when CAVs collect data about travel routes, control sensor data, and their owners and passengers. Such data become a honeypot for malicious attackers. According to the National Institute of Standards and Technology (NIST), the CAV security framework should provide device security, data security, and individuals' privacy.

Especially when public transport modes such as trains, flights, and buses are used, protecting individual privacy while delivering 6G services such as XR holographic telepresence will be challenging. Therefore, the 6G security framework for CAVs has to consider security convergence by combining physical security and cybersecurity and the concept of privacy by design.

10.5.5 Smart Grid 2.0

With the development of smart devices and advanced data analytical techniques, the grid networks are getting smarter and evolving from Smart grid 1.0 to Smart grid 2.0. Smart grid 2.0 may offer features such as automated meter data analysis, intelligent dynamic pricing, intelligent line loss analysis, distribution grid management automation, and reliable electric power delivery with self-healing capabilities [343]. In smart grid 2.0, it is crucial to offer network information and cybersecurity to ensure confidentiality, integrity, and availability of the energy network. The most common security vulnerabilities may include different type of attacks such as physical attacks, software-related threats, threats targeting control elements, network-based attacks, and AI/ML-related attacks [344]. The critical components and services such as data access points, control elements (supervisory control and data acquisition [SCADA]) [345], and the EMS of the cyber physical system [346], metering, billing, and information exchange are heavily targeted in these attacks.

Moreover, the improvement of trust management of trading mechanisms is a critical requirement of smart grid 2.0. One of the key features envisaged by Smart grid 2.0 is trading energy between unknown parties in a peer-to-peer (P2P) manner. Such trading could occur in variations of prosumer-to-prosumer and prosumer-to-consumer due to the popularity of solar photovoltaic (PV)-based, small-scale energy production and electrical cars [347]. Due to the scale of number of such occurrences, the trust should be established with minimal intervention of an intermediary. Moreover, the radical shift in smart grid management from centralized to distributed mode has also created the necessity of instating trust between the buyer and the seller, which has been the role of the third-party intermediary (i.e. distribution systems Operator) in a vertical grid arrangement [348].

10.5.6 Industry 5.0

Industry 5.0 is identified as the next innovation in the industrial revolution, which means people working alongside robots and smart machines need to add a personal human touch to the Industry 4.0 pillars of automation and efficiency [349]. 6G plays a vital role in enabling the advancements of an automated industrial environment. Similar to other 6G-enabled applications, Industry 5.0 will also face critical security threats and also they may need to provide basic security needs such as integrity, availability, authentication, and audit aspects. Factors such as reduced operational cost, diversity of devices, high scalability have to be considered while developing the security mechanisms for Industry 5.0. 6G will mainly be responsible for the data security, and integrity protection [350] in Industry 5.0 as controlling commands and monitoring data will be transferred over the 6G networks. Therefore, 6G era should also provide highly scalable and automated access control mechanisms and audit systems to restrict access to sensitive resources such as intellectual properties related to Industry 5.0.

10.5.7 Digital Twin

The digital twin is a novel industrial control and automation systems concept identified as a key 6G application. A digital twin is defined as a digital or virtual copy of a physical object, an asset, or a product [351, 352]. Digital twin interconnects virtual and physical worlds by collecting real-time data using IoT devices connected to the physical system. These collected data will be stored in locally decentralized servers or centralized cloud servers, and then, the collected data will be analyzed and evaluated in the virtual copy of the assets. After obtaining the results from the simulations, the parameters are applied to the real systems. The integration of data in real and virtual representations will help in optimizing the performance of the physical assets. The digital twin can be used in other use cases such as Industry 5.0, Automation, healthcare, utility management, and contractions.

The biggest security challenge in the digital twin system is that an attacker can intercept, modify, and replay all communication messages between the physical and digital domains. With the popularity of digital twin systems in future, 6G should support highly scalable secure communication channels. Another issue in digital twin systems is that the attacker can modify or alter the IoT data and make privacy attacks. When 6G is used to enable digital twin system, IoT data integrity and privacy protection mechanisms should be utilized. For instance, blockchain can be a candidate technology to enable such features in 6G networks.

10.6 Security Impact on New 6G Technologies

Considering the security requirements and application-specific aspects of the future 6G networks presented in the previous sections, we discuss the threat landscape and possible security solutions related to a few 6G technologies that have already gained the most attention. Although many other emerging technologies show their potential of relevance to 6G, their security and privacy considerations are not yet discovered in the state-of-the-art. In contrast, specific topics such as network softwarization and cloudification are already discussed with respect to 5G security. Based on the current literature, we identified that technologies such as Distributed Ledger Technology (DLT), distributed, and scalable AI/ML and quantum computing, and some PLS-related topics (THz, VLC, reconfigurable intelligent surface [RIS], MC) are quite relevant and have a substantial amount of work and new research directions related to security and privacy in 6G. Therefore, we extensively discuss those listed topics in the remainder of the section. In brief, we discuss the possible security solutions for the key security issues in 6G networks, how the available and evolving technologies can mitigate such security threats, state-of-the-art of security mitigation techniques for the given technologies, and beyond the state-of-the-art vision.

10.6.1 Distributed Ledger Technology (DLT)

Today, among DLTs, blockchain technology has gained the most attention in the telecommunication industry. The advantages of blockchain such as disintermediation, immutability, nonrepudiation, proof of provenance, integrity, and pseudonymity are significant to enable different services in trusted and secure manner in the 6G networks [353].

In addition to the advantages of AI in 6G, AI/ML and other data analytic technologies can be a source for new attack vectors in 6G. It has been proven that ML techniques are vulnerable to several attacks [354] targeting both the training phase (i.e. poisoning attacks) and the testing phase (i.e. evasion attacks). Since data are the fuel for AI algorithms, it is crucial to ensure their integrity and their provenance from trusted sources [355]. DLT can achieve the trust dimensions, such as protecting the integrity of AI data via immutable records and distributed trust between different stakeholders, which will enable confidence in AI-driven systems in a multitenant/multidomain environment.

Furthermore, DLT/blockchain shows the potential of using as a facilitating technology to evolve the 5G service models to support 6G. These services may include, however not limited to, secure VNF management, secure slice brokering, automated Security SLA management, scalable IoT PKI management, secure roaming and offloading handling, and user privacy protection to comply with 6G requirements [356].

10.6.1.1 Threat Landscape

Due to the foreseen alliance of DLT and 6G, the security vulnerabilities of blockchain and smart contracts may also implicitly impact the 6G networks [357]. Most of these attacks occurred due to the reasons such as software programming errors, restrictions in the programming languages, and security loopholes in network connectivity [358]. Moreover, these security issues can occur in both public and private blockchain platforms. They lead to complications such as loss of accuracy, financial losses in terms of cryptocurrency, and reduced system availability. Some of the critical security attacks in blockchain and smart contract systems are listed in Figure 10.5.

Majority attack/51% attack: If malicious users capture the 51% or more nodes in the blockchain, they could take over the control of the blockchain. In a majority attack, the attackers could alter the transaction history and prevent the confirmation of new legitimate transactions from confirming [359]. Blockchain systems which use majority voting consensus [360] are usually vulnerable for majority attacks.

Double Spending Attacks: The spending of the cryptographic token is a key feature of most of the blockchain platforms [361]. However, there is a risk that a user can spend a single token multiple times [362] due to lack of physical notes. Such attacks are called the double-spending attacks [363] and blockchain systems should have a mechanism to prevent such double-spending attacks.

Re-Entrency Attack: The re-entrancy vulnerability can occurred when a smart contract invokes another smart contract iterative. Here, the secondary smart contract which has been invoked can be malicious. For instance such attacks were performed to hack Decentralized Autonomous Organization (DAO) in 2016 [364]. An anonymous hacker stole USD50M worth Ethers.

Sybil Attacks: Here, an attacker or a group of attackers are trying to hijack the blockchain peer network by conceiving fake identities [365]. The blockchain

Key security vulnerabilities of blockchanized 6G services

1 **Majority attack / 51% attack**
A group of malicious users could capture the 51% or more nodes and take over the control of the blockchain.

2 **Double spending attack**
A user spends a single token multiple times.

3 **Re-entrency attack**
A smart contract invokes another iteratively and the invocation of the secondary contract is malicious.

4 **Sybil attacks**
An attacker attempts to take over the peer network by conceiving fake identites explicitly.

5 **Broken authentication and access control**
Potential vulnerabilities and issues in the implementation of authentication and access control mechanisms.

6 **Security misconfiguration**
Use of insecure security configurations or outdated configurations that make the system vulnerable to attack.

7 **Privacy leakages**
Vulnerable to leakage privacy of transaction data, smart contract logics and user privacy.

8 **Other vulnerabilities**
Other security threats such as destroyable contracts, exception disorder, call stack vulnerability, bad randomness, underflow/ overflow errors and unbounded computational power intensive operations.

Figure 10.5 Key security vulnerabilities of blockchanized 6G services.

systems which have minimal and automated member addition systems are typically prone to sybil attacks [366].

Privacy Leakages: Blockchains and smart contracts are vulnerable to several privacy threats such as leakage of transaction data privacy [367], leakage of smart contract logic privacy [368], leakage of user privacy [369], and privacy leakages while execution of smart contracts [370]. Some of the blockchain nodes may follow the strict privacy roles and support too much transparency which may lead to reveal some sensitive information such as trade secrets and pricing information [367]. Moreover, business logic of the organization needs required to be incorporated in the blockchain. The sensitive business logic information such as commissions and bonuses may need be included smart contracts and these information can be revealed to the competitors [368].

Other Attacks: Apart from the above, blockchains and smart contracts are vulnerable to several other security threats such as destroyable contracts [371], exception disorder [372], call stack vulnerability [373], bad randomness [374], underflow/overflow errors [375, 376], broken authentication [377], broken access control [378], security misconfiguration [379], and unbounded computational power intensive operations [380].

10.6.1.2 Possible Solutions

When the DLT/blockchain solutions are adopted in 6G networks, they should always comply with possible mechanisms to mitigate the above security attacks. However, the deployment of some of the security mechanisms can be more momentous in public blockchains than in private blockchains. For instance, the debugging or correcting smart contracts might be a cumbersome process [381] since all the nodes adopt the smart contracts in a blockchain network. Since smart contracts play a vital role in DLT/blockchain systems to enable automation, ensuring the accuracy of the smart contract is necessary. Moreover, the proper validation of correct functionality of the smart contract is required before deploying it in thousands of blockchain nodes. The accurate functionality of smart contacts can be checked by ***identifying semantic flaws*** [382, 383], using ***security check tools*** [384–386] and performing ***formal verification*** [387–390].

Moreover, proper access control and authentication mechanisms should be utilized to identify the malicious bots and AI-agent-based blockchain nodes. Such mechanisms can prevent the majority and sybil attacks. The additional privacy preservation mechanisms such as privacy by design [391] and TEE [392, 393] can be integrated to prevent privacy leakages in blockchain-based 6G services [394, 395].

Moreover, blockchain/DLT support different architecture types such as (i) public, (ii) private, (iii) consortium, and (iv) hybrid blockchain [396]. The impact of the above security attacks naturally varies for different architectures, for example the

51% attacks are highly impacting on public blockchains. In such cases, a consortium of private blockchains can be suitable for certain 6G services (e.g. spectrum management, roaming), which has less number of miners [356]. Therefore, selecting the proper blockchain/DLT type according to the 6G application and services can eliminate the impact of certain attacks.

10.6.2 Quantum Computing

In the next couple of years, it is expected that quantum computing will be commercially available and will impose a significant threat on the current cryptographic schemes. As stated in the current state-of-the-art, quantum computing is envisioned to be used in 6G communication networks to detect, mitigate, and prevent security vulnerabilities. Quantum computing-assisted communication is a novel research area that investigates the possibilities of replacing quantum channels with noiseless classical communication channels to achieve extremely high reliability in 6G. Moreover, with the advancements of quantum computing, it is foreseen by security researchers that quantum-safe cryptography should be introduced in the postquantum world. The discrete logarithmic problem, which is the basis of current asymmetric cryptography, may become solvable in polynomial time with the development of quantum algorithms (e.g. Shor) [397].

Since quantum computing tends to use the quantum nature of information, it may intrinsically provide absolute randomness and security to improve the transmission quality [51]. Integrating postquantum cryptography schemes with physical-layer security schemes may ensure secure 6G communication links [398]. Moreover, new eras may open up by introducing ML-based cyber-security and quantum encryption in communication links in 6G networks. Quantum ML algorithms may enhance security and privacy in communication networks with quantum improvements in unsupervised and supervised learning for classification and clustering tasks. There are promising 6G applications with the potential to apply quantum security mechanisms. For instance, many 6G applications such as ocean communication, satellite communication, terrestrial wireless networks, and THz communications systems have the potential of using quantum communication protocols such as quantum key distribution (QKD) [399]. QKD is applicable in conventional key distribution schemes by providing quantum mechanics to establish a secret key between two legitimate parties. Figure 10.6 demonstrates the envisioned roles of quantum computing and quantum security in the 6G era.

10.6.2.1 Threat Landscape

Within the threat landscape in quantum-based attacks, the adversaries are also considered to have quantum powers. Although quantum computers are yet to

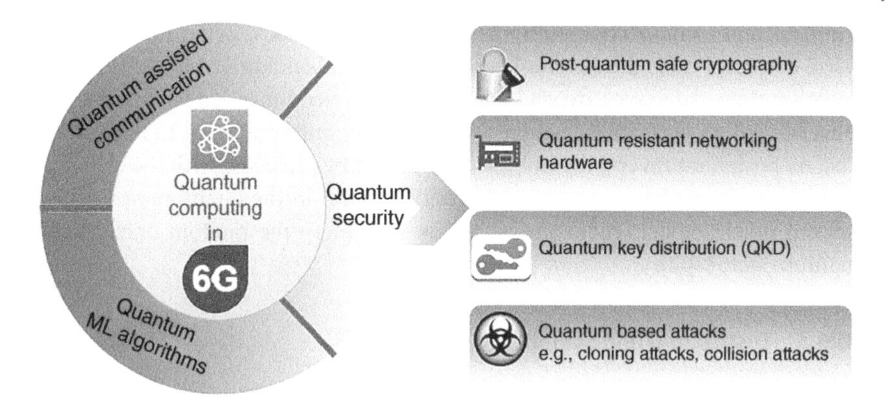

Figure 10.6 Role of quantum computing in 6G.

evolve in the long run, the threat they may generate on IoT devices needs to be carefully considered. Since cryptography is the key security factor in IoT networks and IoT devices, they require light-weight cryptographic solutions. It is always challenging to incorporate postquantum crypto solutions resisting quantum-based attacks in IoT devices. Therefore, device-independent quantum cryptography is a challenge in the postquantum era in the 6G paradigm.

The oblivious transfer (OT) in classical information sharing allows the sender to transfer one of the potentially many pieces of information to a receiver while remaining oblivious as to which piece has been transferred. However, this feature is unable to maintain quantum information since any leakage may create huge damage to whole two-party communication.

As a fundamental law, quantum computers have no-cloning property, which makes it impossible to maintain the exact copy of the quantum state (i.e. rewinding is not achievable). In quantum cloning attacks, an adversary has to take a random quantum state of information and make an exact copy without altering the original state of the data. Although perfect quantum state copies are prohibited, in [400], it is proven that a quantum state can be copied with maximal accuracy via various optimal cloning schemes. Quantum cloning attacks may even occur in high-dimensional QKD schemes as quantum hacking in a secure quantum channel. Moreover, quantum collision attacks can also occur when two different inputs of a hash function provide the same output in a quantum setting.

10.6.2.2 Possible Solutions

Scientists have already started investigating quantum-resistant hardware and encryption solutions to be ready for the threat due to quantum computing in the future 6G era. There are a few postquantum cryptographic primitives identified as lattice-based, code-based, hash-based, and multivariate-based cryptography [401].

In the current context, computational lattice problems show better performance in IoT devices. Due to the smaller key length, they fit better in 32-bit architecture. However, these categories are yet to evolve and are recommended for IoT devices concerning their performance, memory constraints, and communication capabilities. As postquantum cryptography will no longer be protected with the classical random oracle model, it may need to verify security in the quantum-accessible random oracle model where the adversary can query the random oracle with quantum state [402].

10.6.3 Distributed and Scalable AI/ML

6G envisions autonomous networks that can perform *Self-X* functions (self-monitoring, self-configuration, self-optimization, and self-healing) without any human involvement [247]. The ongoing ZSM architecture specifications entailing intent-based interfaces, closed-loop operation, and AI/ML techniques to empower full-automation of network management operations including security are steps toward that goal. Since the pervasive use of AI/ML will be realized in a distributed and large-scale system for various use cases including network management, distributed AI/ML techniques are supposed to enforce rapid control and analytics on the extremely large amount of generated data in these networks. As demonstrated in Figure 10.7, 6G security is mainly revolving around AI in two aspects "AI for security" and "Security for AI."

Distributed AI/ML can be used for security for different phases of cybersecurity protection and defense in 6G. The utility of AI/ML-driven cybersecurity lies

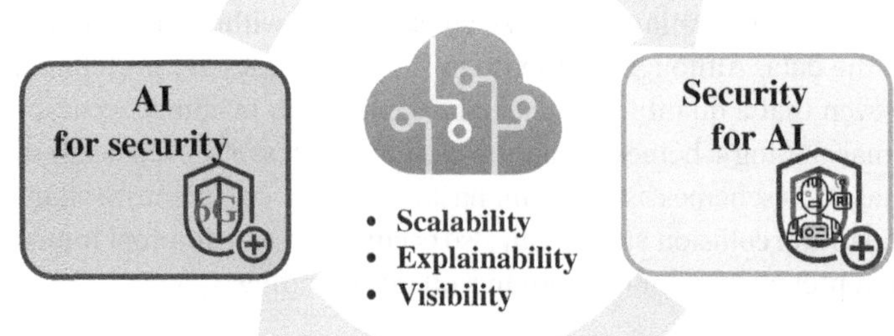

Figure 10.7 6G security and AI.

in the advantages of autonomy, higher accuracy, and predictive capabilities for security analytics. Following are some of the challenges relating to the AI/ML in 6G systems as defined in [403]:

- *Trustworthiness*: Are ML components trustworthy? This is a more important question when critical network functions including security are AI-controlled.
- *Visibility*: For controllability and accountability, visibility is crucial. A research question is how to monitor timely for security-violating AI incidents.
- *AI Ethics and Liability*: Could some AI-based optimization starve some users or applications? Do AI-driven security solutions protect all users the same? Who is liable if AI controlled security functions fail?
- *Scalability and Feasibility*: For federated learning, data transmissions should be secured and preserve privacy. For AI/ML controlled security functions, scalability in terms of required computation, communication, and storage resources is challenging. For instance, further enhanced mobile broadband (FeMMB) leads to huge data flows. Integrated with AI/ML, these flows may cause significant overhead.
- *Model and Data Resilience*: Models should be secured and robust in the learning and inference phases (e.g. against poisoning attacks). However, more attacks are being developed with increasing variety and proficiency in recent years [404], e.g. on federated learning [405].

10.6.3.1 Threat Landscape

It is expected that 6G will heavily rely on AI and ML technologies. However, AI and ML will lead to the 6G intelligence network management system to become a victim of AI/ML-related attacks. Such attacks can target the training phase (poisoning attacks) as well as the test phase (evasion attacks) [406, 407]. During a poisoning attack on the training phase, the attacker can tamper the training data by injecting carefully crafted malicious samples, to influence the outcome of the learning method [408]. Such injection of crafted samples may lead to intelligence services supporting the E2E services to mispredict the resource requirements and misclassify the services. Evasion attacks during the test phases attempts to circumvent the learned model by introducing disorders to the test data. Moreover, model inversion aims to derive the training data, utilizing the outputs of the targeted ML model while model extraction attacks steal the model parameters to replicate (near-)equivalent models. Infrastructure-targeting physical attacks essentially strive for communication tampering, and intentional outages and impairments in the communication and computational infrastructure for decision-making/data processing impairments and may even put entire AI systems offline.

At the AI middleware layer, a significant threat is the compromise of AI frameworks to exploit vulnerabilities in those artifacts or traditional attack vectors toward their software, firmware, and hardware elements. For another type of attack, API-based attacks, an adversary queries and attacks an API of an ML model to obtain predictions on input feature vectors. This may lead to model inversion (recover training data), model extraction (reveal model architecture compromising model confidentiality), and membership inference (exploit model output to predict on training data and ML model) attacks.

10.6.3.2 Possible Solutions

There are different solutions against these threats for AI/ML. Adversarial training injects perturbed examples similar to attacks into training data to increase robustness [409]. Defensive distillation is another defensive strategy that is based on the concept of knowledge transfer from one neural network to another via soft labels, which are the output of a previously trained network and represent the probability of different classes. They are used for the training instead of using hard labels mapping every data to exactly one class) [410]. These two solutions are both effective ones against evasion attacks and adversarial attacks.

Against poisoning attacks in the training phase, protection of data integrity and authentication of the data origin is instrumental. In that regard, blockchain provides a distributed, transparent, and secure data sharing framework perspective [151]. Similarly, moving target defense [411, 412] and input validation [413] are used. The latter is also beneficial against adversarial attacks. To mitigate model inversion attacks, an effective defense is to control the information provided by ML APIs to the algorithms to prevent them. This approach is also effective against adversarial attacks. Another countermeasure against model inversion attacks is to add noise to ML prediction [414]. Noise injection, but to the execution time of the ML model, is also used against model extraction attacks.

10.6.4 Physical-Layer Security

Physical-layer security (PLS) mechanisms rely on the unique physical properties of the random and noisy wireless channels to enhance confidentiality and perform lightweight authentication and key exchange. The flexibility and adaptability of PLS mechanisms, specially for resource-constrained scenarios, joint with the opportunities provided by disruptive 6G technologies may open a new horizon for PLS in the time frame of 6G. Figure 10.8 shows illustrative scenarios for PLS regarding key technologies expected for 6G, which are described next.

10.6.4.1 TeraHertz Technology

In 6G, it is expected to move further to higher carrier frequencies, in the THz range (1 GHz to 10 THz), to improve spectral efficiency and capacity of future wireless networks as well as provide ubiquitous, high-speed Internet access. In those frequencies, the transmitted signals are highly directional, and the propagation

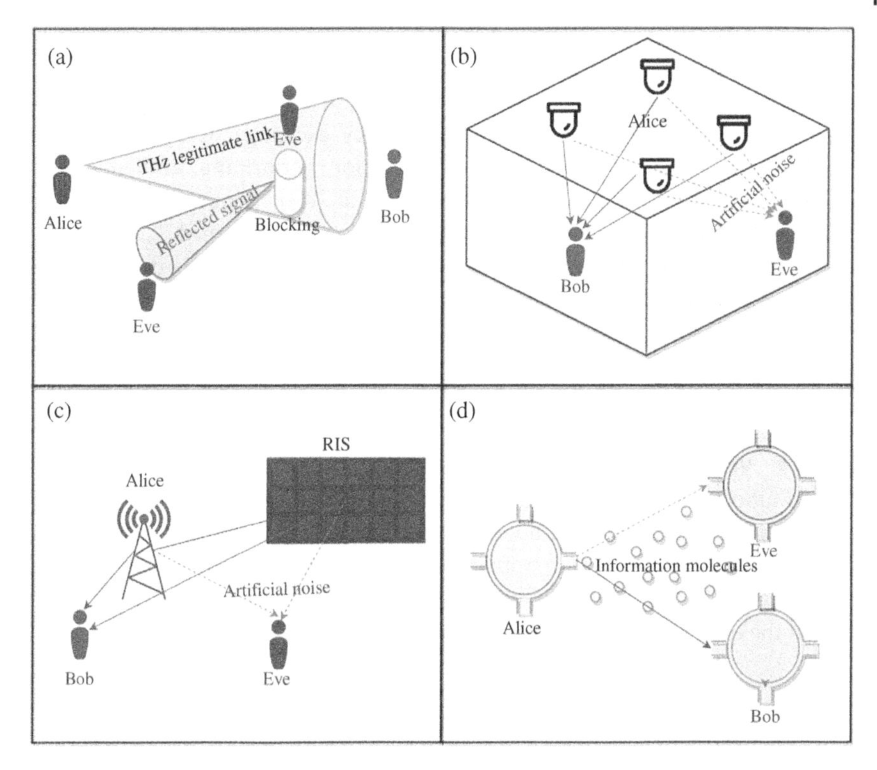

Figure 10.8 Illustrative PLS scenarios in 6G era: (a) THz communications in the presence of eavesdroppers, (b) secure MIMO VLC systems with artificial noise, (c) RIS-aided secure wireless communication, (d) eavesdropping in molecular communications.

environment is harsh, thus the interception of signals is mostly limited to illegitimate users that are located in the same narrow beam of the legitimate user.

10.6.4.2 Threat Landscape

Even with the use of extremely narrow beams, an illegitimate receiver can intercept signals in line-of-sight (LoS) transmissions. Thus, THz communications are prone to data transmission exposure, eavesdropping, and access control attacks.

10.6.4.3 Possible Solutions

In [415], the authors prove that an illegitimate user can intercept signals by placing an object in the path of the transmission so that the radiation is scattered toward him. In that paper, it is proposed to perform a characterization of the backscatter of the channel to detect some, although not all, eavesdroppers. Moreover, in [416], the authors proposed to explore the multipath nature of THz propagation links to enhance the information-theoretic security. By sharing data transmission over multiple paths, the authors showed that the message eavesdropping probability can be significantly reduced, even when several eavesdroppers are cooperating, at a cost of a slight decrease in link capacity. That solution can be explored for transmitting sensitive data or performing secure key exchange in THz networks. Moreover, in [417], a study is conducted for performing authentication at the physical layer in vivo nanonetworks at THz frequencies, where a distance-dependent-path loss based authentication is performed. The authors showed that pathloss can be used as a device fingerprint from a THz time-domain spectroscopy setup. All in all, new PLS solutions, as electromagnetic signature of THz frequencies for performing authentication at the physical layer [39], would benefit THz wireless joint with the incorporation of new countermeasures on the transceiver designs.

10.6.4.4 Visible Light Communication Technology

VLC is an optical wireless technology that has gained significant attention due to its advantages compared with radio frequency (RF) systems, such as high-data rates, large available spectrum, robustness against interference, and low-cost for deployment. VLC also has great potential to complement RF systems in order to exploit the benefits of both networks [418].

10.6.4.5 Threat Landscape

VLC systems are intrinsically more secure than RF systems because light cannot penetrate walls. However, due to the broadcast nature of VLC systems (as in RF), when communication takes place on public zones or with large windows in the coverage, VLC systems are prone to eavesdropping attacks, thus confidentiality may be potentially compromised [419]. Moreover, VLC systems present different characteristics than RF systems that should be considered for the design of PLS mechanisms. For instance, VLC channels are quasistatic and real-valued channels, and VLC systems present a peak-power constraint that impedes unbounded inputs, e.g. Gaussian inputs. Therefore, these operating constraints should be revisited for the performance evaluation and the optimization of PLS strategies in VLC systems [420]. Besides, according to the study conducted in [421], VLC systems are more vulnerable at locations that present strong reflections.

10.6.4.6 Possible Solutions

In [419], the enhancement of the secrecy performance, in terms of the achievable secrecy rate, of a multiple-input multiple-output (MIMO) VLC system is demonstrated by using linear precoding. Therein, the peak-power constraint is considered for the transmitted signal, and only discrete input signaling schemes are used. Also, in [422], a scheme of watermark-based blind PLS was investigated, where red, green, and blue LEDs and three color-tuned photo-diodes are employed to enhance the secrecy of a VLC system by implementing a jamming receiver joint with the spread spectrum watermarking technique.

10.6.4.7 Reconfigurable Intelligent Surface

With the evolution of metamaterials and micro-electro-mechanical systems, RIS has emerged as a promising option to tackle the challenges of intelligent environments regarding security, energy, and spectral efficiency. RIS is a software-controlled metasurface composed of a planar array of a large number of passive and low-cost reflecting elements capable of dynamically adjusting their reflective coefficients, thus controlling the amplitude and/or phase shift of reflected signals to enhance the wireless propagation performance.

10.6.4.8 Threat Landscape

Traditional PLS techniques, such as deploying active relays or friendly jammers that use artificial noise (AN) for security provisioning, may incur increased hardware costs and energy consumption. Moreover, in adverse wireless propagation environments, an adequate secrecy performance cannot be permanently guaranteed even with AN's use. Therefore, it would be desirable to adaptively control the propagation properties of wireless channels to ensure secure wireless communications, which is impossible to be attained with traditional communication technologies.

10.6.4.9 Possible Solutions

By controlling the phase shifts of RIS intelligently, the reflected signals can either be added coherently at the intended receiver to enhance the quality of the received signal or be added destructively at a nondesired receiver to enhance security [423]. In this sense, RIS-assisted PLS has become a promising technology for secure and low-cost 6G networks. For instance, in [424], it is shown that the importance of RIS technology for enhancing security, even if the eavesdropping link is in better conditions than the legitimate link. Moreover, the secret key generation problem for RIS-assisted wireless networks has also been investigated, where each element of the RIS is an individual scatter to enhance the secret key capacity [425].

10.6.4.10 Molecular Communication (MC)

In MC, bionanomachines communicate using chemical signals or molecules in an aqueous environment [426]. This technology is appealing for enabling important applications and use cases related to healthcare innovations in the context of 6G.

10.6.4.11 Threat Landscape

This kind of communication will handle highly sensitive information with several security and privacy challenges on the communication, authentication, and encryption process.

10.6.4.12 Possible Solutions

It is extremely important to tackle security issues in MC from the very early stages of its practical development in order to guarantee the promising benefits of this technology, thus PLS mechanisms would have an impact on providing security for MC. For instance, the notion of biochemical cryptography was introduced in [427], where a biological macromolecule composition and structure are used as a medium to achieve information integrity. Moreover, in [428], the fundamental benefits and limits of PLS are investigated for diffusion-based channels, where the secrecy capacity is derived to obtain insights on the number of secure symbols that can be transmitted over a diffusion-based channel.

10.7 Privacy

The faster the world is moving toward a digital reality, the higher the risk people may put their privacy, which is more precisely called digital privacy. The data is collected for many applications to improve their service performance. Such processed data or information leakage always create huge privacy issues requiring well-balanced privacy-preserving techniques. When more and more end devices tend to share local data with centralized entities, the storage and processing of this data pile with the added privacy protection mechanisms will be difficult. As 6G systems may have simultaneous connectivity up to about 1000 times greater than in 5G, privacy protection should be considered an important performance requirement and a key feature in wireless communication in the envisioned era of 6G [39]. However, in the current data collection and analysis process, privacy protection has not received enough attention and priority level. Therefore, there are many research opportunities for finding the correct balance between increasing data privacy and maintaining them with a lower computation load, which may reduce the speed and accuracy of the computation. In Figure 10.9, we describe illustrating a summary of 6G privacy with respect to privacy types, privacy violation, privacy protection, and related technologies.

The issue in 6G with data privacy will be more challenging when the number of smart devices is increasing and tracking every move of a person with a lack of transparency about what is exactly collected. Especially, in the big data era of decentralized systems, adding privacy protection mechanisms will further increase the communication and computational costs which already show a rapid growth [429].The current European Union's General Data Protection Regulation (GDPR) for privacy assurance should also be subject to change with the evolving 6G applications and specifications. Mainly, there are three key challenges that encounter while protecting privacy in 6G:

- The extremely large amounts of data exchange required in 6G may impose a greater threat on peoples' privacy with extensive attention attracted by the

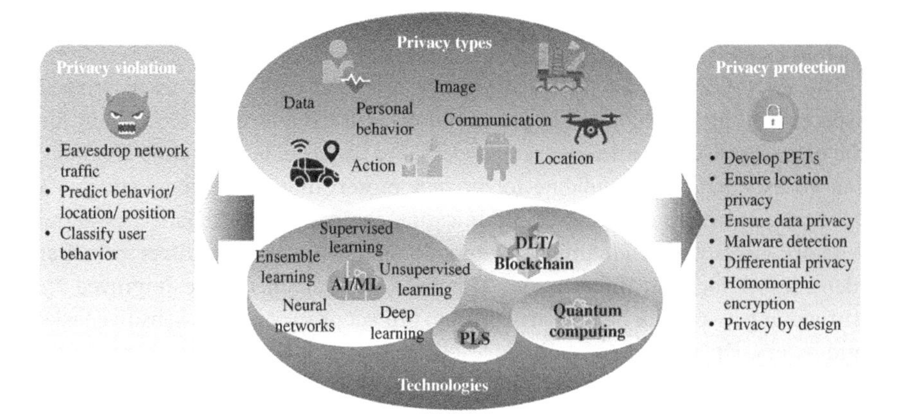

Figure 10.9 Summary of 6G privacy.

governmental and other business entities. This may occur as a large number of small chunks of data accumulation. The easier the data is accessible and collectible in the 6G era, the greater risk they may impose on protecting user privacy and causing regulatory difficulties.

- When the intelligence is moving to the edge of the network, more sophisticated applications will run on mobile devices and there is an increase in the threats of attacks. However, incorporating privacy-protecting mechanisms in resource-constrained devices at the edge of the network will be again challenging. This gives rise to the requirement of introducing lightweight privacy-preserving mechanisms.
- Keeping the correct balance between maintaining the performance of high-accurate services and the protection of user privacy is also noteworthy. Location information and identities are required to realize many smart applications. Therefore, it is necessary to carefully consider data access rights and ownership, supervision, and regulations for protecting privacy.

Considering privacy in the context of statistical and machine learning analysis, differential privacy (DP) is another budding privacy-preserving technology that is also likely to appear in future 6G wireless applications [331, 430]. DP may provide mathematically provable privacy protection against certain privacy attacks such as differencing, linkage, and reconstruction attacks. As stated in [430], DP has interesting properties to enhance privacy protection while analyzing personal information: quantification of privacy loss permits comparisons among different computation techniques; composition allows the design and analysis of complex privacy-enhancing algorithms starting from simple building blocks; allow group privacy; immunity to postprocessing of the privacy concerning algorithms. Rather than using conventional data encryption methods, novel mechanisms can be incorporated with the development of lightweight privacy-preserving techniques such as using HE [431].

The role of blockchain in 6G may have pros and cons in terms of privacy aspects. On the one hand, data privacy in 6G will likely involve blockchain for the ultra-massive and ultra-dense networks. For instance, blockchain technology can be used as a key candidate for privacy preservation in content-centric 6G networks. Having a common communication channel in blockchain may allow network users to be identified by pseudo names instead of direct personal identities or location information. Moreover, blockchain can be improved by introducing new block header structures to protect privacy in high sensitive tasks and actors. On the other hand, since blockchain is a DLT that is intrinsically transparent, it may disclose private information to all participants by creating privacy violations. When the 6G is expected to host a zero-trust architecture that assures embedded trust in the devices and the network, Blockchain is gaining a higher reputation to ensure trust among highly decentralized and distributed applications, and it also brings the biggest issue on data privacy and advanced connectivity. As pointed out in [432], such privacy risks can be addressed by solutions including risk signatures, zero-knowledge augments, and coin mixing.

The fast-growing AI technology in the 6G vision has a close associative with ML technology where privacy is showing a greater impact in two ways [433]. In one way, the correct application of AI/ML can protect privacy in 6G. In another way, privacy violations may occur as AI/ML attacks. Different ML types (e.g. neural network, deep learning, supervised learning) can be applied for privacy protection in terms of data, image, location, and communication (e.g. Android, intelligent vehicles, IoT). As summarized in [433], privacy attacks can occur in ML models during training (e.g. poisoning attack) and testing phases (e.g. reverse, membership interference, adversarial attacks). When AI is used to emulate human brain capability with collaborative/cooperative robots (cobots), they use learning tools to train those digital entities. However, the question is whether the cobots will be ethical, transparent, and accountable for preserving privacy concerns while using data sets during this constant learning and real-time, decision-making process.

While developing more robust and efficient privacy preservation solutions, the properties of quantum mechanics can also be exploited for high security and high-efficiency levels. Such approaches will be very much useful in a postquantum era of 6G networks in the long run. For instance, in [434] the authors propose an encryption mechanism based on controlled alternate quantum walks for privacy-preserving of healthcare images in IoT. Moreover, the work in [435] presents a lattice-based conditional privacy-preserving authentication mechanism for postquantum vehicular communication. Adding quantum noise to protect quantum data will lead the security concept of DP toward quantum. In [436] the author demonstrates this by including depolarization noise in quantum circuits for classification.

On the other hand, critical applications and massive scenarios expected in 5G/6G have raised the importance of novel privacy-related requirements, such as anonymity, unlinkability, and unobservability of the nodes in a network. Thus, from the information-theoretic point of view, a common approach to guarantee privacy is based on the perturbation of data attained by means of a privacy mechanism that performs a randomized mapping to control private information leakage. Quantifying this information leakage is important in order to limit this. Different notions of privacy leakage have been proposed to capture the capacity of adversaries to estimate private information, for example, Shannon's mutual information DP, among others [437], as well as different leakage measures. In that sense, under careful control, privacy can tolerate some leakage to get some utility. There is no a general privacy vs. utility trade-off. Thus, the amount of leakage required to get some utility depends on the application [438].

11

Resource Efficient Networks

6G networks are envisaged to have greater energy efficiency and green computing capabilities compared to its predecessors. After reading this chapter, you should be able to

- Understand energy-efficient network management with 6G.
- Understand energy-efficient security.
- Understand energy-efficient resource management.

The 6G networks will not only improve the way we live or work today but will also impact how we take care of our planet. In this regard, researchers (i.e. from academia and industries) envisioned that the 6G will put significant pressure on energy efficient or green computing (also known as environment sustainable networks) paradigm.

In contrast, currently we are on the verge of 5G that has been rolling out worldwide in an unprecedented way to provide users with high quality of experiences (QoEs). The main objectives of 5G communication towers are to make radical advances, for instance high bandwidth, super high data rate, through put, latency, reliability, and massive connectivity, as shown in Figure 11.1.

To achieve all these objectives, a 5G new radio (NR) is proposed and designed that will enable dense network in cities, as shown in Figure 11.1. The NR is integrated with the new microwave band radio at 3.3–4.2 GHz, and with millimeter-wave for first ever time to greatly improve the data rates up to 10 GPs. In addition, it is integrated with several new network access technologies, such as Beam Division Multiple Access (BDMA) and Filter Bank Multi Carrier (FBMC), Massive Multi-Input Multi-Output (MIMO) for capacity increase, Software-Defined Networks (SDN), and so on, to achieve the fullest network flexibility. Though, the 5G will bring a plethora of features, it will also bring various challenges. Recent research pointed out that 5G base-station (can be attributed to NR) consumes up to twice or more the power than of a 4G base-station. More precisely, in a typical

6G Frontiers: Towards Future Wireless Systems, First Edition.
Chamitha de Alwis, Quoc-Viet Pham, and Madhusanka Liyanage.
© 2023 The Institute of Electrical and Electronics Engineers, Inc. Published 2023 by John Wiley & Sons, Inc.

Figure 11.1 5G communication tower in cities. Source: Pål Frenger/Richard Tano/Telefonaktiebolaget LM Ericsson.

setting, the 5G base-station with higher frequencies need MIMO antennas may have the energy consumption of more than 11 KW, whereas a typical 4G base-station requires less than 7 KW of energy. There may by many more unexplored 5G scenarios, use-cases, or components, where energy consumption demands will be high, such as dense network deployments, faster data converters. Moreover, it is worth to note that in the recent years, the Information and Communication Technology industry is nearly consuming 20% of total power consumption.

11.1 Energy-Efficient 6G Network Management

Exploiting machine-learning algorithm, an energy-efficient mechanism, is suggested for 6G-based industrial network. The main objective of this research is to use the concept of Network in Box (NIB) and to achieve the quality of service (QoS) and QoE for multimedia contents, while reducing the energy-consumption at portable devices. Here, the NIB is a key technology that will reshape the remote industrial automation networking via enabling self-organization, seamless connection for mobility management with less overhead, fast content delivery, and energy-efficient computation. A novel ML-driven mechanism is being used for the mobility management to achieve the efficient and fast communication in industrial 6G-NIB settings, as shown in Figure 11.2. This model allocates the resources in a more efficient way in NIB using wireless links. In the experimental

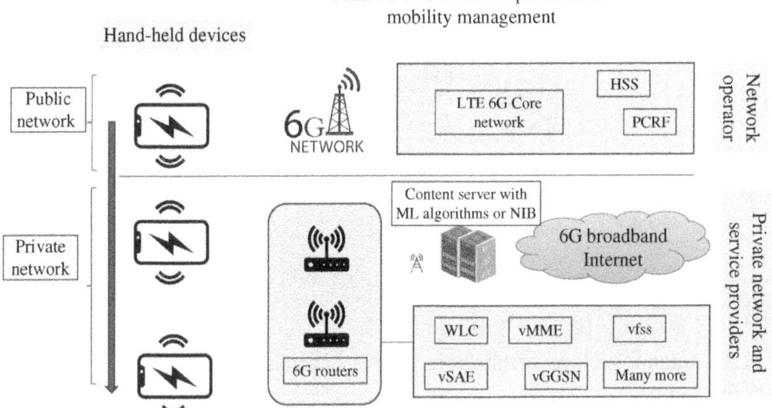

Figure 11.2 NIB-enabled mobility architecture.

settings, real-time industrial dataset is being used and selected parameters are passed through the ML-based algorithm to analyze the 6G-enabled NIB for mobility management. The authors claimed that their algorithm has achieved best performance results in 6G-based NIB system.

Similarly, in [1], the authors have proposed another energy-efficient communication architecture for IoT-assisted smart automation systems. In the automation networks, the IoT systems monitor the extensive mechanical systems those exchange complex and sensitive data (i.e. multimedia contents) and request the desired and useful information from other systems. As the multimedia contents broadcast over AI-enabled 6G networks, the QoE is restricted over the end-user equipment which is running of battery power. However, how to achieve the QoE and efficiency at end user's level are big concerns – especially when multimedia contents are being transmitted in a smart automation network. To enhance the QoE, (i) a QoS-based joint energy and entropy optimization approach (also called QJEEO) is suggested. Three layers have been used, mainly intelligent radio network, data receiving model QoS mapping, and QoE to enable energy efficiency in smart automation. (ii) QoE structure model is suggested that will transmit multimedia contents in automation network via AI-enabled 6G network. A layered architecture is suggested that include multimedia IoT devices, 6G-assisted visionary layers, and automation systems layer, as shown in Figure 11.3. Multimedia data are fed/received to/from video encoder to reduce the redundancy. Then the encoder data is transmitted over AI-enabled 6G base-station. Finally, the server

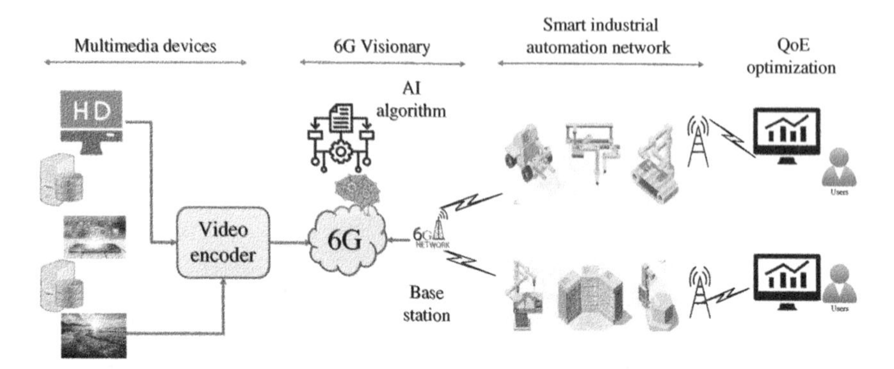

Figure 11.3 Automation layer architecture. Source: Scanrail/Adobe Stock, zozzzzo/Adobe Stock.

will analyze the contents, manage data traffic and allocate resources (in terms of QoE) to the users.

However, to realize the energy-efficiency with QoE in 6G network – the authors have simulated their network in MATLAB with following parameters, such as variable data rate, 250 m radio range, 15 m/s speed, simulated area 1000 m × 1000 m. Nevertheless, their result show that the proposed AI-based algorithm is an efficient-aware, but more work needs to be done to achieve the big vision on 6G green communication.

Energy-consumption in radio network is one of most promising research areas. A sleep mode integrated with machine learning mechanism for radio traffic forecast has been investigated in [4]. In this research, a radio network (i.e. base-station) is being connected with the photo-voltaic cell (i.e. green energy) with energy storage unit, and connected to the power-grid (i.e. brown energy), as shown in Figure 11.4.

Each LTE-A RAN offer various services to an area, and it consists of one macrobase-station and few small cells. The energy is supplied through various resources (such as PV, battery storage) to the cluster. The authors examined various prediction models on energy using the machine learning algorithms. The proposed solution includes real data modeling to forecast the future traffic load and renewable energy sources. The model is consisting of two phases (training phase and run-time phase), as shown in Figure 11.5. Training phase forecasts the renewable energy projection and the energy demand. Based on the training model, in run-time phase, the RAN network is operated and saved the energy. The results discussed that many ML algorithms may succeed in achieving a best trade-off between energy-consumption and QoS, for the details please refer to [4].

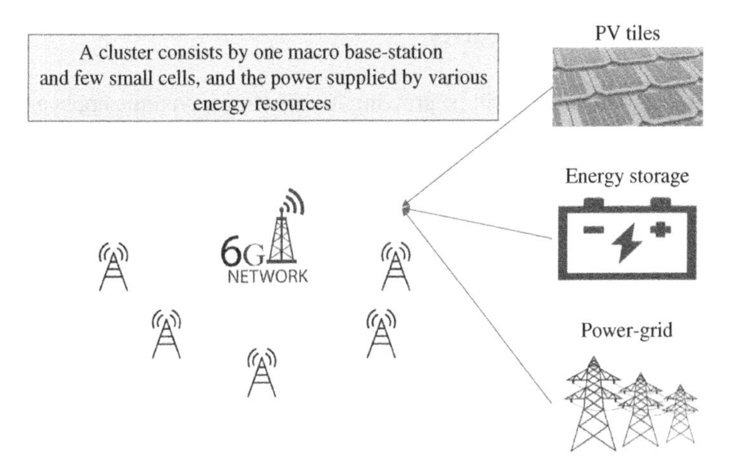

Figure 11.4 6G RAN network cluster powered by distributed energy sources.

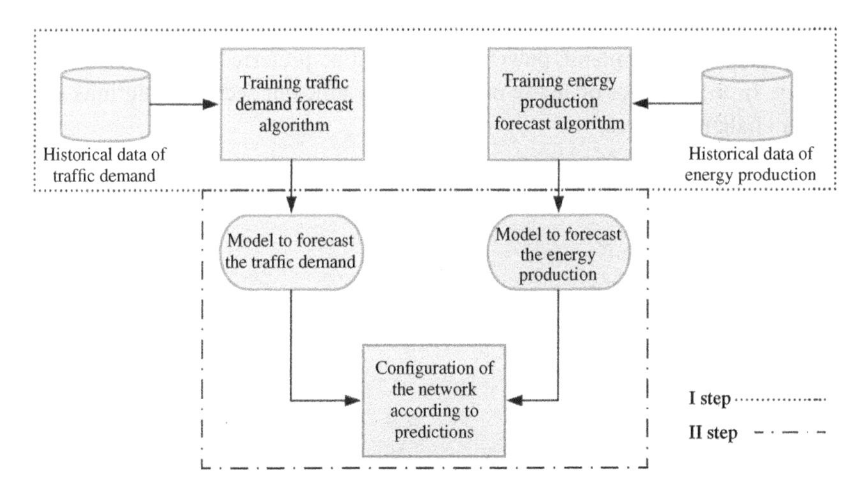

Figure 11.5 A RAN network operations: training and run-time phases. Source: Adapted from Pereira and Sousa [4].

In summary, the NIB would be able to reshape industrial automation through self-organization, seamless mobility, and fast computation. Likewise, several other architectures that are focusing on energy-efficient QoSs are a good start of rethinking of energy-efficient management. However, these solutions are at very early stage, need more testing, prototyping, and validations. Moving forward, more research needs to be done how a mathematical model would fit to the 6G applications.

11.2 Energy-efficient Security

As the 6G systems and networks will be growing in real-world dynamic applications, they would require different security mechanisms to be implemented in several devices in the universal settings. Many of these heterogeneous devices (including base stations) have different energy sources, some may run on batteries and others may run on the solar, or other source of energies. However, one security solution can fit to all strategy will no longer be practical in 6G due to the heterogeneity of the devices, dynamic services, energy constraints, threat vulnerabilities, and so on. This may lead to challenge how to implement bulky security solutions (e.g. symmetric or asymmetric cryptography, protocols, quantum-based security solutions) to the devices, in particular resource-constrained devices, in energy-efficient manner. For instance, smart devices in the setting of 6G network may have different hardware capabilities and those may be distributed or deployed in more adverse environments, which may not be covered by 5G, and security requirement must be tailored accordingly. Another example, as shown in Figure 11.6 undersea communication equipment usually are battery-powered devices than devices on land, power/energy must be preferred for many months or years. In this respect or many others, energy efficient security solutions are another challenging area.

Figure 11.6 Under water wireless communication.

A security-energy optimization framework has been suggested in [5]. The framework uses nonadditive measure to select the various attribute and then applies function approximation for model training. The nonadditive modeling can reduce the computational complexity significantly while providing model accuracy. A supervised learning is being used to generate the optimization model. Several attributes (e.g. application, AI utilization, battery capacity) in model belong to the subset those have largest interaction with the AI-driven optimization model training – so that the complexity of model can be significantly reduced.

Another research proposed a virtual mobile smart cell framework with high efficiency [6]. The proposed framework supports several services, efficient software-defined networking (SDN) virtualization, energy-aware wireless security, and energy-efficient handover control mechanisms to enhance the power consumption trade-off. AS shown in Figure 11.7 several data centers are deployed to support the mobile small cells through the L2 Link. Such framework offers a practical cell offloading solution that reduces the holistic energy consumption, where mobile cell head (MCH) offload and distribute data with the MSC. Such offloading can also exploit the intercell cooperation among different MCHs. In addition, the framework also supports efficient and secure key management, and all of these entities may use their self-signed certificates that would provide high level of trust among the mobile nodes in a MSC or on the move. A high-level key management diagram is shown in Figure 11.8.

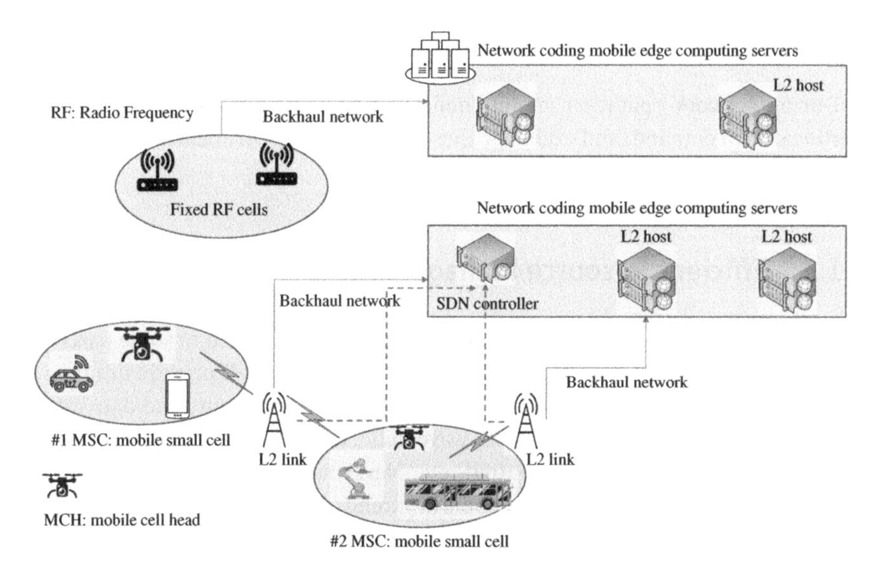

Figure 11.7 Network-coded cooperative mobile edge computing in mobile small cell management.

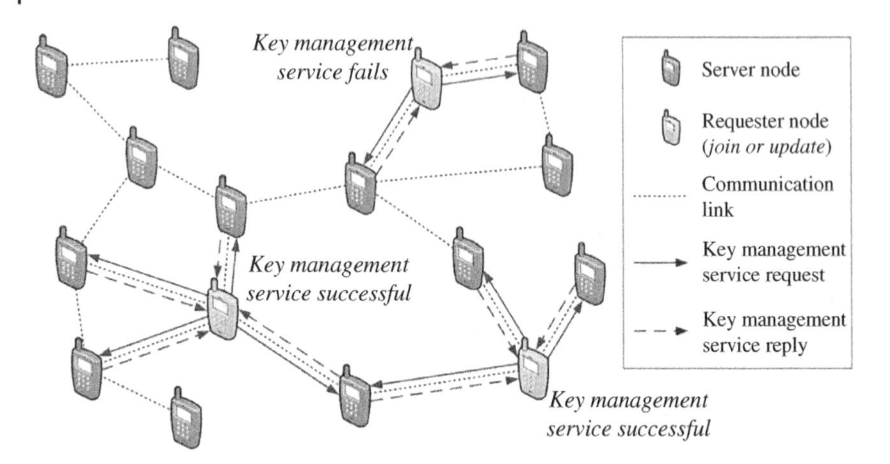

Figure 11.8 Secure and energy-efficient key management [6].

In the network initialization phase, a Trusted third party (TTP) initiates an initial set of network users by providing them with a share of the master private key. This enables a threshold amount of them to provide KM services during network operation:

- Providing a requester node with its proxy key pair, enables it to sign its self-generated certificates as if this was signed by the TTP.
- Providing a requester node with its unique share of the master private key and joining the MSC.

During network operation, mobile nodes can join the network, self-generate certificates on demand, and exchange these self-generated certificates to establish secure communications channels.

11.3 Efficient Resource Management

G mobile network is at the forefront and already being deployed in some parts of the world. It is anticipated that current 5G infrastructure will provide ultrahigh reliability, latency as low as 1 ms, and increased network capacity and data rates using several technologies, millimeter wave radio, utilization of higher frequency bands, multiple-input-multiple-output, etc. Nevertheless, the vision of 5G network has shifted to the 6G network. Some key trends that are envisioned to emerge in 6G are the following: virtual and augmented reality (or mixed reality), 8K high definition videos, holograms, remote surgery, the industry 4.0, super intelligent homes, artificial intelligence integrated services, mobile edge AI, Unmanned Aerial Vehicles (UAV), and autonomous vehicles, to name a few. However, all abovementioned trends will demand much more from mobile networks in terms of reliability, latency, and data rates than 5G. It is widely known that resource management is one of the major challenges in 5G, and due to the ever-growing scale of devices, ubiquitous connectivity and communication, future applications and many more, the resource management will be more challenging in 6G networks. Several proposals have been suggested in the literature, and each proposal has its own advantages and disadvantages. Most of them are deployed as a centralized system. Therefore, the centralized system may not be practical where diverse cooperation and coordination required among several data hungry applications in 6G.

A blockchain-enabled resource management (i.e. spectrum, computation, and energy sharing) framework is suggested in [7]. All these resources are dynamically allocated via network slicing and virtualization and are controlled by the blockchain. As shown in Figure 11.9, the dynamics of blockchain can be deployed to manage resources in a distributed manner that can organize customers and producers in energy market, and transparent information flow can accelerate the transactions. In summary, the spectrum auction or trading and computing platforms are incorporated in this resource bucket, where spectrum is efficiently distributed, and network slices are controlled and managed by the network operators. Consequently, the blockchain can enable efficient and secure resource management in 6G networks.

Achieving energy-efficiency in full-spectrum sharing from sub-6 GHz to THz is a big challenge. One solution fit-to-all is not practical in diverse 6G domain. Several challenges has been discussed in various research, but most of them point

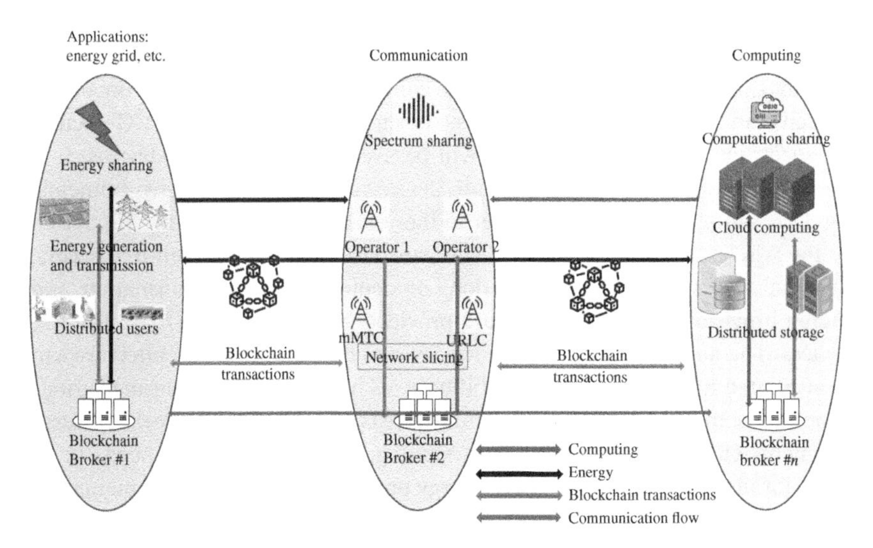

Figure 11.9 Resource management using blockchain.

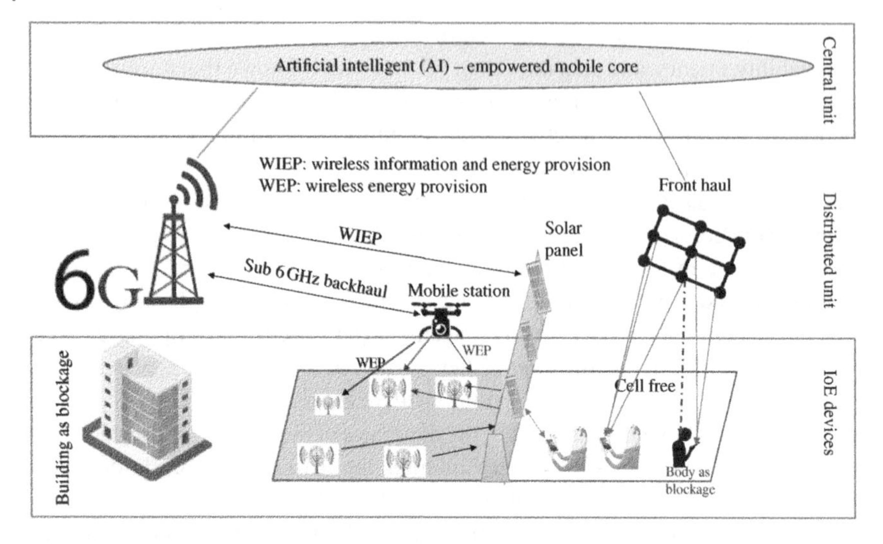

Figure 11.10 Energy-efficient self-sustainability architecture in 6G.

toward the self-sustainability using self-powered, green-powered, etc. However, the research proposed in [8] studied the intelligent reflecting surfaces, visible light communication, and proposed an architecture self-sustainable architecture for Internet of everything (IoE), as shown in Figure 11.10. In this, the traditional base station is broken into two units: central unit (CU) – it is analogous but super intelligent; and distributed unit (DU) – it is a cell-free network that resides very close to the end devices. Here, the DU is an employment of much higher frequency in 6G. The authors have divided their architecture into three layers: CUs, for example, AI-empowered mobile cores; DUs, for example, self-organized attocells; and zero-energy IoE devices. As shown in Figure 11.10, CU decides the spectrum as well as which DU will be used to cater the IoE devices. It is worth to mention that through the IoE devices and surrounding environments, few DUs setup the exact cell to cater to these IoE devices, which are supported by the wireless energy provision. These surrounding could be cell-free access, airborne access, IRS, etc. In addition, on-demand wireless information and energy provision (WIEP) services are provided by the CU and DU to the IoE devices. The authors claimed that each layer of their proposed architecture can be attributed to energy self-sustainability in 6G applications considering various dimensions. Indeed, the proposed architecture is one of earliest work that focuses on energy-efficiency in full-spectrum sharing, but more research needs to be done, for instance what if Wireless Energy provision (WEP) energy remains low due to various environmental reasons that may cause path loss due to channel attenuation, etc.

It is apparent that 6G will pave the way for deploying several billions of devices and handing them intelligently using advanced solutions (e.g. Edge AI, blockchain) in smart future applications. These applications will be facing a lot of technical and environmental challenges, including energy-efficiency. The dynamic nature of 6G will undoubtedly lead to an increase in power consumption in various devices, particularly the resource-constrained devices those will be using ML or AI complex algorithms. From the perspective of energy-efficiency, several energy-efficiency techniques in network management [1–4], energy-efficient security solutions [5, 6] and energy-efficient resource management [7–9] architectures have been proposed.

Acknowledgement

This chapter is contributed by Pardeep Kumar (Department of Computer Science, Swansea University, United Kingdom).

12

Harmonized Mobile Networks and Extreme Global Network Coverage

Harmonized mobile networks refer to the smooth interconnection of multiple communication technologies, data storage, and processing platforms at different scales. Extreme global network coverage refers to the digital inclusion through seamless global service coverage by connecting remote, rural, deep-sea, and even space locations. After reading this chapter, you should be able to

- Understand harmonized mobile networks in 6G.
- Understand extreme global coverage supported by 6G wireless systems.
- Explore limitations and research questions that need to be addressed in the future to enable network harmonization and extreme global coverage.

12.1 Harmonized Mobile Networks

Similar to 5G networks, future 6G networks will need to utilize both stand-alone and nonstand-alone deployments by leveraging both legacy networks and future new radio to enable new use-cases, satisfy new network requirements, provide extreme global network coverage, and support flexible deployments of network resources. Thus, future 6G networks will interconnect the different types of networks ranging from visible light communication to conventional terrestrial communications and satellite communications by aggregating various resources and technologies from Internet of Things (IoT) devices to core-network/cloud infrastructures. In terms of bandwidth, various frequency bands are adopted in 6G for different network deployments and purposes, ranging from sub-6G GHz to THz, free-space optics (FSO), and visible light communication (VLC). THz signals have many distinct features, for example THz radiation can penetrate many materials (e.g. clothing, glass, plastic, masonry, and wood) and are strongly

6G Frontiers: Towards Future Wireless Systems, First Edition.
Chamitha de Alwis, Quoc-Viet Pham, and Madhusanka Liyanage.
© 2023 The Institute of Electrical and Electronics Engineers, Inc. Published 2023 by John Wiley & Sons, Inc.

reflected by metals. Thus, THz bands have been used in many popular sensing, imaging, and localization applications. In addition, THz bands have found many promising use-cases in future 6G wireless systems, such as vehicular and drone communications, wireless data centers, and space communication networks. For illustration, we present three scenarios of 6G enabled by THz communications in Section 4.1, which include high speed transmissions, integrated backhaul and access networks, and high speed satellite communications.

The use of FSO is critical in beyond 5G and 6G, especially when high speed fronthaul and backhaul links in the range of multi-Gbps are needed to be implemented for remote and isolated areas. Generally, FSO systems are promising for many applications, e.g. campus connectivity, video surveillance and monitoring, disaster recovery, backhaul/fronthaul for wireless networks, security, and broadcasting. For example FSO backhaul links are implemented between the satellite and ground station and between the satellites located at the highest tier in aerial radio access network (ARAN) systems, as presented in Section 9.2 and illustrated in Figure 12.1. In addition, VLC using light-emitting diodes (LEDs) has

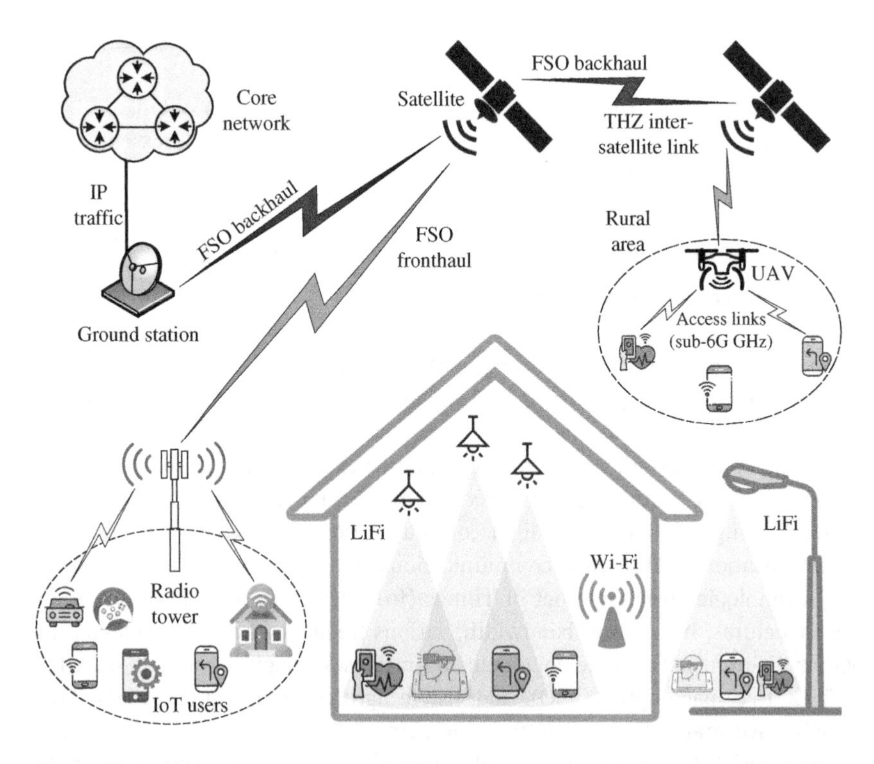

Figure 12.1 Network convergence of different frequency bands, including, Sub-6 GHz, FSO, VLC, and Wi-Fi, in future 6G wireless networks.

emerged as a promising technology in optical wireless communications. VLC has recently received much attention from both industry and academia due to lots of offered advantages, such as easy installation, high data rate communication, low cost and power consumption, and high security. Because of these features, VLC is considered a key-enabling technology in 6G networks and can be integrated with the other technologies to bring significant improvements. For instance, the integration of NOMA (non-orthogonal multiple access)-VLC is promising and has attracted significant attention. In fact, NOMA performs well at a high signal-to-noise ratio (SNR), which can be guaranteed in VLC systems thanks to the short distance from the LED transmitter to users and Line of Sight (LoS) propagation. Moreover, the channels stay almost constant in VLC systems, which is vital to NOMA functionalities in performing successive interference cancellation operation. Besides, each LED as a small access point in VLC systems and can provide services to a few users, while NOMA typically multiplexes a few number of users in a cluster. In aerial access networks, UAVs using radio frequency (RF) resources for communications are energy consuming, which is opposed to the fact that UAVs typically have a finite battery. This challenge can be addressed by equipping UAVs with VLC capabilities. Many experiments show that VLC systems, also known as light fidelity (LiFi), can achieve the peak data rate on the order of tens of Gbps. Moreover, many research centers and tech companies focus on developing commercial products using VLC. For example, LiFi-based access points can be deployed through the integration with existing light infrastructure, and thus they can be used for both simultaneous illumination and communication purposes. For the purposes of illustration, a 6G wireless system using THz, VLC, and FSO is shown in Figure 12.1.

In harmonized networks, it is important to consider heterogeneous networks, which utilize both licensed bands, such as sub-6 GHz, mmWave, and THz, and unlicensed bands, such as FSO, VLC, and Wi-Fi. Hybrid RF and VLC systems are promising for the implementation of VLC networks, where the downlink and uplink transmissions are carried over VLC and RF links, respectively. By capitalizing on the advantages of the two technologies, hybrid VLC-RF networks can achieve high capacity in VLC spectrum while ensuring network coverage in the RF spectrum. The coexistence of FSO and RF is necessary to overcome critical challenges of pure FSO systems in which FSO links are severely affected by turbulence channels and strong winds [118]. As RF links are more robust to turbulence and point errors than FSO links, they reserve backup transmission channels once FSO links become outage or unavailable. As an example of mixed RF-FSO networks, the work in [439] considers the use of RF spectrum for transmissions between massive ground users and a HAP through spatial division multiple access, and FSO is adopted between the HAP and satellite.

Due to the limited spectrum in RF networks, various Wi-Fi standards have been developed so far, for example 802.11ax for dense deployments, 802.11az for indoor location, 802.11ah for IoT, 802.11p for automotive, and 802.11ba for lower-power applications. In this regard, the coexistence of Wi-Fi and RF networks has the potential to bring significant improvements in terms of throughput, latency, security, coverage, and positioning. An illustration of a heterogeneous LiFi–Wi-Fi network is shown in Figure 12.2, adapted from [440]. In such scenarios, data traffic can be offloaded from Wi-Fi to local LiFi networks. As a result, Wi-Fi resources can be saved and used by users who are outside of the coverage of LiFi networks. In the network, illustrated in Figure 12.2, the software dened networking (SDN)-based switch can collect context and quality of service (QoS) information from both networks before being forwarded to the SDN controller, which is responsible for making control decisions based on the global view of the network.

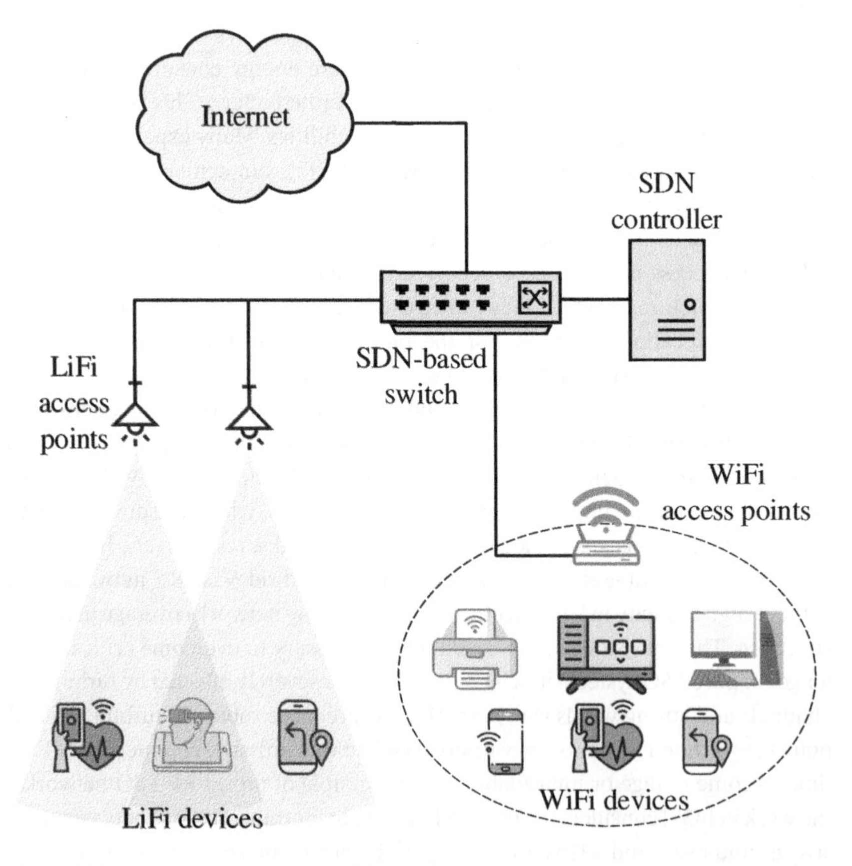

Figure 12.2 Network convergence of Sub-6 GHz, FSO, VLC, and Wi-Fi in 6G.

Because many applications and services will be available in 6G, it is expected that these technologies, along with many potential ones as presented in Chapter 4, will be efficiently harmonized in future 6G wireless networks. It is noted that since optimizing these heterogeneous networks leads to complex problems and requires a huge amount of data for learning models, AI and ML have found many applications in improving network performance. In Wi-Fi networks, a recent survey in [441] shows that numerous ML approaches have been developed for optimizing Wi-Fi features, e.g. channel access and sharing, traffic prediction, beamforming, spatial reuse, resource management, and signal classification.

The previous network generations (i.e. from 1G to 4G) focus mainly on providing wireless communications, while advanced services are mostly ignored. The 5G networks and beyond 5G are responsible for performing four major functions that are communication, computation, control, and content delivery (4C), as illustrated in Figure 12.3. One example of control functions is MEC coordination and collaboration. In other words, multiple nearby MEC servers can coordinate and collaborate with each other to tackle the limited resource limitation of each individual MEC server. This is similar to the concept of cloud radio access network (C-RAN), where a resource pool is created by virtually grouping resources from individual servers within the collaboration space. Due to the fact that (i) joint 4C optimization is needed for improving the network performance, and (ii) conventional

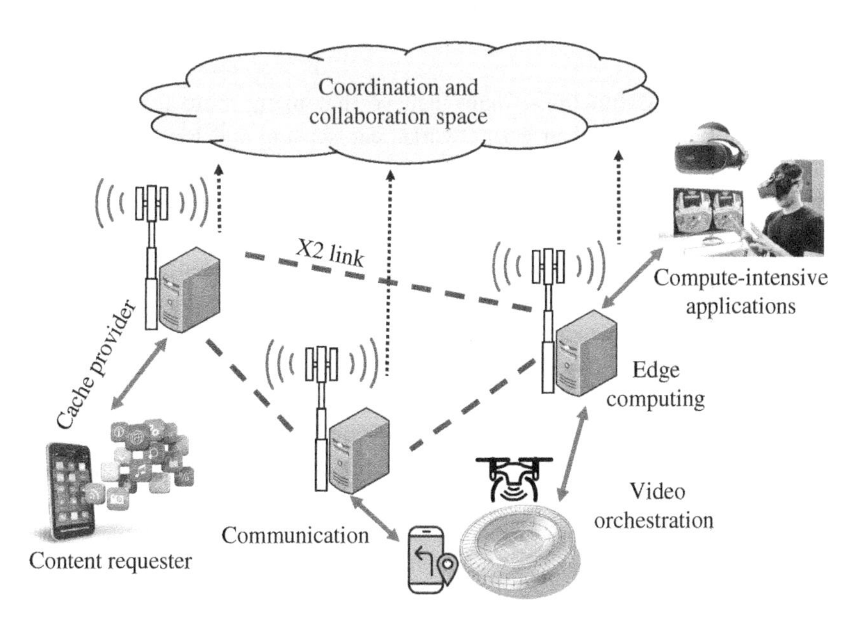

Figure 12.3 Illustration of 4C functions provided in edge computing systems. Source: Scanrail/Adobe Stock, Indo-Asian News Service/Red Pixels Ventures Limited, Techyuga.

approaches cannot efficiently solve complex problems with large action and state spaces, recent studies have addressed various problems pertaining to joint 4C optimization. For example, the work in [442] investigates two deep Q-learning models for mobile edge caching and computing in vehicular networks. To reduce the computational complexity of the original problem and circumvent the high mobility constraint of vehicles, the authors further proposed deploying two deep Q-network (DQN) models at two distinct timescales. In particular, each epoch is divided into several time slots and then the large timescale deep Q-learning model is executed at every epoch while the small timescale model is performed at every time slot. The work in [443] observes that existing works on deep reinforcement learning (DRL)-based offloading for edge networks consider discretized channel gains as the input data, thus suffering from the curse of dimensionality and slow convergence in the case of high quantization accuracy. Therefore, a continuous control with a DRL-based framework of computation offloading and resource allocation in wireless-powered MEC systems is proposed. The proposed algorithm is composed of two alternating phases: offloading action generation and offloading policy update. In particular, relaxed actions are quantized into a number of binary offloading actions in the former, while the best offloading action among quantized ones is used to update the deep neural network. Besides the integration of edge computing, in-network caching, and device-to-device (D2D) communications, the work in [444] also takes into consideration the social relationships among mobile users to improve the reliability and efficiency of resource sharing and delivery in mobile social networks.

In addition to 4C functions, many believe that many more functions and services will be converged in 6G networks, e.g. sensing and localization. It is since massive IoT and mobile devices are equipped with advanced sensors and navigation capabilities that facilitate the deployment of sensing and localization services in future networks, these devices may collect a huge amount of data but may not have enough storage and computing resources to perform the collected data successfully. As an attempt, the work in [445] investigates two problems in MEC systems, and the sum throughput is maximized by optimizing the time allocation for sensing, offloading, and computing and the resource allocation at the edge server. The two multiple access schemes, time-division multiple access (TDMA) and NOMA, show their superiority under different conditions, e.g. TDMA (NOMA) offers higher sum throughput under heterogeneous (homogeneous) sensing rates, while better fairness can be achieved by NOMA at the sacrifice of the throughput performance.

More recently, joint communication and radar sensing (JCAS) is an emerging research theme, which integrates the functionalities of the two individual systems, communications and radar, into one system only, as illustrated in Figure 12.4. Along with the emergence of remote sensing applications, radar has been recently

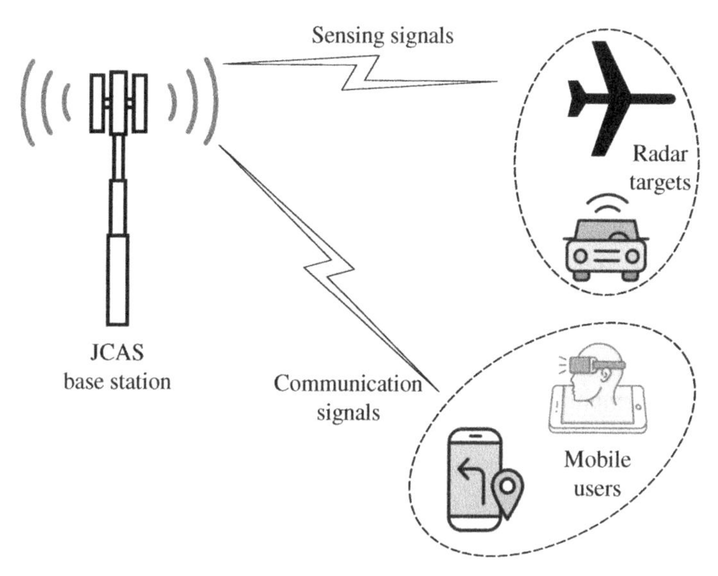

Figure 12.4 An example of JCAS systems, where a JCAS node transmits a single signal for both communication and sensing purposes.

used in a number of civil applications, such as traffic control, autonomous driving, and activity recognition. Therefore, JCAS makes existing mobile networks evolve into perceptive mobile networks, i.e. future 6G networks have eyes to see and understand the network environments. In the meanwhile, JCAS systems also expose some issues relating to spectrum management, and waveform recognition plays a key functionality in the physical layer of JCAS systems. More recent studies using deep learning have been investigated for JCAS waveform recognition due to the capability of processing high-dimensional, unstructured, and massive data. For instance, the work in [266] proposes a novel network with deep convolutional neural network architecture to learn radar-communication waveform patterns in the training stage and to classify the waveform of incoming signals in the inference (also known as prediction) stage precisely. Experimental results show that the accuracy improves further as the SNR increases, where all the eight waveforms attain over 90% at 30 dB SNR.

To enable harmonized networks, 6G should utilize new, advanced networking technologies, such as autonomous mesh networks [446], IoBNT, IoNT, and Internet of Space Things [447], scalable D2D communications [448] in addition to legacy wireless technologies. New 6G applications with enhanced human–machine interaction and human cyber–physical interaction will be needed through the exploitation of biosensors and brain-computer interfaces. Moreover, novel and alternative computing paradigms, e.g. serverless computing [449], in conjunction with the ultraflexible splitting of functionality between end devices and edge nodes, are also needed to satisfy the demands and requirements of new 6G applications. Moreover, 6G harmonized networks should be utilized for both AI and network programmability concepts to automate the harmonization. In this regard, there are limitations related to the 5G system architecture. For instance, the 5G architecture is not optimally designed for integrating new features, such as managing network nodes operating at above 100 GHz frequencies, interconnecting millions of (sub-) mesh networks, and network procedures exploitation by using AI. Therefore, it is needed in 6G to integrate and leverage these trends at the 6G concept development phase to offer significant improvements in flexibility, cost efficiency, and complexity. Moreover, 6G network harmonization will cover various new environments, such as modular flexible production cells, zero-energy sensors, solutions for on-demand connectivity in remote areas with different stakeholders. Since different networks have different network characteristics, network harmonization in 6G will allow to provide the required coverage, connectivity, and dependability and also satisfy the requirement of heterogeneous networking environments. AI-enabled network harmonization will allow 6G to deploy ultraflexible resource allocation procedures in challenging environments, such as those populated by mobile devices with special requirements (e.g. reliability, energy efficiency, and security) and in need of coverage (e.g. emergency scenarios and remote areas).

To realize harmonized mobile networks in 6G, a key enabler is the enhancement of radio access virtualization and softwarized methods, e.g. network function virtualization (NFV), SDN, network slicing (NS), and Zero touch network and Service Management (ZSM). As presented in Chapter 6, SDN offers the ability to control the network traffic routing centrally and intelligently using software applications, and NFV is a concept that is used for packaging network functions so that they can run on commodity and general hardware devices. NS is a technology that divides the physical network into separate logical networks known as slices, each of which can be configured to offer specific network capabilities and network characteristics. Meanwhile, ZSM allows the accommodation of new services and the modules to be scaled and deployed independently. As an example, key features of SDN, e.g. centralized control, network programmability, and abstraction, can be exploited to decide which one of the networks should be selected in case a mobile device is under the coverage of both LiFi and Wi-Fi as in Figure 12.2. SDN can be used in conjunction with NFV to provide flexibility, security, and scalability to the dynamic design of 6G networks with heterogeneous resources and technologies. In fact, facilitating heterogeneous QoS requirements of 6G IoT applications through the same network infrastructure is challenging. NS ensures facilitating QoS requirements of different 6G IoT applications through dedicated slices for each use case and dynamic resource allocation between slices.

12.2 Extreme Global Network Coverage

Connectivities vary among different regions and can be divided into four levels, including not connected, underconnected, connected, and hyperconnected. On the one hand, the connectivity can be connected/hyperconnected in urban areas, where various networks are deployed to serve a large number of populations. On the other hand, people living in underconnected/not connected areas may have no Internet access or have limited access with very low data rates and discontinuous services. Since the first deployment in South Korea in April 2019, 5G networks have been deployed in many areas and countries in the world. However, more than 50% of the world population (i.e. more than four billion people) are still not connected or underconnected. The digital divide issue is further emphasized by the percentage difference between developed countries (22%) and less-developed countries (85%) [450]. 6G has been conceptualized to go much beyond the maximum coverage that can be achieved by the full-fledged deployment of 5G. By 2030, 6G energized telecommunication industry is expected to offer global services encompassing remote places (e.g. rural and underconnected areas), oceans, vast-land masses with low inhabitants, air, and space [53]. The obvious advantages of connecting the unconnected will be the minimization of the digital divide by connecting everyone and everything, higher business opportunities in every part of the world supported by the continuous growth in local operators, and easy roll-out of essential services like safety and basic governmental benefits. 6G shall put digital inclusion as one of the top priorities and encompass efficient and affordable solutions for global service coverage, connecting remote places, e.g. in rural areas, transport over oceans or vast landmasses, enabling new services and businesses that will promote economic growth and reduce digital divide as well as improving safety and operation efficiency in those currently under-/uncovered areas.

5G networks were originally conceived as terrestrial networks and presumed that a large population of the world would not be connected [333]. The unconnected areas mostly include oceans, deserts, deep forests, rural areas, mountains, and airspace. There are many critical challenges in providing Internet access to unconnected areas, such as low potential returns, inaccessibility, spectrum availability, maintenance, infrastructure, and power grid. However, 6G is believed to overcome this shortfall of predecessor generations and emerge as universal communication systems offering ubiquitous global coverage. Potential technological solutions that have been identified to play a vital role in driving the extreme global network coverage are UAVs as mobile and agile aerial base stations, satellites, in particular, low earth orbit (LEO) and very low earth orbit (VLEO) for broader coverage, cell-free architecture based on distributed multiinput-multioutput (MIMO) for multiple simultaneous connectivities to access points, THz and VLC for new resources to boost capacity, blockchain-integrated AI for efficient spectrum management, and decentralized business ecosystem for incentivizing smaller players to participate in extending the radio footprints. Thus, 6G, with its extreme global coverage, is envisaged to be a key enabler for never-seen-before types of applications and will also minimize the digital divide. The latter is especially important in an emergency or global pandemic situation like Coronavirus disease (COVID-19). Besides technological factors, providing global network coverage requires considering the other important factors (e.g. politics, economics, societies, environments, and legalities) and needs the involvement of multiple stakeholders (e.g. governments, mobile network operators, and infrastructure providers).

Among potential technological solutions, nonterrestrial networks are needed in 6G to achieve global service coverage by offering full digital inclusion in the areas, such as the deep sea and space. To be more specific, a nonterrestrial network is composed of different platforms, from low-altitude platform (e.g. UAVs and drones) and high-altitude platform (e.g. aircraft) to satellites (e.g. LEO satellites). This network architecture has some distinct features, which make it distinguished from other architectures:

- *Ubiquity*: The LEO communication platform guarantees service continuity across the globe with three advantages, including wide coverage, networking backup/resilience, and emergency broadcast, for which existing terrestrial communication systems have limited capacity and cannot provide ubiquitous wireless connectivity.
- *Mobility*: The dynamicity of aerial low-altitude platform (LAP)/high-altitude platform (HAP) topology implementation and the networking overlay among the LAP, HAP, and LEO communication systems help ARANs to adapt flexibly to the requirements of end users anywhere on the ground and in the air.
- *Availability*: Operating in the air at various altitudes, ARANs are not commonly affected by natural and man-made disasters capable of rendering terrestrial communication infrastructures vulnerable and interrupting service. In other words, better service availability is provided regardless of the recipient's terrain, such as mountains, seas, and deserts.
- *Simultaneity*: Multitier LAP/HAP/LEO communication systems can adaptively self-organize to forward information-centric services effectively across discrete (aerial and terrestrial) locations on simultaneous multicast and broadcast streams using various wireless access technologies.
- *Scalability*: Because aerial base stations interlink to each other using aerial wireless ad hoc technologies and there are hierarchical networking overlays among LAP, HAP, and LEO communication platforms, ARANs can quickly establish scalable topologies for local sites without service interruptions.

This nonterrestrial network, referred to as ARAN, is presented in Section 9.2, when we explain different RAN concepts for future 6G wireless networks. ARAN, along with terrestrial communication systems, can provide extreme global network coverage and has full support for emerging applications with various applications and diverse deployment scenarios. More details of the ARAN architecture and its network analyses and enabling technologies can be found in [122].

The hierarchical architecture of 6G ARANs provides heterogeneous communication capabilities from multiple platforms with distinct functionality and features. As a result, ARAN can support a broad range of emerging applications and services, such as wireless coverage expansion, aerial surveillance, precision agriculture, and commercial delivery. Based on the networking requirements, applications of 6G ARANs can be divided into three categories, including event-based communications, scheduled communications, and permanent communications. The first category, event-based communications, defines application scenarios in which networking infrastructures are temporarily required to provide and boost communication services for short-duration events, for example disaster and search and rescue scenarios. An example of the first category is [451, 452], where HAP stations are considered as super macrobase stations to provide various wireless and computing services for not only disaster and remote applications but also advanced network scenarios. The second category is scheduled communications in which ARAN components are deployed to fly on a predefined path to provide users with networking services of a given duration. Aerial surveillance and smart agriculture are two well-known examples of this category. For example, the work [453] reviews many kinds of UAVs equipped with suitable sensors for smart farming scenarios, such as weed management, crop health monitoring, plant counting and numbering, pest management, and assessing plant quality. The third category, permanent communications, involves application scenarios where networking services are required continuously over a long period, e.g. urban monitoring, healthcare, intelligent transportation systems, and networking in underserved areas. For example UAVs are assigned as intermediate aerial nodes to improve coverage as well as boost the system capacity in [454]. This work can improve the efficiency by up to 38% and reduce the delays by up to 37.5% when it is compared with the systems without UAV assistance. Therefore, such a network model is suitable for areas with high demand for connections.

Besides the above application scenarios, each platform in the ARAN network has distinct applications and carries its complement to the other platforms and terrestrial communications. From the architectural viewpoint, HAPs are located in the middle tier of ARAN systems and have considerable advantages when compared with other platforms, including large-area coverage with the sell size up to 10–20 km, adaptability to traffic demands, rapid deployment compared to terrestrial and satellite systems, high endurance compared with LAP platforms, and green operation with solar power [122]. HAP stations have the potential for various application domains, such as IoT applications, backhauling for small and isolated base stations, temporary unpredicted events, agile computation offloading, flying data centers, coverage holes, massive UAVs and aerial users, and intelligent vertical domain applications [454]. In the following, we explain the two representative use cases of LAP/HAP systems for IoT applications and wireless backhauling. More details on the applications of ARAN and HAP systems can be found in [122, 451, 452].

The basic concept of IoT is that anything can be interconnected with the global information and communication infrastructure at any time and any place. Although IoT can potentially benefit modern society, many technical issues and requirements remain to be addressed, such as automatic networking, massive connectivity, manageability, interoperability, and autonomic services provisioning [71]. Thanks to the quasi-stationary feature and wide footprint, HAP stations have the potential to enable IoT applications. For example, IoT devices with limited computing resources can offload their computation tasks to the HAP-based edge servers for remote computation. Moreover, to support a large number of IoT devices, HAP stations can be deployed to form a collaborative edge computing server with more powerful capabilities and higher time-endurance, which helps to execute the computation tasks offloaded from IoT devices on the ground and aerial components (e.g. UAVs and LEO satellites). An illustration of collaborative computation between the LAP and HAP platforms in a two-tier computing system is shown in Figure 12.5. In this scenario, the HAP station is empowered with powerful computing capabilities to perform computation tasks offloaded from UAVs, which are mainly responsible for data collection from ground IoT users.

It is noted that the HAP/LAP can also be considered as aerial users in edge computing systems and thus offload their computational tasks to terrestrial computing servers for remote processing. There are challenges that need to be considered before HAP stations can be used for IoT applications in practice. The first challenge is the power consumption of IoT devices required to transmit over a long distance (i.e. 10–20 km). Since IoT applications typically have low data rates, e.g. around 50 kbps for LoRa and 1–10 Mbps for NB-IoT, IoT devices of these applications may only need to spend a low amount of transmit power in order to communicate with HAP stations. Moreover, due to large coverage, HAP stations may receive simultaneous transmissions and computation requests from massive IoT devices. This is a new challenge, namely mx-MTC (machine type communication) with $x \gg 1$, which needs to handle many more IoT connections than that of the massive MTC (mMTC) use-case in 5G networks [452]. Therefore, a significant challenge is how HAP systems can keep a balance between packet

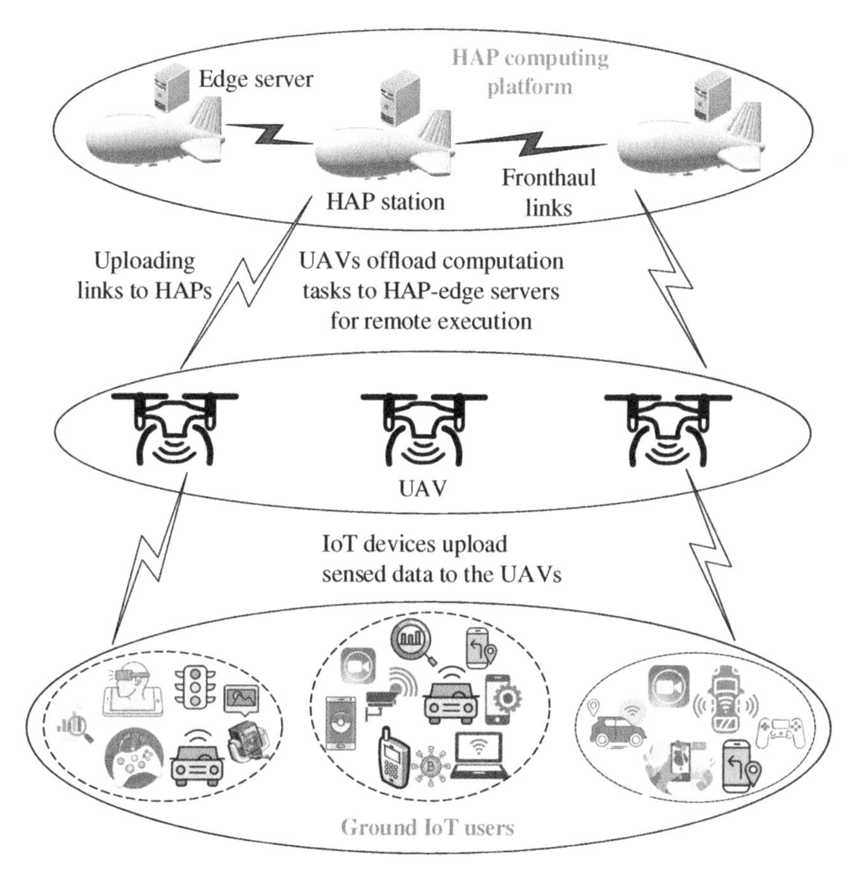

Figure 12.5 A collaborative computation system, where HAP-edge servers execute computation tasks received from UAVs and IoT devices. Source: Adapted from Pham et al. [165].

collision and reliability requirements of mission-critical IoT applications. HAPs in 6G networks are expected to address these challenges in order to enable future IoT applications.

To fulfill the key requirements of future beyond 5G and 6G networks, network densification will be one of the key solutions. A fundamental question is how to design efficient backhaul solutions with the capability to forward and receive massive traffic from/to small cell users [455]. With the increasing network densification, implementing wired backhaul for a great number of small cells would not be affordable or even feasible. The reasons for this are as follows:

- It is costly and time-consuming to implement wired backhaul for a large number of small cells [455]. In addition, deployment of wired backhaul depends on various factors, such as the location of small cells, quality of service (QoS) requirements of mobile/IoT users.

- Different frequency bands have been proposed for wireless backhaul [455, 456], such as cellular frequency band, mmWave band, sub-6 GHz band, satellite frequency band, FSO, and TV white space band. Therefore, wireless backhauling provides a practical solution for dense small cells in the emerging 5G networks.
- Providing wireless access to rural/remote areas and some urban areas requires carefully considering the deployment cost. In such scenarios, wireless backhaul is a practical and affordable solution, which can simplify the deployment and drive down the maintenance cost.
- In heterogeneous networks with all wired backhaul links, broken links may not be recovered instantly, and thus emergency services can be severely impacted due to the slow network recovery and low reliability. Deployment of wireless backhaul can enable to mitigate the aforementioned issue.

Due to many advantages, wireless backhauling has been considered a well-accepted and cost-effective solution and received enormous attention from the communication community. Motivated by recent advancements in aerial access communications, HAP/LAP stations can be exploited to provide backhauling solutions for small cells, as illustrated in Figure 12.6. One option for wireless backhauling in HAP systems is using FSO links. Although FSO communication links are much affected by unfavorable weather conditions (e.g. cloudy, rainy, and foggy), many studies have shown that the data rate of FSO links can be very high and in the range of (multi-)Gbps in favorable conditions (e.g. clear sky and foggy) [451]. Another solution candidate is using mmWave communications. To compensate for high signal attenuation and absorption losses of mmWave communications, several promising approaches have been proposed, such as mmWave-MIMO combination, hybrid mmWave/THz/FSO network, intelligent surfaces, resource management, and reflectarrays [80]. In addition to the above merits, HAP stations are quasistationary and have a large footprint; therefore, backhauling small cells in metropolitan areas and megacities can be realized via the deployment of HAP/LAP stations. Deploying HAP/LAP stations becomes more appealing when one needs to implement the backhaul links for base stations in hard-to-reach or isolated areas (e.g. deep seas and deserts).

There are a number of enabling technologies to realize ARANs and extreme global coverage in 6G networks, including energy refills, operational management, and data delivery. Energy refilling technologies (e.g. energy harvesting and wireless power transfer) are necessary as many IoT devices and typical aerial components in the ARAN architecture face battery storage capacity limitations. On the operational management plane, key enablers are network softwarization, mobile cloudization, and data mining. For example, SDN, as a

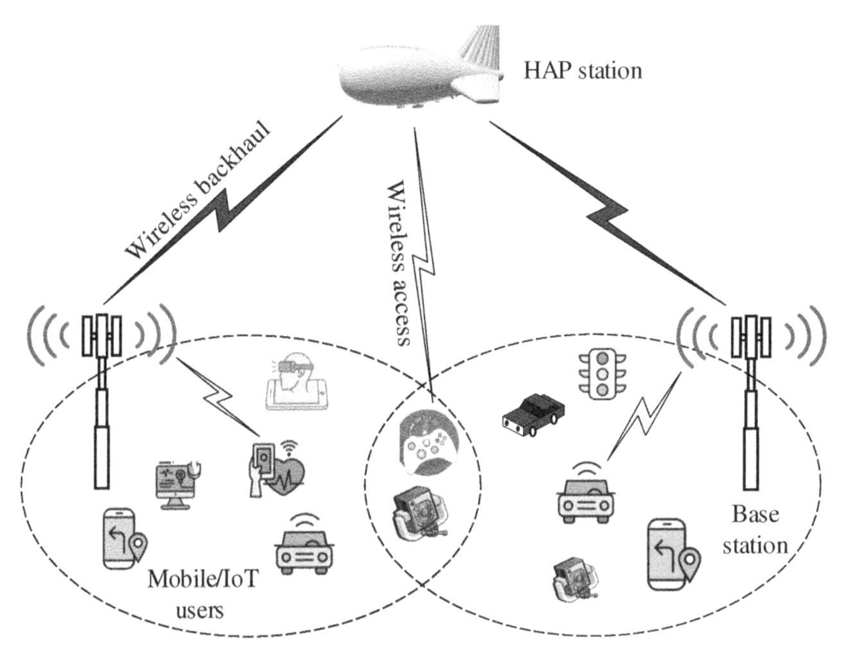

Figure 12.6 Illustration of HAP/LAP deployment for backhauling solutions in beyond 5G/6G networks, where HAP stations can provide direct access to users and backhaul links with terrestrial base stations.

network softwarization technology, empowers ARANs with enhanced control, situational awareness, and flexibility, coordinates interference avoidance, and facilitates interoperability between network nodes. Mobile cloudization distributes computing resources at all the tiers of the ARAN and thus enables the deployment of compute-intensive applications and facilities via the flexibility of network resource allocation. Moreover, data mining techniques can be applied to extract useful patterns from collected data and context information, thus reducing energy consumption, improving network security, sharing the workload across an entire network, increasing bandwidth, self-organizing networking configurations, achieving autonomous training operations, handling mobile big data, and analyzing mobility [122]. Further, by exploiting the value of data mining, AI and ML can provide solutions for simultaneous massive numbers of user connections in a dynamic, heterogeneous, and unpredictable network resource such as ARANs in the context of 6G. Finally, various frequency bands, as presented in Section 12.1, and multiple access technologies, such as cell-free massive MIMO, NOMA, and intelligent surfaces, can be used to support extreme global coverage in 6G networks.

12.3 Limitations and Challenges

In Section 12.1, we present the vision of harmonized mobile networks in 6G. The above discussions reveal the following limitations and research questions that are required to be addressed to enable harmonized networks in 6G.

- What is the new 6G architecture in which harmonized networks enable flexible network typologies and backward compatibility for legacy networks?
- How to harmonize different networks to achieve the convergence of the biological, digital, and physical world?
- What are the architectural options to enable the allocation, instantiation, and operation of AI functions over harmonized network deployments?
- How to tackle the large variations of available capacity at short timescales in 6G applications at architecture level?
- What are the possible global-level optimization mechanism to increase the collaborative utilization of available resources at different networks?
- How to enable the network harmonization in a trustworthy manner?
- How to create new services, such as dynamic function placement, network programmability, and processing of offloading at a global level in harmonized networks?
- What are the possible supporting protocols which can enable integrated and distributed AI to automate the network harmonization?
- How to design an automated and intelligent harmonized network equipped with distributed computing capabilities?
- How to develop technologies for new harmonized network requirements, such as network function refactoring, run-time scheduling of network resources and predictive orchestration?
- How to enable the integration in wireless networks closely interacting with brain-computer interfaces and human-machine interfaces?

Based on the above discussions, limitations, and challenges, the future 6G research should focus on network harmonization by focusing on the integration of nodes using above 100 GHz spectrum (including THz, FSO, and VLC), nonterrestrial networks, nanoscale networks, autonomous D2D communications. One of the key enablers for such harmonization in the 6G era will be the enhancement of radio access virtualization methods. It will enable abstraction and programmability, which are essential to develop harmonized networks in 6G. Moreover, 6G should evolve as new service-based networks which can be enabled via network harmonization. Service-based networks have optimal service independency, flexibility, and re-usability. In fact, 6G can achieve this by evolving from fully cloud-native network functions in 5G to alternative computing paradigms, such as edge AI and serverless computing. This approach will need to redesign the signaling procedures and interfaces in legacy mobile architectures. For a better realization, network harmonization should support crucial factors, such as high reliability, accountability, availability, and liability. The cross-layer technologies, such as communication-control co-design and spatiotemporal network design [457] and also novel computing techniques, such as blockchain [264] can be used to achieve the above requirements. Moreover, the end results of the network harmonization can be further improved by the optimization of multifunction resources, such as 4C, sensing, and localization.

There are also interesting research questions that need to be answered to realize the vision of extreme global network coverage for 6G wireless systems.

- How to intelligently manage the seamless three-dimensional mobility while moving vertically up (in air or space) or down the ground level?
- How to efficiently manage and control a swarm of LEO and very-LEO satellites as well as enable secure and cost-efficient intercommunications?
- What could be the time-efficient solutions that can help overcome Doppler shift and Doppler variation issues that occur due to the relative motion of the earth and LEO satellite [333]?
- How to optimize power radiation and improve system efficiency while achieving extreme global network coverage?
- What are the channel models that can be used for heterogeneous scenarios (e.g. UAV, maritime, satellite, and vehicle-to-X) that will use different frequency bands such as sub-6 GHz, mmWave, and THz bands?
- How to develop an automated, open, and decentralized wireless market that motivates business entities of all sizes to participate in realizing extreme global network coverage?

Considering these limitations and challenges, some of the important areas need dedicated research inputs for the vision of extreme global network coverage in 6G. In the context of flying relays and aerial base stations, we need to optimize three-dimensional localization and network planning [458]. These areas of research become even more important since UAVs have resource constraints and are characterized by heterogeneous processing and storage capacities. Another area of future work concerning global network coverage would be secure interoperability among various operators and smart three-dimensional roaming mechanisms applicable for heterogeneous networks utilizing different technologies. Federated learning has the potential to globally optimize the 6G ecosystem while ensuring data privacy and ubiquitous network coverage. Despite being a key enabler of extreme global coverage, security and privacy are critical issues in 6G ARANs due to LoS communication links, high mobility, and limited resources of aerial components. Therefore, designing secure and privacy-preserving communication protocols for 6G ARANs that consider limited resources and mobility features of aerial components is crucial and worth further investigation. Yet another area of future work that can drive extreme global coverage is channel estimation and dynamic link optimization. Advanced techniques, such as data mining and AI, are necessary to provide efficient and cost-effective solutions for ubiquitous coverage. Finally, it would be highly beneficial to design simulation tools for 6G networks that can fully characterize design features, constraints, environmental factors, allowing precise simulations of various 6G technologies, protocols, and network deployments.

13

Legal Aspects and Standardization of 6G Networks

6G introduces a new landscape for mobile communication. This requires new legal framework and standardization efforts. This chapter focuses on 6G legal aspects and standardization approaches. After reading this chapter, you should be able to

- Understand the improvements needed in legal framework to fit the 6G era.
- Understand the key standardization efforts related 6G networks.

13.1 Legal Aspects

The evolving 6G landscape will also demand changes in legal and regulatory domains. For instance, 6G mobile networks will heavily depend on artificial intelligence (AI) technologies such as machine learning and deep learning, to optimize network performance while enhancing efficient resource utilization and user experience. These AI technologies will utilize the data obtained through numerous network and user equipment. The data obtained through numerous connected devices may be sensitive, obtained in such a way that users are not perfectly aware of and cause privacy attacks. Hence, legal frameworks should evolve to identify who has the ownership of data and who will decide the technologies, rules, and regulations on how the data will be processed and used.

In addition, spectrum management in THz frequency range and different rights in the context of spectrum sharing are areas that need to be thoroughly explored [272]. Furthermore, 6G will require denser network architectures with small cells. However, existing regulatory barriers may hinder the deployment of small cells. Also, there can be regulatory challenges when 6G evolves as a network-of-networks, as different networks may be considered under the purview of various regulatory and legal frameworks.

6G Frontiers: Towards Future Wireless Systems, First Edition.
Chamitha de Alwis, Quoc-Viet Pham, and Madhusanka Liyanage.
© 2023 The Institute of Electrical and Electronics Engineers, Inc. Published 2023 by John Wiley & Sons, Inc.

Furthermore, emerging technologies, such as Explainable AI (XAI), which interprets the outcomes of AI models, may increase privacy concerns of users [459, 460]. XAI models are able to discover strong nonlinear mappings using data. Thus, XAI models can understand the following:

- Data types and features vital to make decisions.
- Data that is required to make better predictions.
- How to improve algorithms to improve the accuracy of predictions.
- Unravel hidden bias between data and algorithms.
- How to assist and train humans to discover new patterns.

However, the existing legal framework needs to be updated considering the advancements of AI technologies. Also, it should be noted that humans have the capability of explaining their decisions, if required, in the eyes of law. Similarly, machines that utilize AI techniques, such as ML and DL, are also required to explain their decisions to ensure their accuracy and reliability.

13.1.1 Recent Developments of Legal Frameworks

Considering the growing challenges to govern and regulate AI and other 6G technologies, legal frameworks are being developed to be more suitable for the future society. Some of the recent developments of legal frameworks to accommodate the advancements of AI and other 6G technologies are briefed below.

- *General Data Protection Regulation (GDPR)* [461]: The GDPR is an European Union (EU) regulation law to ensure data protection and privacy in the EU and the European Economic Area (EEA). The GDPR came into effect from May 2018. Under GDPR, businesses require data protection compliance to ensure user privacy. GDPR also protects the fair usage of data collection, processing, and usage and also maintains an accurate an updated reflection of data. GDPR also declares that a data subject should have the right to not be a part of a decision taken by automated evaluations and processing that may lead to legal effects. This also includes user profiling, which makes decisions or performs actions considering personal aspects, in situations having legal concerns. For instance, companies are required to obtain permission from the user prior to installing a cookie in the user device to track user behavior. However, in exceptional circumstances, such data processing can be used with provisions for right to human intervention, right to obtain human intervention to challenge the predictions and opportunity to obtain an explanation on how the predictions were made.
- *French Digital Republic Act* [462]: This act amends the French Data Protection Act and modifies French law concerning several areas, such as intellectual property, medical research, and consumer protection. This also marks the initiative to implement the GDPR in all EU member states. This act empowers the

French data protection authority while providing new rights for individuals. These new rights include right of self-determination (right to control use of his personal data), right of access rectification (individuals can make a request when accessing and using their personal data), right of self-determination (decide and control the use of his or her personal data).

- *Equal Credit Opportunity Act (Regulation B)* [463]: This ensures that credit scores of applicants are not affected by factors, such as their race, color, religion, or sex. Creditors should also notify applicants on actions taken including collecting information of spouses, applicant's race, and other personal characteristics.
- *Algorithmic Accountability Act of 2019* [464]: Similar to GDPR, this mandates companies to provide a detailed description on how their automated decision-making works. For instance, companies should justify when particular adverts are displayed to avoid any bias or discrimination.
- *Other approaches* [459]: The United States has also chosen to implement specific privacy and data protection policies in various sectors including healthcare, finance, consumer rights, and federal agencies.

Many countries in the world have also started implementing laws and regulations to secure user privacy. One such example is the Personal Data Privacy law in China [465]. This law mandates tech companies to ensure that user data is neither misused nor mismanaged while user data is stored securely to avoid any privacy violations. In addition, Russian Federal Law No. 152-FZ also regulates the processing of personal data [466]. Additionally, Brazil's Lei Geral de Proteção de Dados (LGPD), Australia's Privacy Amendment (Notifiable Data Breaches), and Japan's Act on Protection of Personal Information are also important steps toward improving legal frameworks, such that it protect user privacy, in the AI-based landscape facilitated by 6G communication networks.

13.2 6G Standardization Efforts

Standardization is important to define the technological requirements of 6G networks and also to select the suitable technologies to deploy 6G network. Thus, the global telecommunication markets are shaped by standards. Many Standards Developing Organizations (SDOs) are working or at least planning to work on 6G standardization (Figure 13.1).

13.2.1 European Telecommunications Standards Institute

European Telecommunications Standards Institute (ETSI) [467] is a large telecommunication SDO with over 900 member organizations from 65 different countries. Since 6G-related research are still is at very early stages, ESTI is currently focusing

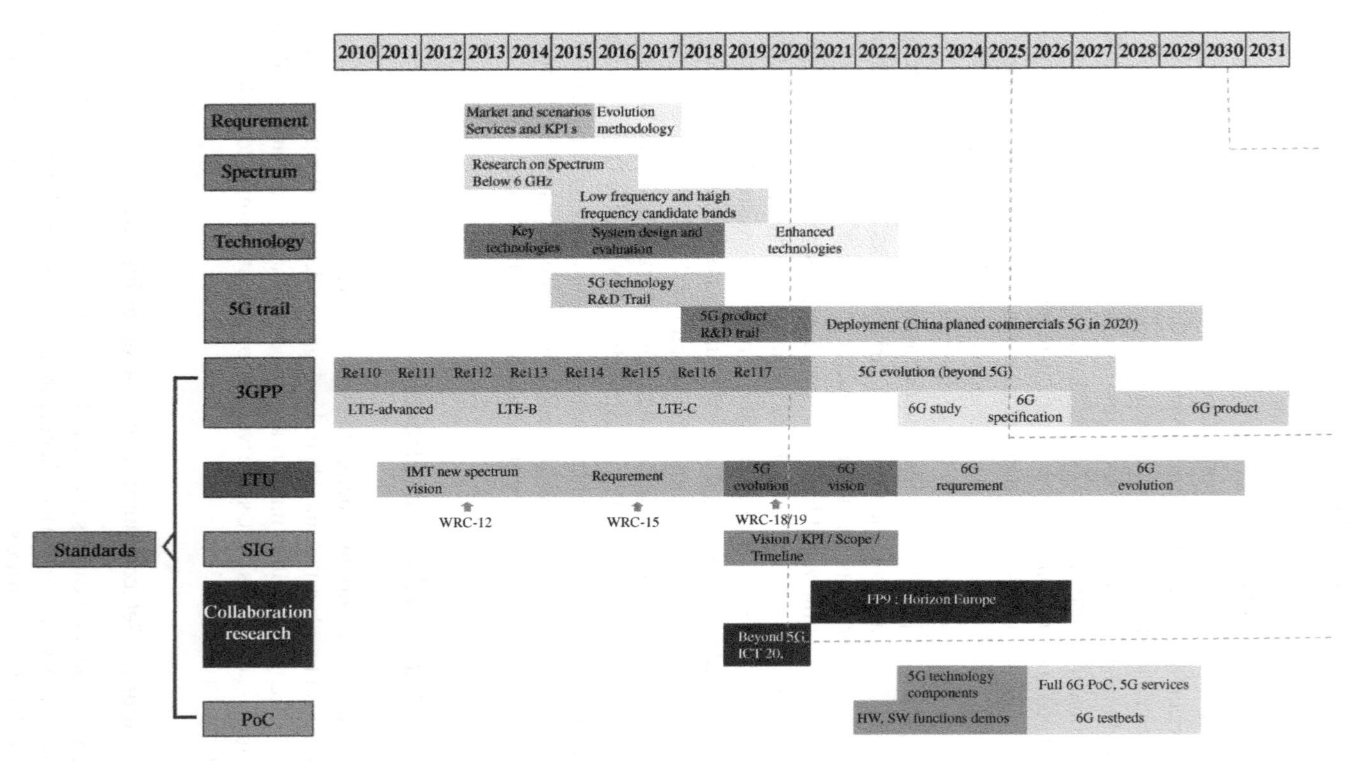

Figure 13.1 Global 6G Standards.

on 5G and 5G advanced standardization activities. ETSI is also expecting the first 6G services only appearing after 2030. Moreover, ETSI supports the 6G corresponding European Funding program, i.e. (Horizon Europe 2020–2027) which could generate impact research on standards. Preliminary funding suggests that technologies such as beyond mmW or THz Communications, smart surfaces, large intelligent surfaces, AI, SSN, energy harvesting/transfer and nanophononics (glass to radio) will be interested in 6G standardization.

13.2.2 Next Generation Mobile Networks (NGMN) Alliance

Next Generation Mobile Networks (NGMN) [468] is an association which is focusing on mobile telecommunications standard development. It has members from different telecommunication stakeholders such as mobile operators, vendors, manufacturers, and research institutes.

The NGMN alliance kicked off a new project called "6G Vision and Drivers." This project is focusing on providing early and timely direction for global 6G research and development activities. NGMN will work with all NGMN partners, i.e. Mobile Network Operators, Vendors/Manufacturers and Research/Academia across the globe on development of the mobile network technologies for the NGMNs beyond 5G.

13.2.3 Alliance for Telecommunications Industry Solutions (ATIS)

Alliance for Telecommunications Industry Solutions (ATIS) [469] is a leading ICT solution development organization with 150 member companies globally. AITS is focusing on different technologies including 6G, 5G, IoT, Smart Cities, AI-enabled networks, DLT/blockchain technology, and cybersecurity.

ATIS launched a call to action promoting US leadership on the path to 6G. It promotes innovative research and relies on 6G to position the United States as the global leader in 6G services and technologies during the next decade and beyond. This call to action has later outgrown as the Next G Alliance.

13.2.4 Next G Alliance

Next G Alliance [470] is an organization under AITS which is focusing on the supporting 5G evolutionary paths and 6G development in North American region. This is private sector-led novel initiative which is focusing on rapid commercialization of 6G technologies by supporting the complete 6G commercialization life-cycle including research and development, manufacturing, standardization, and market readiness.

13.2.5 5G Automotive Association

The 5G Automotive Association (5GAA) [471] is a global alliance which is working on better utilization of 5G for automotive industry. 5GAA is developing concepts called the Cooperative Intelligent Transportation Systems (C-ITS). The key enabling technology of C-ITS is wireless communication which is collectively referred to as Cellular vehicle-to-everything (C-V2X) communication.

The 5GAA is working on identifying the superior standards-based cost-effective and scalable 5G era and beyond wireless communication technologies for C-ITS and Connected Vehicle applications.

13.2.6 Association of Radio Industries and Businesses (ARIB)

The Association of Radio Industries and Businesses (ARIB) [472] is a SDO based in Japan. ARIB promotes R&D of new radio systems to achieve the efficient use of the radio spectrum.

ARIB is currently working on early nationwide deployment of 5G, promotion of Post-5G networks, and promotion of 6G R&D activities. ARIB is also supporting the Japanese government's vision on "Society 5.0," by developing wireless technologies which are expected to be the infrastructure to enable the Society 5.0 services.

13.2.7 5G Alliance for Connected Industries and Automation (5G-ACIA)

5G Alliance for Connected Industries and Automation (5G-ACIA) [473] is an alliance which is mainly focusing on applying 5G technologies within industries. It supports and guides the development of 5G technology to address the industrial requirements.

5G will managed to address many important aspects of Industrial Internet such as uRLLC, native support for LAN services, time-sensitive communication, and nonpublic networks. Therefore, 5G-ACIA believes that next level of research should increasingly focus on the further evolution of 5G toward 6G.

13.2.8 Third-Generation Partnership Project (3GPP)

3-Generation Partnership Project (3GPP) [474] is united alliance of seven telecommunication SDOs, i.e. ARIB, ATIS, CCSA, ETSI, TSDSI, TTA, TTC. These SDOs are known as "Organizational Partners" of 3GPP. The main objective of 3GPP is to offer a platform for its organizational partners to define specifications and report that define 3GPP telecommunication technologies.

3GPP is continuously working toward the standardization of beyond 5G network and 3GPP Technical Specification Groups (TSG) are currently working on Release 17. From a standardization point of view, 3GPP will focus on concrete 6G standardization work from Release 20 onward. This is expected to happen around 2025.

13.2.9 International Telecommunication Union-Telecommunication (ITU-T)

International Telecommunication Union-Telecommunication (ITU-T) [475] standardization sector assemble telecommunication experts across the globe to develop telecommunication standards. ITU-T has launched a Focus Group on Technologies for Network 2030 (FG NET-2030) which will be studying about the networks for the year 2030 and beyond. It will be focusing on different types of networks including 6G. FG-NET-2030 is also organizing a series of workshops to highlight the requirements for future networks (FN).

13.2.10 Institute of Electrical and Electronics Engineers

Institute of Electrical and Electronics Engineers (IEEE) [476] is the largest professional association for electronic engineering and electrical engineering. IEEE FN initiative is specially focusing on the development and deployment of 5G and beyond networks. It is working on establishing a 5G and beyond technology road map to highlight the short term (1–3 years), midterm (4–5 years), and long-term (6–10 years) research and technological trends. FN also organizes technical conferences and workshops to promote beyond 5G research activities. Moreover, FN will work in collaboration with IEEE Standards Association to contribute IEEE standards related to beyond 5G networks.

13.2.11 Other SDOs

13.2.11.1 Inter-American Telecommunication Commission (CITEL)

is an entity of the Organization of American States (OAS) which is focusing on the development of telecommunication and ICT solution in the American region [477]. CITEL is working together with ATIS and Next G Alliance on 6G development.

13.2.11.2 Canadian Communication Systems Alliance (CCSA)

CCSA is a SDO in Canada which is focusing on Internet, television, and telephone technologies [478]. CCSA promotes the beyond 5G activities within Canada and participates the standardization activities with 3GPP.

13.2.11.3 Telecommunications Standards Development Society, India (TSDSI)

TSDSI is a SDO in India [479]. It is developing standards for telecommunication systems to satisfy the India-specific needs.

TSDSI is developing Roadmap 2.0 of India's National Digital Communications Policy (NDCP) with telecommunication companies, mobile operators, academic partners, government R&D agencies, and vertical operators. Roadmap 2.0 is mainly targeting beyond 5G and 6G networks by focusing on spectral efficiency, new air interface technologies, and new radio architectures.

13.2.11.4 Telecommunications Technology Association (TTA)

TTA is the South Korean SDO which is focusing on the advancement of telecommunication technologies and ICT services. TTA launched a "Mobile Communication Technical Committee (TC11)" which will be focusing on the standard development for beyond 5G and 6G networks together with other global SDOs [480].

13.2.11.5 Telecommunication Technology Committee (TTC)

TTC is a nonprofit SDO in Japan. TTC is contributing to the standardization activities related to the ICT domain. Together with other SDOs, TTC contributes to the 3GPP's beyond 5G standardization efforts [481].

Part IV

Applications

14

6G for Healthcare

6G is envisaged to play a key role in emerging healthcare applications. After reading this chapter, you should be able to

- Understand the evolution of telehealth toward intelligent healthcare applications.
- Emerging trends of intelligent healthcare applications.
- Security and privacy aspects of future healthcare applications.

14.1 Evolution of Telehealth

The emergence of 5G networks have boosted the development of telehealth and smart healthcare applications. For instance, remote patient monitoring allows healthcare professionals to monitor conditions of patients at their residence or in a remote facility. Data gathered from different sources such as wearable devices attached to the patient's body [482], patient's own smart mobile device, and sensors placed in the patient's room can be used for monitoring. The aggregated data from different sources are examined by the healthcare professional to make a judgment on the patient's condition to take necessary actions. In a similar vein, telemedicine provides remote clinical services to patients with the use of high-quality audio and video streams. Remote surgery is extremely useful in a pandemic which enables a surgeon to perform surgical procedures from a remote facility with his surgical console. Actions of the surgeon are replicated on a patient residing in a different location. A robotic mechanism executes the surgical procedure on the patient and proper haptic feedback is sent back to the surgeon. The feedback can be enhanced by integrating the data from different sensors at the operating theater to ensure the accuracy. Augmented Reality (AR) technology is useful in telesurgery where experienced surgeons guide other

6G Frontiers: Towards Future Wireless Systems, First Edition.
Chamitha de Alwis, Quoc-Viet Pham, and Madhusanka Liyanage.
© 2023 The Institute of Electrical and Electronics Engineers, Inc. Published 2023 by John Wiley & Sons, Inc.

surgeons who perform the surgery next to the patient. Robots deployed at the hospitals minimize the human involvement in treating the hospitalized patients, distributing essential items, performing periodic monitoring. AR technology can be utilized to increase the productivity of the service by providing remote guidance.

5G enables direct integration of heterogeneous Internet of Things (IoT) devices into the network via massive Machine Type Communication (mMTC) service, without Wi-Fi or additional IoT gateways. It also supports 10× longer battery life for devices in mMTC and 100× higher device density. Since the remote monitoring of patients requires integration of various low power devices, 5G services can be effectively utilized to build a proper remote monitoring infrastructure for patients. A remote clinical service (telemedicine) which requires 4K video streaming at 25 fps (frames per second) needs a data rate of 8–16 Mbps. This is realizable with 5G networks via eMBB which supports an average data rate of 100 Mbps. A Local 5G Operator (L5GO) deployment at the healthcare premise is suitable for catering ultra Reliable Low Latency Communication (URLLC) use case like remote surgery. The ultralow End-to-End (E2E) latency requirement of a use-case like remote surgery [483] is achievable with 5G networks making the application viable. An AR-assisted telesurgery require both high bandwidth and ultralow latency. To avoid cyber-sickness of AR communication, E2E latency should be less than 50 ms [484]. Defining a network slice for AR ensures the service levels and adds extra privacy and security to the data stream. Utilizing robots to assist the patients in hospitals requires precise coordination with a controlling server and between robots, calling for mMTC services of 5G. The coordination and communication between robots happen locally and exact details are mostly irrelevant beyond the premises. This is established through a multi-access edge computing (MEC) server to manage the robots and guarantee the service levels, provide ultralow latency, add extra security and privacy, and reduce congestion in the infrastructure network beyond the MEC server.

14.2 Toward Intelligent Healthcare with 6G

Recent developments in healthcare systems have given rise to Healthcare 5.0 with the emergence of digital wellness. AI-driven intelligent healthcare will be developed based on various new methodologies including Quality of Life (QoL), Intelligent Wearable Devices (IWD), Intelligent Internet of Medical Things (IIoMT), Hospital-to-Home (H2H), and new business models [21, 70]. Thanks to recent advances in wearable sensors and computing devices, it is possible to monitor and measure the health data in real time. The sensing data collected from wearable devices can be preproceeded by the nearby edge node and then sent to the doctors for remote diagnosis. Also, with the realization of holographic communications, tactile Internet, and intelligent robots in 6G, the doctor can remotely do the surgery. Such a telesurgery would remove the need of on-site operations and avoid the risks caused by virus spreading, especially ones, such as, COVID-19 and any other transmissible diseases.

Many studies have considered 6G as a key enabler of intelligent healthcare (i.e. Healthcare 5.0). In particular, promising technologies such as edge intelligence (i.e. cloud/edge computing + AI), holographic communications, tactile Internet, and Internet of Bio-Nano-Things (IoBNT) are expected to play a key role. For example, since mission-critical healthcare applications would have different QoS requirements (e.g. latency and computing resources), an approach, named as ACTION, is investigated in [485] by jointly employed tactile Internet and network function virtualization techniques. Considering that passive optical network has the potential to support bandwidth-hungry healthcare applications, the work in [486] developed a joint dynamic wavelength and bandwidth allocation framework. The experimental results show that the proposed framework can provide better performance than the benchmarks in terms of packet loss and delay.

As the healthcare data are huge in volume and increase significantly, exploiting AI (e.g. deep learning) to develop data-driven healthcare solutions has become one important trend in healthcare research and development [487]. Meanwhile, since most patients do not want to share their personal health data, if not mandatorily requested by the governments and medical staffs, effectively overcoming the security and privacy challenges of healthcare data usage is of great importance. Among many solutions, FL and blockchain are promising. In particular, in FL-based healthcare solutions, the users need not share their raw data, instead they just need to share information of the local model [488]. With its decentralization and security nature, blockchain is also identified as a promising technology toward providing security and privacy for transferring and acquiring of healthcare data [489].

14.3 Personalized Body Area Networks

Body Area Networks (BANs) with integrated mobile Health (mHealth) systems are advancing toward the personalized health monitoring and management. Such personalized BANs can collect health information from multiple sensors, dynamically exchange the such information with the environment and interact with networking services including social networks [490]. Personalized BANs has a wide range of applications, covering both medical to nonmedical domains. For example, personalized BANs can be used to avoid the need of cable wiring in polysomnography tests (also known as sleep disorder diagnosis). Personalized BANs has also found in nonmedical applications such as emotion detection, entertainment, and secure authentication applications [491]. More recently, the Internet of Nano-Things (IoNT) and the IoBNT have been developed as the next generation of IoT for healthcare services. The concept of IoBNT engineers information communication within the biochemical domain, while connecting to the Internet via the electrical domain is a recent development [492].

Despite various applications, IoBNT is still in the infant stage of research and development. In order to realize IoBNT, some efforts have been devoted to the design of nanosensors. Graphene membranes are utilized to develop highly sensitive capacitive nanomechanical sensors which are shown to have a superior responsivity over the commercial nanosensors [493]. From the wireless communication perspective, Akyildiz *et al.* [447] pointed out several significant challenges in IoNT and IoBNT such as energy consumption, interference control, network and routing protocols, coding scheme and modulation technique, experimental validation, and data management. The features of 6G such as high data rate, reliability, new spectrum (visible light and THz) communications, and ultralow latency will help to address these challenges, along with other efforts from existing studies for IoNT and IoBNT.

In [494], the human insulin–glucose system is analyzed in terms of the data rate, channel capacity, and propagation delay, and from the communication perspective. Several characteristics of the correlation between insulin rate/resistance and the derived models are illustrated via the experimental results. Due to the diffusion and attenuation properties of molecular communications in IoBNT, extending the coverage range is of importance. To address this issue, Wang *et al.* [495] proposed to deploy an intermediate nanomachine to relay, amplify, and then forward the received signals between the transmitter and receiver. Various performance metrics such as mean square error, minimum error probability, and maximum probability detection are derived, and more interestingly the performance gain can reach 35 dB when the number of released molecules is 500. For the architecture designs of transmitters and receivers in IoBNT molecular communications, the interested readers are invited to read the review article [496].

14.4 XR for Healthcare Applications

XR is an emerging immersive technology with the fusion of physical and virtual worlds where wearables and computers generate human–machine interactions [59, 340]. XR utilizes different sensors to collect data regarding the location, orientation, and acceleration. This requires strong connectivity, extreme data rates, high resolution, and extreme low latency, that is envisioned to be facilitated 6G. For instance, utilizing XR technologies in Brain-Computing Interfaces will require over 1 Tbps data rates [12]. XR will be instrumental in a wide array of healthcare applications. These applications range from virtual medical visits, rehabilitation therapies, remote medical tests and scans, remote surgeries, and training of healthcare professionals.

14.5 Role of Blockchain in Medical Applications

Blockchain, as explained Chapter, is expected to harness the capabilities of 6G and enable emerging medical applications and services. For instance, blockchain will enable the coordinated operation of large numbers of IIoMT devices. This is due to the distributed nature of blockchain and its inherent support for smart contracts. Smart contracts can perform tasks, such as the decentralized validation of IIoMT devices, facilitating large-scale decentralized healthcare platforms. Furthermore, the decentralized operation also minimizes the operational costs of such medical applications by eliminating the requirement of centralized computing and storage resources, and minimizing transmission overheads in B5G and 6G networks [497].

In addition, the operation of widespread decentralized medical applications will also require to perform a large number of transactions. This can also be facilitated through blockchain-enabled cryptocurrencies. Through the utilization of lightweight and scalable security mechanisms and encrypted data transmissions, blockchain can be utilized as a secure and efficient platform to store medical records and medical history. Blockchain can also facilitate the sharing of such sensitive data with healthcare services and medical research, in a secure, transparent, and accountable fashion. Additionally, blockchain can be efficiently utilized to facilitate seamless supply chain management and operation [498] (Figure 14.1).

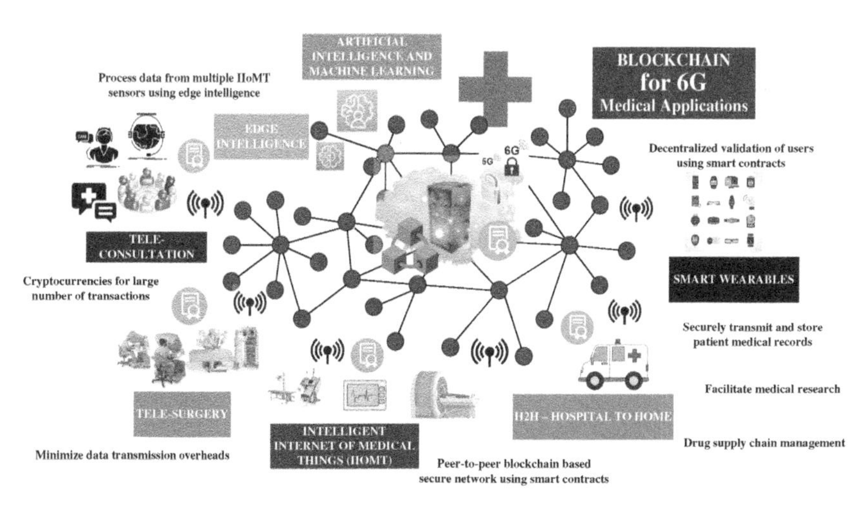

Figure 14.1 Blockchain for 6G medical applications. Source: Tamás Haidegger/Zoltán Benyó/IntechOpen, V. B. Chernetsov/Zhuravlev Yu.Yu, macrovector/Adobe stock.

14.6 Security and Privacy Aspects of 6G Healthcare Applications

Digital healthcare or e-health care services are evolving for new dimensions. Within few years, AI-driven intelligent healthcare will be developed based on various new methodologies including QoL, IWD, and novel business models [21, 70]. The growth of elderly population may create an increase in the importance of e-health than ever before. BANs with the integrated intelligent health systems are advancing toward personalized health monitoring and management. Such personalized BANs can collect health information from multiple sensors, dynamically exchange the collected information with the environment and interact with networking services including social networks [490].

6G will be the main communication platform to interconnect the intelligent healthcare services in the future. Thus, enabling the secure communication, device authentication, and access control for billions of IoMT and wearable devices will be critical security challenges to solve in 6G era.

Privacy protection and ensuring of the ethical aspects of user data or electronic health records will be a critical issue in future healthcare system. As explained above, the utilization of AI is mandatory to manage billions of IoMT devices and process the health-related information. However, current AI model are mainly focused on performance optimization rather than the ethical aspects. Specially, AI models should follow strict ethical rules on data collection and use of user data for the model training [499]. Moreover, AI models should comply with privacy rules and regulations enforced by the regulation bodies. As the main communication infrastructure for future healthcare systems, 6G networks should protect both privacy and integrity aspects of the patient information and records.

15

Smart Cities and Society 5.0

In recent years, the rising urbanization of the world population, the proliferation of Internet of Things (IoT), and the use of artificial intelligence (AI) and big data have substantially impacted our living environments, conditions, and quality of life. In this context, the concept of the "smart city" provides an opportunity to address these urban issues. As the future of wireless and communication networks, 6G is expected to help create various smart city services and facilitate the deployment of smart cities. After reading this chapter, you should be able to

---

- Understand the fundamentals of smart cities and society 5.0, which will help you explore the role of future 6G networks more fully.
- Understand how future 6G networks help improve the performance and services on offer by smart cities and society 5.0.
- Explore potential applications of 6G for smart citizens, smart transportation, smart grid, supply chain management, and miscellaneous scenarios.

15.1 Preliminaries of Smart Cities

In a smart city, the administration/government uses Information and Communications Technology (ICTs) for efficient administration and governance to provide quality and better services to the urban citizens [500]. Digital technologies, social capital, and infrastructure are combined in a smart city that can stimulate sustainable economic development and offer an environment that is attractive for all the residents in a smart city. Some of the key enabling technologies of smart cities are communication infrastructure, edge computing, AI, big data analytics, and IoT. Several services can be provided in a smart city by combining

6G Frontiers: Towards Future Wireless Systems, First Edition.
Chamitha de Alwis, Quoc-Viet Pham, and Madhusanka Liyanage.
© 2023 The Institute of Electrical and Electronics Engineers, Inc. Published 2023 by John Wiley & Sons, Inc.

the technologies mentioned above. For example, smart parking, smart garbage collection, traffic prediction, accident prevention, smart healthcare, smart energy distribution through smart grid, and crime detection and prevention can improve the quality of life of the residents [501]. The real-time data collected through sensors, citizens, assets, and buildings in a smart city can be fed to AI-based algorithms and big data analytics to understand the changing demand patterns and requirements. They would help governments/administrations make better decisions and respond quickly with better and lost cost solutions. In a smart city, the officials can interact directly with the infrastructure providers and the communities to monitor the events and their evolution. Cities will become more responsive and more livable when they get smarter. A smart city enables the citizens to participate in the planning/improvements of cities. The citizens can provide their thoughts/feedback on projects, such as road and traffic conditions. Citizens also can help the governments in constantly mentoring the facilities/infrastructure. For example the citizens can take a photograph of potholes and bad roads and send them to the municipality officials for their immediate actions.

Recent advances and innovations in technologies and devices, such as smart phones, cloud computing, edge computing, AI and allied technologies, and ICT convergence, are driving the development of smart cities. For example sensors can be placed in several key locations in the city, such as streetlights, electricity poles, traffic lights, vehicles, bins, wastewater management systems, to collect the data in real time through which the status of the assets in a city can be monitored regularly. Moreover, AI techniques (e.g. deep learning and federated learning) and big data analytics can be applied on the real-time data to uncover the patterns that help in making predictions. These predictions help the governing bodies or concerned authorities to make decisions in real time to improve the quality of life of the residents in smart cities.

The key enabling technologies for smart cities are discussed as follows:

- *Internet of Things*: Several IoT devices use sensors to collect and analyze the data that can be used to provide better services and enhance public utilities and infrastructure. As a result, the cost of operations can be reduced, and the quality of life of citizens can be enhanced by IoT [502].
- *Artificial Intelligence*: AI plays a key role in extracting useful information from the data collected from several sensors that can be used to make predictions that the governments/concerned authorities can use to make decisions to enhance the quality of life of the citizens in a smart city. Some of the applications that AI can benefit are controlling traffic congestion, disaster management, intelligent transportation systems, and meeting the requirements of power by industries [503].

- *Communication Infrastructure*: Communication infrastructure is one of the important technologies that help create interconnected and modern cities that provide fast, reliable, and comprehensive Internet connections to several applications. The communication infrastructure will allow the acquiring, sharing, and processing of big data generated from several devices in a smart city [504].
- *Big Data Analytics*: In a smart city, millions of IoT devices generate data every second from several applications, such as smart grid, smart transportation, smart homes, and smart buildings. Therefore, big data analytics is an essential technology that can help in processing the big data generated in smart cities for real-time analytics [505].
- *Cloud Computing*: Cloud computing is another key enabling technology for smart cities. It is essential to meet the dynamic storage requirements of the big data generated from the smart city applications [506].

Communication infrastructure plays a vital role in the realization of a smart city. As more and more data are generated from several sources, communication technology has to process and transfer the data to the storage infrastructure so that meaningful insights from the data generated from different smart city applications can be extracted in real-time. The existing 5G infrastructure faces several challenges in addressing issues, such as reducing latency, providing the required bandwidth, and handling big data in real time for applications embedded with augmented and virtual reality, upcoming metaverse-based applications [507]. Some of the applications, such as environment monitoring, virtual navigation, smart traffic, full high definition video transmissions in connected robots and drones, require very high bandwidth and ultrareliable and low latency communications [508]. 6G is a potential solution for several applications of smart cities. 6G can meet the bandwidth, ultrareliable and low latency, fully autonomous network management requirements of several smart city applications [59]. The rest of this chapter discusses the need of 6G for several smart city applications, such as smart citizens, smart transportation, smart grid, supply chain management, and other applications.

15.2 6G for Smart Citizen

Smart citizens develop and use information using various advanced systems to ensure efficiency and sustainability in a smart city. The role of smart citizens in smart cities is enormous, especially in the development of various solutions that are tailored to fulfill the needs of its residents and other visitors. The primary goal and objective of any effort contributed to a smart city setup is to increase the quality of life and ensure prosperity for all its dwellers. Therefore, the residents play an immensely significant role in the implementation of the most appropriate processes and solutions. The key enablers in this regard are technology and data, the absence of which makes people un-smart and hinders the ability to engage in any smart city project leading to failures.

A smart city setup utilizes humans for leveraging information to develop a more advanced smart city setup and makes humans empowered. Hence, smart citizens are individuals who use the technologies of smart cities and help in leveraging information to the smart devices, which enable the devices to provide better output. The role of 6G in enhancing the life of smart citizens is discussed in the section. Some applications of 6G for smart citizens in smart cities, such as smart communication, smart safety security, smart resource and waste management, and tracing during pandemic, are depicted in Figure 15.1.

The role of 6G wireless networks is much more promising in supporting smart city architecture and infrastructure [509]. For example, in the case of security in a smart city setup, any information needs to be sent at the earliest so that the mitigation measures can be implemented at an accelerated rate. During the pandemic, 6G technologies can be used to track human movement and transmit information using the 6G wireless spectrum, which enables faster data transmission. Even in extremely high-populated setups, the use of 6G enables higher performance in delivering a huge amount of data relevant to mass human movement from various city locations with minimum air latency. This helps immensely in curbing human movement and identifying potential COVID-19 patients at a much faster

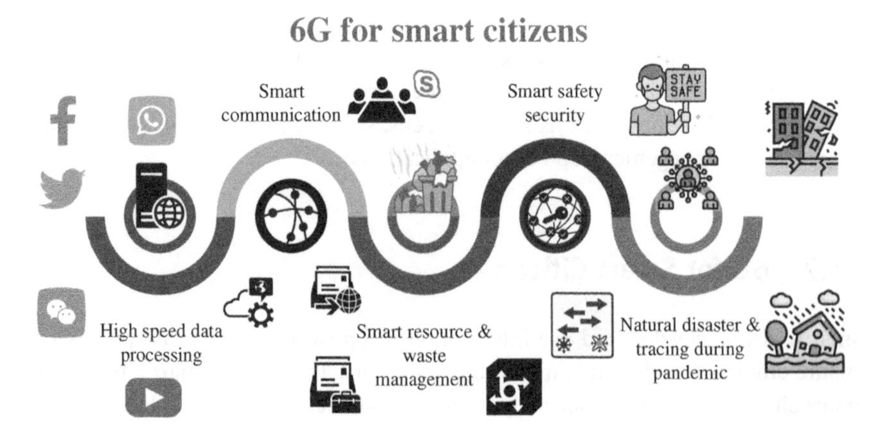

Figure 15.1 6G for smart citizen.

speed. In countries like India, where awareness of safety during a pandemic is still a concern, such technologies help identify individuals not abiding by COVID-19 protocols, capture their information, detect their identity, and transfer the information to the administration at accelerated speed for necessary action. This would ensure health care safety and containment of the diseases [504]. On the other hand, the incidence of theft and sexual harassment could be curbed by the use of data collection and transmission using 6G technology. The faster data transmission enables administrations to immediately provide the required support and save human lives and assets. Since the past few years, citizens have witnessed various natural disasters costing millions of lives. In such scenarios, tracing victims and rescuing them at the earliest possible time is the most difficult challenge. The faster the rescue, the higher are the chances of saving precious lives.

In the present smart city infrastructure, drones and unmanned aerial vehicles (UAVs) are used to trace humans, cattle, or animals who are victims in such situations. The use of 6G technology enables coverage of larger areas in tracing and help in faster transmission of information if the search process identifies a victim. This enables rescue groups to reach faster and provides the necessary support to restore human lives [510]. On another note, it is a known fact that waste management is an essential service in a smart city setup. But despite the use of IoT and various other advanced technologies, waste management systems often become resource-intensive, inefficient, and extremely outdated. The use of IoT has immensely contributed to optimizing garbage collection services and transforming waste management into a data-driven collection system.

With the growth in population and urbanization, waste generation has increased immensely. Although IoT technologies are used, the transmission delay leads to a lag in garbage collection by the necessary authorities, making bins overflow with trash. 6G technologies have huge potential to help in this regard by keeping up with this fast pace of filling garbage cans and transmitting information faster, enabling garbage collection on time. Furthermore, the performance of 6G enables larger location coverage and provides information from all nooks and corners ensuring the entire city is kept clean and hygiene.

15.3 6G for Smart Transportation

Smart transportation is one of the major applications of IoT, which involves the integration of modern technologies and management strategies into the transportation system. These technologies cater innovative services to various aspects of traffic and transportation management, ensuring that the users are more informed on the smarter usages of transportation networks. Various technologies are used to implement smart transportation, namely in the navigation of cars,

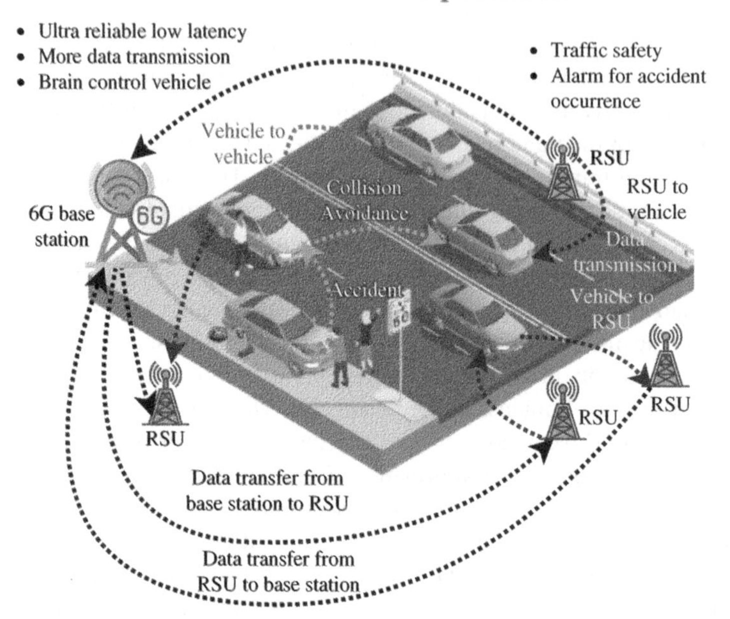

Figure 15.2 Applications of 6G for smart transportation.

controlling the traffic signals, automated number plate recognition, camera-based speed monitoring systems integrating live data, and feedback from various sources. The use of intelligent transportation systems plays a major role here, enabling improvised use of such transportation networks and leading to the development of smart infrastructure capable of meeting the future needs of smart cities in future 6G networks. Some applications of 6G for smart transportation in smart cities are depicted in Figure 15.2.

Intelligent transportation systems in smart transportation ensure communicating and traveling in a city more convenient and safer and at the same time immensely cost-effective. This is achieved by integrating IoT devices and 5G/6G communication technologies wherein IoT devices are equipped with inexpensive sensors that are embedded into physical devices that are managed and controlled remotely without human intervention. 5G technology enables high-speed communication for accurate maneuvering and controlling the transports in real-time with minimum latency. Smart transportation can be categorized into two major types–public infrastructure-oriented smart transportation and automotive industry-based smart transportation. For example, in traditional city life, when pedestrians intend to cross the road while waiting drivers get busy in finding their destinations; the street traffic lights get activated either through timers, pressure plates located under the road, or based on the buttons pressed by the pedestrians. Hence, in such circumstances, it becomes important for both the driver and the pedestrian to concentrate and pay attention, or else the consequences pertaining to safety lags become inevitable. On the contrary, in the case of a smart transportation system, vehicles use a combination of blue tooth and Light Detection and Ranging (LIDAR) system to detect any pedestrians leading to the generation of automobile brakes.

The street traffic lights function based on signals received from the vehicle informing on the number of cars waiting. Moreover, the communication of street lights with vehicles even triggers starting of the car engine and stopping of the same. The use of IoT and 5G communication ensures controlling of all such real-time actions and remote sensing with utmost accuracy and efficiency [511].

ICTs have evolved rapidly since the past decades and have revolutionized traditional domains, such as agriculture, education, health, energy, and transportation. The advent of 5G technologies has further accelerated such advancements as various countries across the globe have taken the initiative to implement 5G infrastructures ensuring processing of high data volume in low latency. This change has been prominent even in the transportation sector, wherein a significant growth in connected and autonomous vehicles has been observed. An increase in logistics efficiency, urban transportation implementing Mobility-as-a-Service (MaaS) platform has also been visualized [512]. The use of 5G technologies has enhanced the efficiency of the operation of autonomous vehicles on land, sea, and water. It has simplified the traditional communication and signaling processes, reducing operation costs. It has further enhanced the capabilities of connected freight, enhancing the ability to track goods through streamlined logistics and supply chain management systems. 5G technologies have paved the way to implement dynamic transport planning, which reduces traffic congestion allocating ample space for cyclists and pedestrians. 5G would also help in the monitoring of smart cards and control systems that would benefit the transport planning by a better understanding of mobility patterns of the different population groups. The real-time monitoring of public transport vehicles and demand management allows coordination between supply and demand. This is achieved by developing origin-destination matrix proxies that help transport drivers manage empty or overloaded vehicles, ensuring the quality of services among users. The increase in multimodal connectivity among different modes of transport integrates varied mobility options under one platform – MaaS. Thus, users can choose among multiple options intelligently and further as part of better transportation, they experience different onboard entertainment and hospitality services. The safety and security aspect is maintained through enhanced video connectivity between the center and police that helps avoid incidents of violence and sexual harassment in the public transport network.

Although 5G networks provide user-based and application-specific QoS and quality of experience (QoE), the future conditions of ultrasmart cities require comparatively much higher data rate, 3D connectivity, localization within 1 cm, and optimized reliability for automated transportation systems. The mobility requirement of smart transportation using 6G generally vary between 240 and 1200 km/h. For example a self-driving car may need to communicate and coordinate with roadside sensors and vehicles moving in the adjacent lanes while both move at extremely high speeds. The delivery drone may need to communicate with the ambulance to collect medical supplies and deliver them to a remote rural setting. These are examples wherein 6G technology has the potential to play a significant role in supporting low latency in autonomous driving, enabling intelligent decision-making. In the case of brain–vehicle interfacing, a brain-controlled vehicle is used, which is controlled by the human mind rather than physical intervention. Although this type of vehicle is a futuristic idea, it has huge potential to make people with disabilities more independent. Here the brain activity signals are converted to motion commands further transported to the vehicle using brain–computer interface [90]. Some of the other applications in transportation systems, such as object detection, need the processing of images/videos captured through vehicles' cameras in real-time. These applications need high bandwidth and ultrareliable and low latency communication mechanisms to make decisions in real time. If the decisions are not taken in real time, it may lead to loss of lives or property [513]. Vehicle-to-vehicle (V2V) communications can play an important role in handling several prevalent issues in cities. In V2V communications, several vehicles will be interconnected, and they can communicate with each other. In case of traffic congestion or accidents, the vehicles in proximity to the locations of the incidents can pass on this information to the other vehicles in the network so that they can take alternate routes to avoid traffic congestion. The vehicles can also pass on the information regarding the accidents to the nearby hospitals and police stations for their necessary actions [514]. 6G plays a very important role in the realization of the full potential of V2V communications, which can help improve the traffic management and immediate response of emergency teams in case of accidents.

15.4 6G for Smart Grid

The smart grid is one of the key technologies that can play a vital role in the sustainable development of a smart city. A smart grid can be used to balance the electricity distribution among household and industrial sectors, effective demand response management, faulty lane detentions, load prediction can be managed effectively in a smart grid. A smart grid uses artificial intelligence and big data analytics that can analyze electricity data generated from household appliances and industrial sectors to extract the patterns that can be used for predictions. Electricity blackouts also can be predicted and necessary preventive measures can be taken by the electricity companies in a smart grid. A smart grid can also be used to detect and prevent electricity thefts, address security issues, and predict and prevent power outages [515]. Furthermore, a smart grid enables two-way communication between the utility and the customers that can be used to sense throughout the transmission lines. The smart grid comprises new equipment, automation, controls, sensors, computers, and new technologies that collaborate to effectively respond to the electricity demands of household and industry customers [516].

The smart grid provides quality and diverse power supply, characterized by interaction, automation, and informatization, to the customers in a secured and efficient manner. IoT provides the core infrastructure for the smart grid. To enable peer-to-peer energy trading, real-time asset supervision, sharing of data, management of the electricity load, and other services in a smart grid, a robust, automated, reliable, secured communication infrastructure is required [517]. To make the smart grid more sustainable, the key is to manage and connect the remote devices through smart meters [518]. The automated, high bandwidth, ultrareliable, and low latency communication services provided by 6G networks can significantly improve several real-time services provided by the smart grid, such as remote monitoring and controlling of the energy sources and demand response automation. 6G networks can also help in the wide deployment and coverage of smart meters in the distributed network, which can help prevent blackouts and enable the self-healing capabilities of smart grids. Some of the applications that require high-speed connectivity such as video surveillance during natural calamities in real time, predictive maintenance will also be benefited by future 6G networks [519].

The massive amount of data generated through constant connectivity and communication in a smart grid can be dealt with by sophisticated AI/machine learning (ML) techniques that can help in the decision-making process. These ML-based techniques can assist in data collection and analyze the existing patterns in the data that can be used to effectively and efficiently run the smart grid operations. An ML-assisted 6G can solve some real-time issues in smart grids, such as predicting electricity consumption price, detecting the intruders in a smart grid, lane maintenance, meeting the electricity demand requirements, and predicting stability in a smart grid [520]. Another key enabling technology for a smart grid is IoT sensors that can monitor, distribute, and control the flow of electricity. Some of the use-cases for 6G-enabled smart grid are discussed below and summarized in Figure 15.3.

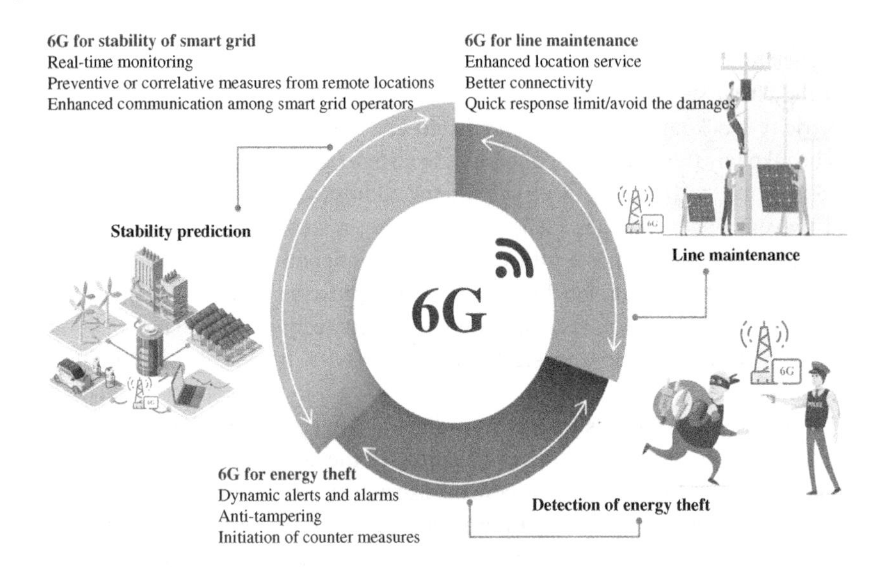

Figure 15.3 Illustration of 6G use-cases for smart grid.

- *Stability Prediction for a Smart Grid*: One of the key challenges in a smart grid is maintaining the stability of the smart grid. The two important criteria in the maintenance of stability in the smart grid are the following:
  - having a reserve of battery storage for meeting the ever-changing electricity demands,
  - providing sufficient capacity for the stability of the voltage at each location throughout the distribution lines.

  Several issues, such as blackouts and power outages, may emerge due to the instability of a smart grid, which may, in turn, lead to huge revenue losses in various industries [521]. To predict the stability of a smart grid, the data generated from several sensors can be analyzed by ML algorithms. Early prediction of smart grid stability enables the smart grid operators to take necessary preventive actions [522].
- *Energy Theft Detection in Smart Grid*: The utilities are losing a great amount of money every year because of energy thefts, which result in the increase of energy prices for the consumers. Several approaches, such as tapping a line between transformer and house and hacking/tampering of meters, are used by energy thieves [523]. 6G-enabled smart grid can be effectively used to monitor the data from the smart meters to enable effective training of AI/ML algorithms to monitor and predict the energy thefts in real time [524].
- *Line Maintenance in Smart Grid*: Properly maintaining the infrastructure improves the reliability of the smart grid. To prevent blackouts due to several

issues, e.g. weather conditions and equipment failures, aging/deterioration of power lines and transformers, have to be detected at early stages [525]. 6G-enabled smart grids can effectively monitor and predict the aging of the lines based on the real-time data and prediction.

15.5 6G for Supply Chain Management

Supply chain management plays a very important role in every industry. It directly impacts the success and customer satisfaction of any organization. In the past few years, international trade disputes and various natural disasters have immensely disrupted supply chain management, wherein the consequences have been expensive and catastrophic. These unwanted circumstances have largely affected various company's ability to deliver its product on time leading to poor revenue generation and loss of brand name. To add to this, COVID-19 has further added fuel to this unexpected setback in supply chain management wherein all the major companies are experiencing supply chain disruptions. The supply chain process, in general, is quite complex, and it involves active coordination among various disbursed and commonly connected stakeholders in the supply chain, namely the producers, the brokers, the transporters, the wholesalers, the retailers, the consumers, and many more. Monitoring and controlling all the people, processes, parts, and products in this supply chain management's lifecycle is extremely important to ensure efficiency and the establish a strong reputation. The pandemic situation has tested the efficiency, resilience, and flexibility of all the leading supply chain management systems across the globe and highlighted the need for a transparent, agile, sustainable, robust, and digitally enabled system.

The role of ICTs and 6G is huge in this regard which enables continuous monitoring, assessment, and optimization of real-time data on all stakeholders and products involved providing unprecedented visibility in the entire supply chain process. Some applications of 6G for supply chain management in smart cities (e.g. synchronized planning, data integration, dynamic fulfillment, data analytics, faster delivery, and shipping management) are depicted in Figure 15.4. For example, in the case of on-road asset tracking, the implementation of curbside pickup, Buy Online, Pick in Store (BOPIS), smart lockers, and Direct to Consumer (D2C) deliveries has become highly popular. In the case of perishable goods and pharmaceutical items, the demand has risen due to the use of advanced cold chain technologies. But all of these systems require on-road asset management to ensure the quality of the devices. IoT in association with 6G technologies has a huge role in this regard. The connected sensors collect a coherent stream of real-time data providing information on the exact location of the item and

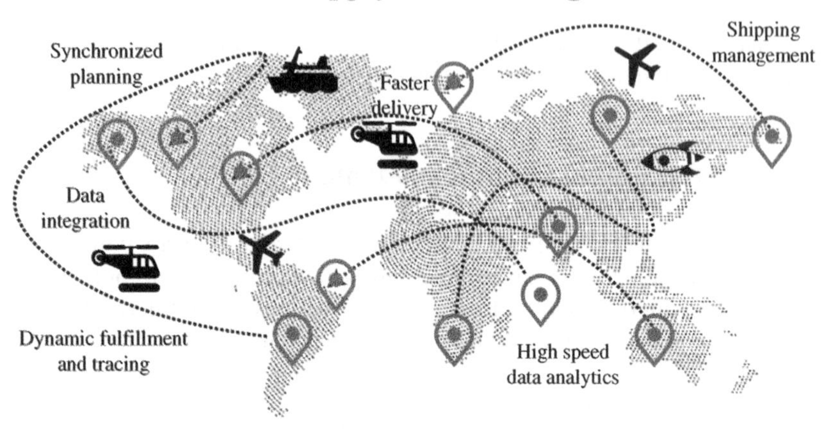

Figure 15.4 Applications of 6G for supply chain management.

the time it would take to move between different locations in the supply chain management's life cycle, including the movement of the trucks.

6G wireless networks have huge potential to help faster data transmission and ensure coverage of large locations during product tracking and data transmission on the same. The bottlenecks will be identified faster, allowing enough time for contingency measures and planning an alternative route for expedited delivery. With the use of 6G-based IoT technology, supplier, manufacturers, and distribution centers would be able to send and receive goods much faster, reducing handling time ensuring efficient processing of materials. With such huge data collection being transmitted at high speed, delivery forecasts will become more accurate for the vendors and consumers.

Moreover, 6G wireless networks also have the potential to contribute immensely to cold chain monitoring. Goods such as pharmaceuticals and food items are often perishable and need to be stored in ideal conditions. The use of smart cold chain enables monitoring, locating, and addressing any issues much faster using IoT. To be more specific, the environment sensors can track suitable conditions relevant to temperature, humidity, air quality, light intensity, and various other factors that impact the quality of such products inside the storage facility, cargo containers, or the delivery vehicle. In case a threshold breach is identified, an alert is generated in real time for prompt mitigation to ensure the integrity of the product is unharmed. 6G technology can help in this regard in transmitting the data faster from various locations to ensure much more accelerated services are rendered in minimum time. Warehouse errors are often expensive in case of the absence of an effective inventory management system. Indoor wireless IoT solutions have been extremely instrumental in the management of activities in warehouse management systems [526].

IoT in association with 6G technology can help in reducing manual labor and human errors and increasing the processing speed to improve overall warehouse efficiency. Moreover, connected sensors can help monitor the movement and use of materials, products, and assets inside various facilities to ensure effective use of materials, prevention of thefts reduction in search time, and avoid out-of-stock situations. Furthermore, the 6G technology can enable the connected sensors placed on shelves to communicate stock levels much faster in real time to identify the consumers' usage patterns. This accelerated yet continuous transmission and feedback of data will streamline coordination between the logistics providers and warehouse operations, ensuring space utility with utmost efficiency [527]. In the case of fleet management, IoT sensor networks enabled with 6G can collect data relevant to vehicle usage, speed, location, emission, and various others. The sensors using 6G can transmit the key health and operational parameters pertinent to tires, hydraulics, engine, and other components to ensure on-time or predictive maintenance of the same. Telematic sensors can use 6G technology and transmit information pinpointing underutilized machinery time and idle time to prevent wastage of fuel allowing operators to make accurate and faster-informed decisions. Similarly, unauthorized usage can also be detected to avoid thefts. 6G-enabled sensors can also provide more accurate estimations pertinent to delivery time, thus reducing unnecessary waiting time at destinations and enhancing the workflow in the supply chain.

15.6 6G for Other Smart Scenarios

Communication infrastructure plays a vital role in some of the other smart city applications, such as disaster management, smart healthcare, smart governance, monitoring the movement of citizens during pandemics/epidemics, monitoring citizens/sensitive structures/locations using drones/unmanned aerial vehicles, as discussed below.

- *Disaster Management*: Disaster management includes taking necessary actions during disasters, such as clearing the citizens in low-lying areas during floods, tracking and monitoring the spread of fire in forests, providing the necessary services to the people affected by cyclones/tornadoes. The governments/administration has to access the data in real time to take the necessary actions/decisions to help the citizens who are affected by the disasters. Future 6G networks can help in drastically improving the disaster management in smart cities.
- *Smart Healthcare*: Due to traffic conditions, several citizens avoid going to the hospitals for their regular checkups. Transportation of the patients to the hospitals in time during emergencies is also a challenge. Smart healthcare can provide solutions to the problems mentioned above. In smart healthcare,

doctors and other healthcare officials can regularly monitor the conditions of patients through wearable devices. The patients also can be treated through telemedicine/remote monitoring. 6G networks can tremendously improve smart healthcare through its high bandwidth and ultrareliable and low latency characteristics [76].

- *Smart Governance*: Smart governance is the process of utilizing modern digital technologies and communication infrastructure by the governments/administrations to provide sustainable and transparent services to the citizens. Through smart governance, the citizens can effectively participate in administrating cities. Citizens can provide their feedback on several policies and planning of cities. Even the industries can participate in the framing of policies regarding the industries, and the government can continuously monitor the industries to check whether they are adhering to several regulations like pollution and waste disposal. 6G networks can play a vital role in providing the required communication infrastructure to fully realize the potential of smart governance.
- *Tracing the Citizens During Pandemics*: During pandemics and epidemics like recent COVID-19, it is very important to continuously monitor and trace the citizens to contain the spread of the virus. Technologies like IoT and drones can be used to continuously monitor the citizens. To get real-time information regarding the movement of citizens in real time so that the administration can take necessary actions, 6G can play a vital role [528].
- *Monitoring the Infrastructures Through Drones*: Drones/Unmanned Aerial Vehicles can be used effectively to monitor several key infrastructures and locations to prevent crimes and terrorism by monitoring the movement of citizens in sensitive areas. 6G communication infrastructure can play an effective role in safeguarding the sensitive areas and protecting the citizens by preventing burglaries [37].

In all the applications mentioned above, huge bandwidth requirements, along with ultrareliable and low-latency communications, necessitates the need of 6G for seamless smart services for making the life of the citizens comfortable, making the cities more livable for its citizens, containing the spread of pandemics, limiting the loss of lives and properties during disasters, and containing criminal activities.

Acknowledgement

This chapter is contributed by Thippa Reddy Gadekallu, Sweta Bhattacharya, and Praveen Kumar Reddy Maddikunta (School of Information Technology and Engineering, Vellore Institute of Technology, India).

16

Industrial Automation

Industrial automation together with the advent of Industry 5.0 is envisaged to be one of the key emerging 6G applications in the coming decade. After reading this chapter, you should be able to

- Understand the network-level requirements to enable Industry 5.0, Collaborative Robots, and Digital Twin.
- Understand the relevance of 6G networks for realization of Industrial Automation.
- Understand the importance of 6G technologies for Industry 5.0.

16.1 Introduction

Vehicles, clothing, houses, and weapons have been designed and manufactured by humans and/or with the help of animals in the past centuries. With the emergence of Industry 1.0 in 1974, industrial production began to change significantly. Figure 16.1 shows an overview of the evolution of Industrial *X*.0 [529]. The development time for the first three revolutions was around 100 years, and it took only 40 years to reach the fourth from the third. In 1800s, Industry 1.0 evolved through the development of mechanical production infrastructures for water and steam-powered machines. There is a massive gain in the economy as production capacity has increased. Industry 2.0 evolved in the year of 1870 with the concept of electric power and assembly line production. Industry 2.0 focused primarily on mass production and distribution of workloads, which increased the productivity of manufacturing companies. Industry 3.0 evolved in 1969 with the concept of electronics, partial automation, and information technologies.

6G Frontiers: Towards Future Wireless Systems, First Edition.
Chamitha de Alwis, Quoc-Viet Pham, and Madhusanka Liyanage.
© 2023 The Institute of Electrical and Electronics Engineers, Inc. Published 2023 by John Wiley & Sons, Inc.

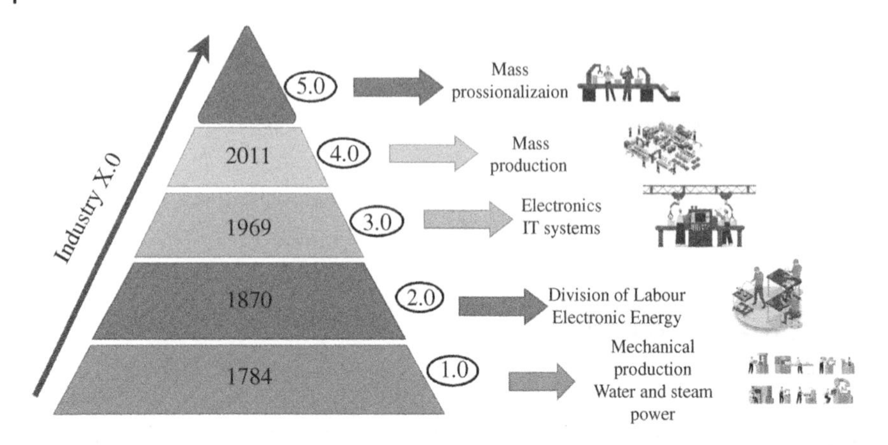

Figure 16.1 Illustration of industrial evolution.

Industry 4.0 evolved in 2011 with the concept of smart manufacturing for the future. The main objective is to maximize productivity and achieve mass production using emerging technologies [530, 531]. Industry 5.0 is a future evolution designed to use the creativity of human experts working together with efficient, intelligent, and accurate machines [532].

16.1.1.1 Motivations Behind the Evolution of Industry 5.0

Industry 4.0 standard has revolutionized the manufacturing sector by integrating several technologies, such as artificial intelligence (AI), the Internet of Things (IoT), cloud computing, cyber physical systems (CPSs), and cognitive computing. The main principle behind Industry 4.0 is to make the manufacturing industry "smart" by interconnecting machines, devices that can control each other throughout the life cycle [129, 533–535]. In Industry 4.0, the main priority is process automation, thereby reducing the intervention of humans in the manufacturing process [536, 537]. Industry 4.0 focuses on improving mass productivity and performance through the provision of intelligence between devices and applications using Machine Learning (ML) [538–540]. Industry 5.0 is currently conceptualized to leverage the unique creativity of human experts to collaborate with powerful, smart, and accurate machinery. Many technical visionaries believe that Industry 5.0 will bring back the human touch to the manufacturing industry [349]. It is expected that Industry 5.0 merges the high speed and accurate machines and critical, cognitive thinking of humans. Mass personalization is another important contribution of Industry 5.0, wherein the customers can prefer personalized and customized products according to their taste and needs. Industry 5.0 will significantly increase manufacturing efficiency and create

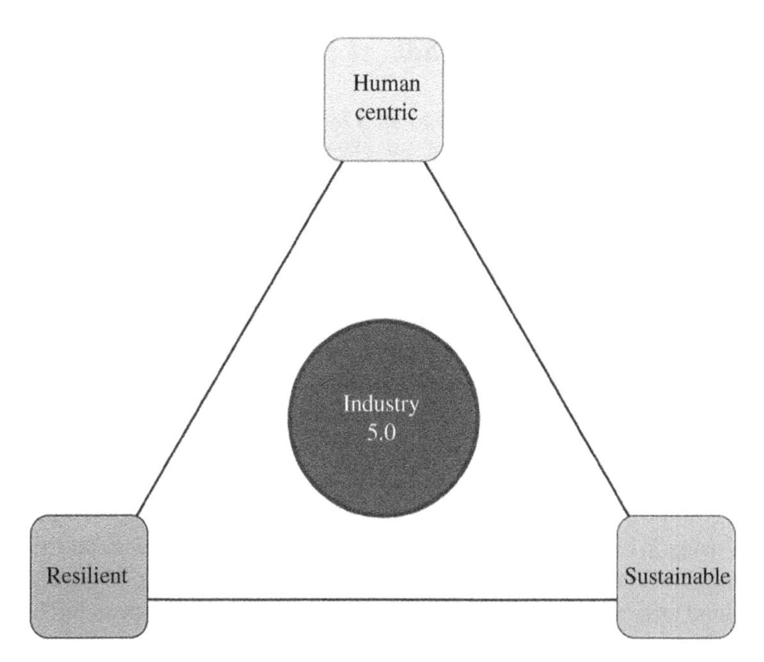

Figure 16.2 The core elements of Industry 5.0.

versatility between humans and machines, enabling responsibility for interaction and constant monitoring. The collaboration between humans and machines aims to increase production at a rapid pace. Industry 5.0 can enhance the quality of the production by assigning repetitive and monotonous tasks to the robots/machines and the tasks which need critical thinking to the humans (Figure 16.2).

Industry 5.0 promotes more skilled jobs compared to Industry 4.0 since intellectual professionals work with machines. Industry 5.0 focuses mainly on mass customization, where humans will be guiding robots. In Industry 4.0, robots are already actively engaged in large-scale production, whereas Industry 5.0 is primarily designed to enhance customer satisfaction. Industry 4.0 focuses on CPS connectivity, while Industry 5.0 links to Industry 4.0 applications and establishes a relationship between collaborative robots (cobots). Another interesting benefit of Industry 5.0 is the provision of greener solutions compared to the existing industrial transformations, neither of which focuses on protecting the natural environment [541]. Industry 5.0 uses predictive analytics and operating intelligence to create models that aim at making more accurate and less unstable decisions. In Industry 5.0, the majority of the production process will be automated, as real-time data will be obtained from machines in combination with highly equipped specialists.

16.2 Background of Industry 5.0

The first industrial revolution has started in 1780 by generating mechanical power from different sources, followed by electrical energy utilization for assembly lines. Information technology has been employed to automate activities in the production industry. For example, the fourth industrial revolution utilized IoT and cloud to connect the virtual and physical space, later named CPS [542, 543]. Though standard 4.0 changed the manufacturing industry, the process optimization ignored the human resources resulting in unemployment. Therefore, the industry pioneers are looking forward to the next revolution where both human intelligence and machines will be integrated for a better solution. The fourth industrial revolution aimed at transforming the manufacturing agents into CPS from complete physical systems through effective integration of business processes and production. This includes integrating all the entities in the manufacturing industry's supply chain, right from suppliers to production lines to end-users using IoT [544]. Industry 4.0 uses CPS to communicate with all the entities through the IoT network. As a result, the significant data accumulation is stored in a cloud environment for effective processing. Industry 4.0 uses technology concepts such as CPS, IoT, AI, Robotics, cloud computing, edge computing, big data analytics, ambient intelligence, virtual reality, and cybersecurity to achieve its central theme of Smart Manufacturing [545, 546]. Industry 4.0 has reduced production, logistics, and quality management costs with increased mass production. Although Industry 4.0 has improved the manufacturing cost, it has ignored the human cost through process optimization. This inadvertently leads to the backward push of employment and will raise resistance from labor unions thereby affecting the full adoption of Industry 4.0 [541]. Industry 5.0 is expected to solve this issue through increased participation of humans.

With the rapid increase in environmental pollution from Industry 2.0, the manufacturing industry is focused on taking adverse effects to manage waste effectively and reduce its impacts on the ecosystem. Industry 4.0 does not ensure any environmental protection. So the need for a technological solution to provide pollution-free manufacturing processes has led to the next industrial revolution [547, 548]. Industry 5.0 ensures the sustainability of civilization by reducing waste generation through bioeconomy leading to a pollution-free environment. The fifth industrial revolution has focused on intelligent manufacturing by bringing back human intelligence to the production floor by enabling the robots to share and collaborate with humans. The Industry 5.0 has brought back the humans to cowork with machines (robots) on factory floors, thus utilizing human intelligence and creativity for intelligent processes [349]. Humans will share and collaborate with the cobots without fear of job insecurity, thus resulting in value-added services. This section provides definitions, features, and state of arts of Industry 5.0.

16.2.1 Definitions

As Industry 5.0 is yet to evolve to the fullest, and various industry practitioners and researchers have provided various definitions. Some of the definitions are discussed here.

Definition 1: Industry 5.0 is a first industrial evolution led by the human based on the 6R (Recognize, Reconsider, Realize, Reduce, Reuse, and Recycle) principles of industrial upcycling, a systematic waste prevention technique and logistics efficiency design to valuate life standard, innovative creations, and produce high-quality custom products [549] by Michael Rada, founder and Leader, Industry 5.0.

Definition 2: Industry 5.0 brings back the human workforce to the factory, where human and machine are paired to increase the process efficiency by utilizing the human brainpower and creativity through the integration of workflows with intelligent systems [349].

Definition 3: European Economic and Social committee states that the new revolutionary wave, Industry 5.0, integrates the swerving strengths of cyber-physical production systems (CPPS) and human intelligence to create synergetic factories [550]. Furthermore, to address the manpower weakening by Industry 4.0, the policymakers are looking for innovative, ethical, and human-centred design.

Definition 4: Friedman and Hendry suggest that Industry 5.0 compels the various industry practitioners, information technologists, and philosophers to focus on the consideration of human factors with the technologies in the industrial systems [551].

Definition 5: Industry 5.0 is the age of Social Smart factory where cobots communicate with the humans [552]. Social Smart Factory uses enterprise social networks for enabling seamless communication between human and CPPS components.

Definition 6: Industry 5.0, a symmetrical innovation and the next-generation global governance, is an incremental advancement of Industry 4.0 (asymmetrical innovation). It aims to design orthogonal safe exits by segregating the hyperconnected automation systems for manufacturing and production [553].

Definition 7: Industry 5.0 is a human-centric design solution where the ideal human companion and cobots collaborate with human resources to enable personalizable autonomous manufacturing through enterprise social networks. This, in turn, enables human and machine to work hand in hand. Cobots are not programmable machines, but they can sense and understand the human presence. In this context, the cobots will be used for repetitive tasks and labor-intensive work, whereas human will take care of customization and critical thinking (thinking out of the box).

16.2.2 Additional Features of Industry 5.0

Industry 5.0 is the enhanced version of the fourth industrial revolution. The added features of Industry 5.0 are discussed in this subsection.

16.2.2.1 Smart Additive Manufacturing

The most popular cost-effective approach for current manufacturing industries, which support producers to execute development plans, reduce pollution and resource utilization throughout the development lifecyle, is sustainable manufacturing [554]. Additive manufacturing is the sustainable approach adopted for industrial production, which builds the product part layer by layer instead of a solid block, thereby developing lighter but more robust parts one layer by layer. It adds up material layer by layer on the 3D objects. Smart additive manufacturing (SAM) applies AI algorithms, computer vision to add more accuracy and better graphical representation of product design in 3D printing. Now, 5D printing, a new subset of additive manufacturing, is employed for better compositions. The recent enterprises and researchers are focusing on deploying smart manufacturing products in their research and industrial domains. With the recent advancement of technologies such as AI, IoT, Cloud computing, Big Data, CPS, 5G, digital twin (DT), edge computing (EC), and manufacturing, smart empowering technologies are becoming popular and remarkably strengthened the development of smart manufacturing. Sustainability, profitability, and productivity are the main advantages of smart manufacturing industry. From the last decade, SAM has become emerging technology in smart manufacturing domain [555]. One of the prominent features of Industry 5.0 is additive manufacturing referred to as 3D printing which is applied to make manufacturing products more sustainable. Additive manufacturing in Industry 4.0 focused on customer satisfaction by including benefits in products and other services. It also facilitates transparency, interoperability, automation, and practicable insights [556]. SAM defines the various processes in which the component to be manufactured is developed by adding materials, and the development is executed in various layers. SAM has capability to save energy resources, helps to reduce material and resource consumption which leads to pollution-free environmental production. To obtain the complete benefits of Industry 5.0, SAM is merged with integrated automation capability to streamline the processes involved in supply chain management (SCM) and reduces the delivery time of the products.

16.2.2.2 Predictive Maintenance

As the economy of the world is moving toward globalization, the industries are facing many challenges. This is forcing the manufacturing units move to upcoming transformation such as predictive maintenance (PdM). To enhance the productivity and efficiency, the manufacturers started utilizing evolving technologies, such as CPS approaches and advanced analytical methods [557]. Transparency is the capability of industry to uncover and assess the uncertainties in order to estimate the manufacturing ability and availability. Basically, most of the manufacturing schemes assume the availability of equipment continuously. However, it never practically happens in the real industries. Thus, the manufacturing units should transform themselves to the PdM to acquire transparency. This transformation needs application of state-of-the-art prediction tools in which the data are processed to information systematically and defines the uncertainties to allow the manpower in taking smart decisions. The implementation of IoT provides the basic framework for PdM with the utilization of smart machines and smart sensor networks. Enabling self-conscious capability for systems and machines is the main goal of PdM. Smart computational agent is the key technology for PdM which includes smart software to provide functionalities for predictive modeling. In Industry 5.0, PdM helps to perform maintenance activity for avoiding problems instead of performing planned and scheduled maintenance and when a problem arises [558].

16.2.2.3 Hyper-Customization

Industry 4.0 targeted linking machines, created intelligent supply chains, promoted the production of smart products, and isolated the manpower from automated industries. But Industry 4.0 has failed to manage the growing demand for customization, whereas Industry 5.0 does it using hyper-customization. Hyper-customization is a personalized marketing strategy which applies cutting-edge technologies such as AI, ML, cognitive systems, and computer vision to real-time data in order to provide more specific product, service, and content to every customer. The integration of human intelligence with robots helps manufacturers to customize the products in bulk. In order to achieve this, many variants of the functional material is shared with other personnel with the motive of customizing the product with different variants for customers choice. Industry 4.0 aimed at huge production with low wastage and maximum efficiency, whereas Industry 5.0 aims at mass customization with minimum cost and maximum accuracy. The collaboration between manpower and robots along with cognitive systems enable the industries to coordinate the processes in the manufacturing to implement the customer needs and market changes. The first step in hyper-personalization is transition to agile manufacturing process and supply chain. This also needs human intervention, production team, and customer preferences. Also, the applicability of hyper-customization depends strongly on the cost-effectiveness of the developed products [559].

16.2.2.4 Cyber Physical Cognitive Systems

Due to the advancement of technologies such as smart wearable devices, IoT, cloud computing and big data analytics, CPS has become popular nowadays. The fourth industrial revolution has transformed the manufacturing process from complete manual systems to CPS [560, 561]. The framework for Industry 4.0 is established on the communication between CPS with the help of IoT. Cloud technology is used for huge amount of efficient, secure data storage and exchange [562]. Also, cognitive methods are used in several applications such as surveillance, industrial automation, smart grid, vehicular networks, and environment monitoring to increase the performance of the system and thus called as cyber physical cognitive system (CPCS) [563, 564]. Cognitive capabilities such as observe/study the environment and take actions accordingly are contained in the nodes of CPCS. Learning and knowledge are the primary components of decision-making in CPCS. The CPCS has been introduced for human robot collaborative (HRC) manufacturing. The HRC executes the assembly of components in manufacturing division in collaboration with a robot and human. The integration of machine–human cognition is modeled and applied for this collaboration work in real time. The fifth industrial revolution confined the merits of fourth industrial revolution and brings back the human labor for production. The fifth revolution facilitates the robots and skilled labor to work together in order to produce customized products and services in Industry 5.0 [565].

16.3 Applications in Industry 5.0

This section discusses some of the potential applications of Industry 5.0.

16.3.1 Cloud Manufacturing

Cloud manufacturing is a novel way to revolutionize the traditional manufacturing paradigm in to an advanced manufacturing process by integrating latest technologies such as cloud and EC, IoT, virtualization, and service-oriented technologies. In a cloud manufacturing process, multinational stakeholders will collaborate together to operate efficient and low-cost manufacturing process. The distinguishing features of cloud manufacturing include reliability, high quality, cost-effectiveness, and on-demand capabilities. In addition, it has positive impact on environment as cloud manufacturing can eliminate the long haul delivery requirements of raw materials for the manufacturing process. Moreover, cloud manufacturing is leading advanced manufacturing models such as additive manufacturing and manufacturing grid as well. Figure 16.3

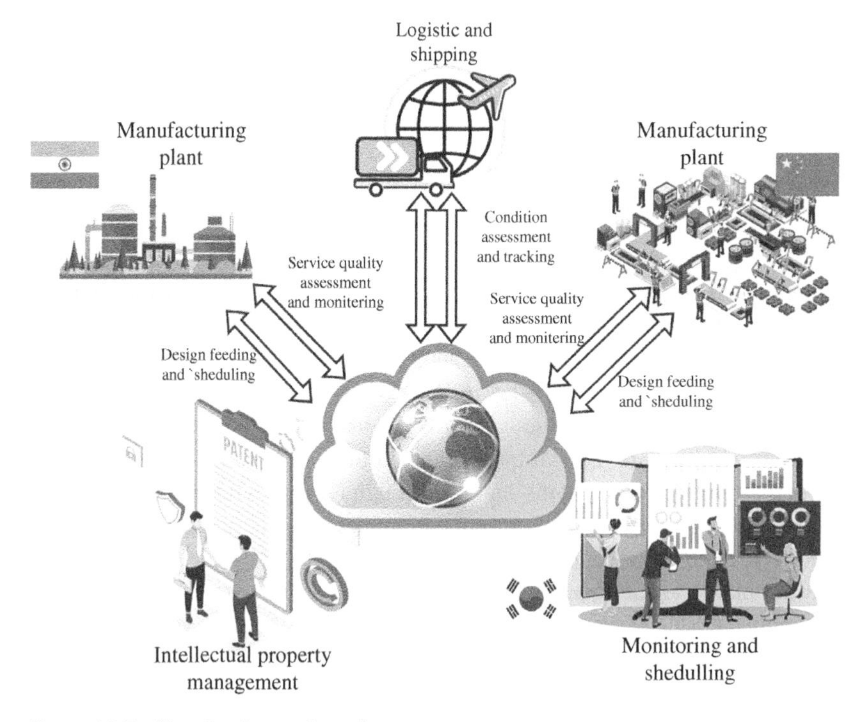

Figure 16.3 The cloud manufacturing ecosystem.

illustrates the multinational collaborative nature of typical cloud manufacturing ecosystem.

Cloud manufacturing allows the designers to protect their intellectual components such as design files of manufacturing items by storing in the cloud with robust access control and utilize the manufacturing resources dispersed across different geographical regions [566]. In this manner, the designers are allowed to place their manufacturing plants closer to the raw material and also countries where manufacturing cost is cheaper. Here, the control of the machines in the plant and the operations of the manufacturing life cycle, such as service composition [567] and scheduling [568] are handled by the cloud. The working condition information of the manufacturing process can be collected via IoT sensors and analyzed in the cloud [569]. Li *et al.* [570] and Tao *et al.* [571] presented how cloud manufacturing can be deployed as a service-oriented manufacturing model. Xu *et al.* [350] explained the potential business models in cloud manufacturing, including pay-as-you-go business model.

With Industry 5.0, the next generation of cloud manufacturing systems is expected to cater to different and complex requirements in the engineering, production, and logistics contexts. The technological evolution of AI/ML technologies, EC features, and 5G-based telecommunication networks open up different avenues to expand the capabilities of future cloud manufacturing systems exponentially.

16.3.2 Digital Twins

A digital replication of a physical system or an object is called a DT. Real-world objects such as wind farms, factories, jet engines, buildings, or even larger systems such as smart cities can be represented digitally through DT [572]. Even though the concept of DT has been proposed in 2002, it has become a reality only in the last few years due to the surge of IoT. IoT made DT cost-effective thus making it accessible and affordable for many industries [573]. Through IoT devices, the data from the physical objects are fed to their digital counterpart for simulation. This mapping of real-time objects/systems digitally through DT makes it possible to analyze, monitor the digital version, and prevent the problems before they occur in the real world. The rapid advancement of AI, ML, and big data analytics has enabled DT to reduce maintenance costs and improve performance of system [574].

In Industry 5.0, DT can offer significant value for the development of customized products on the market, enhanced business functions, reduced defects, and rapidly growing innovative business models to achieve profits. The DT can enable Industry 5.0 to overcome technical issues by identifying them at a faster speed, identifying items that can be reconfigured or renewed on the basis of their productivity, making predictions at a higher accuracy rate, predicting future errors, and avoiding huge financial losses. This type of smart architecture design enables organizations to realize economic advantages successively and more quickly than ever before. In Industry 5.0, DT can be used to generate simulation models, access real-time computational data so that companies can remotely modify and update physical objects [575]. In Industry 5.0, DT is used for customization that can improve the user's experience of their product needs, a purchasing process that enables clients to build virtual environments to see the results.

16.3.3 Cobots (Collaborative Robots)

Recent trends in automation and robotics have made it increasingly important for people to work with robots. Due to the massive rapid changes in AI, smart technology, it is clear that all devices with computational capabilities have become more intelligent and have introduced a new technology called cobots. Collaborative robots are robots designed to work collaboratively with humans, and this collaboration helps to make human capabilities more efficient, extremely easy to automate for individuals and small businesses than ever before. The first cobots were developed by professor Edward Colgate and Michael Peshkin of Northwestern University in 1996 [576]. The first generation of cobots did not have motors and was also very passive in operation and had brakes during operation. Although today's cobots are very different from traditional industrial robots that have the ability to work with humans without enclosures. Cobots are usually embedded with sensors and are highly responsive to the detection of unpredictable impact, which gives them the ability to stop spontaneously when human workers detect any misplaced objects in their path. This tends to make them extremely reliable when it comes to safety at work compared to standard industrial robots [577].

Robots are extremely good in the manufacturing process of high-volume products and are much more compatible than humans. In comparison to human beings, robots are inefficient in critical thinking. Customization or personalizing of products may be a major challenge where robots require guidance. Managing human connections within production processes is therefore crucial. Cobots can offer significant value in Industry 5.0. Working with humans, robots can achieve their intended goal, thereby helping to deliver mass customized and personalized products to customers with high speed and accuracy. Personalizing of cobots can take many forms throughout Industry 5.0 by providing medical treatments, smart applications that efficiently summarize a patient's healthy life, and medical requirements to create a fully customized health fitness routine [577]. Surgery is one of the applications of cobots where a highly qualified doctor and a robotic assistant work together to perform surgery. Medical centers already enjoy the medical benefits of collaborative robotic-supported industrial processes. The Davinci surgical system is an innovation in cobot technology as it enhances the operating capability of surgeons in the operation theater. Davinci cobot is widely used in urology and gynecology surgeries, as well as in other surgeries. The role of cobots in Industry 5.0 is used to increase productivity and helps to build a new relationship between humans and machines. In Industry 5.0 applications, cobots help to improve safety and performance while at the same time facilitating more interesting responsibilities for human workers and increasing productivity growth [578]. Industries must realize that cobots offer not only the ability to improve business performance but also the potential to reduce rising labor costs in highly competitive markets.

16.3.4 Supply Chain Management

Disruptive technologies that enable Industry 5.0 such as DT, cobots, 5G, and beyond, ML, IoT, EC, are aligned with the smartness, and innovation of humans can help the industries in meeting the demand and delivering the personalized and customized products at a faster pace [579]. This helps SCM in integrating mass customization, which is a key concept in Industry 5.0, into their production systems.

DT can be used to create a digital replica of the SCM that consists of warehouses, inventory positions, assets, and logistics. The DT encapsulates factories, suppliers, contract manufacturers, factories, transportation lanes, distribution facilities, and customer locations. DT supports in the entire life cycle of the SCM, right from the design phase, to the construction and commissioning, to the operations [580, 581]. Through simulating the real-time SCM systems, DT can sense the real-world data through IoT sensors. ML, big data, etc., and can use these data to predict the difficulties faced during several phases of SCM. The industries can hence take preemptive corrective measures to minimize the losses and errors during several phases of SCM and can help in delivering customized products to the customers in quick amount of time [582]. With DT, businesses can evaluate the complex interconnected trade-offs in capacity, service, inventory, and total landed cost. DT can also help industries in increasing their margins, reduce operational costs during several phases of the SCM. Some of the state-of-the-art works that present DT as a solution for SCM are discussed below.

Cobots can play a very important role in SCM. Tasks which are routine/dangerous, such as packaging, routine quality checks, carrying of heavy goods, that humans hesitate to do can be performed by robots, whereas the expertise of the humans can be used in more complex jobs in the SCM life cycle [577]. Cobots can be used in applications throughout the life cylce of SCM, such as material handling, assembly of the materials, packing, performing quality checks, transportation, delivery of the products to the customers, and picking the return of the products from the customers. Through cobots, the SCM industries can reduce their total cost of ownership. Hence, cobots streamline all the processes in SCM, such as systematic inventory management, tracking of stocks, order fulfillment, and return of the products [583].

16.3.5 Manufacturing/Production

It is generally acknowledged that in past technological revolutions, the introduction of robotics and automation brought about paradigm changes in the manufacturing industry globally. Robots have historically done risky, monotonous, or physically demanding work in manufacturing settings, such as welding and painting in car factories and loading and unloading heavy consignments in warehouses [584]. Industry 5.0 is aimed at combining these cognitive computing skills with human intelligence and resourcefulness in collaborative operations as machines in the workplace grow smarter and more connected. It is therefore conceivable that the fifth industrial revolution will bring

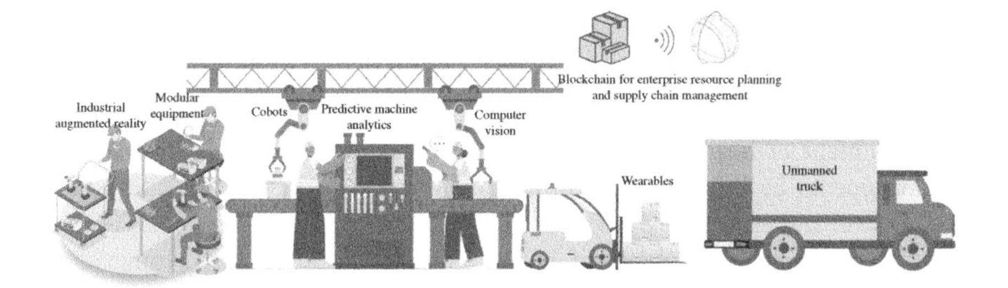

Figure 16.4 Industry 5.0 Technologies.

shifts in norms and make fundamental changes in our approach to industry and manufacturing (Figure 16.4).

16.4 Role of 6G in Industry 5.0

In the future, 6G can offer significant value-added services to Industry 5.0. Radio infrastructure with a very dense chain of thousands or millions of sensors, hardware elements, and robots is a challenge. With the vigorous growth of smart infrastructure and potential applications with current networks (e.g. 4G and 5G networks), it will not be possible to meet rapidly increasing bandwidth requirements. The use of 6G and beyond in the Industry 5.0 revolution makes it possible to deliver better latency, support high-quality services, as well as extensive IoT infrastructure, and integrated AI capabilities [25]. In Industry 5.0 applications, 6G networks help improve application performance efficiently and effectively by providing smart spectrum management, AI-powered mobile EC, and smart mobility [14, 59]. For Industry 5.0 applications, 6G networks are expected to meet the standards of an intelligent information society that can deliver ultra-high data rates, ultra-low latency, ultra-high reliability, high-energy efficiency, traffic capacity, etc. Mobility and handover management are the most significant challenges for 6G networks in Industry 5.0. The 6G networks will be large-scale, highly dynamic, multilayer networks that lead to frequent handovers [76, 585]. AI techniques can be used to obtain optimal mobility predictions and optimal handover solutions to ensure efficient connectivity [90]. The greatest challenge for Industry 5.0 applications is to provide a high data rate for different applications. Quantum communication and free-space optical communication in 6G can fix these issues. In Industry 5.0 applications, a large number of smart devices are connected and an excess amount of energy is consumed therefore energy management is a critical challenge in Industry 5.0. 6G networks optimize energy management through the use of advanced energy consumption strategies and energy harvesting methods.

16.4.1 Internet of Everything

Internet of Everything (IoE) is an interconnected link between people, processes, information, and things [586, 587]. IoE can provide significant value for the establishment of new opportunities for Industry 5.0 applications [588]. IoE's advancements in Industry 5.0 can create new functionalities, provide better experience, and expected benefits for industries and nations. The role of IoE in Industry 5.0 enhancing customer loyalty and delight, building customization experience based on IoE-generated data. The use of IoE in Industry 5.0 provides an opportunity to minimize operating costs by eliminating bottlenecks on communication channels and reduces latency. Supply Chain and Logistics Efficiency is a challenging issue for Industry 5.0. IoE to minimize supply chain waste and optimize production processes. Due to the immense development of IoE, information sharing between humans takes place in wireless mode, essentially with the help of wireless sensors. For example, in the Internet of medical things, sensors are fixed to the patient. These sensors detect abnormalities in patients and transmit the sensed data to the doctor or nurse concerned. The physicians will take appropriate steps on the basis of the information obtained.

16.4.1.1 Big Data Analytics

Big data have recently become a major focus of discussion in both industry and academia [589–591]. It represents a large and diverse set of data collected from all types of sources. Many data analysis techniques include Big Data technologies, such as ML, AI, social networking, data mining, data fusion, and so on [592, 593]. Big Data Analytics tends to play an important role in the field of Industry 5.0. In Industry 5.0, some companies can use Big Data Analytics to better understand consumer behavior in order to optimize product prices, focus on improving production efficiency, and help reduce overhead costs [594]. Understanding the current behavior of the user, social relations and human behavior rules are a critical challenge. Big Data Analytics is used by certain companies, such as Facebook, Twitter, and Linkedin, which can help to promote products and increase sales on the basis of consumer satisfaction. Data infusion, massive customized manufacturing processes, and smart automation in the production process are essential for the resolution of the Industry 5.0 ecosystem. Big Data Analytics can be used to make real-time decisions to enhance the competitive advantage of industries, with a focus on providing recommendations on predictive discoveries for major events in Industry 5.0 applications. In Industry 5.0, Big Data Analytics helps with mass customization processes with zero-fail integration with the available resources [595]. Real-time analytical data shared with smart systems and data centers helps manufacturers to produce and handle high-data volumes. Continuous process improvement is another critical challenge in Industry 5.0, which often requires the collection of detailed information on the entire manufacturing cycle. Big Data Analytics techniques are used to recognize and eliminate nonessentialities to maximize predictability and explore new possibilities.

16.4.1.2 Blockchain

Blockchain technology can offer significant value additions in future Industry 5.0. Centralized management of a large number of heterogeneous connected devices in Industry 5.0 is a critical challenge. Blockchain can be used to design decentralized and distributed management platforms by enabling distributed trust [596]. Blockchain-enabled secure peer-to-peer communications offer an immutable ledger to keep records [489, 597, 598]. Moreover, the immutable ledger supports operational transparency and accountability for the significant events in Industry 5.0 applications. Especially, transparency is important for the dispute resolution in Industry 5.0 ecosystem [599]. The smart contracts can be used for security enforcement, such as authentication as well as automated service-oriented actions of the future Industry 5.0 applications. Also, a higher level of protection for data and transactions can be offered by using a compartmentalized and distributed approach using blockchains [600, 601]. Data receiving and gathering [602] can also be enabled via blockchain.

Blockchain can be used to create digital identities for different people and entities in Industry 5.0 for efficient subscriber management. It is needed for access control and authenticating the stakeholders in any industrial activities over a public network [600]. Moreover, these digital identities can be further expanded to manage properties, possessions, objects, and also services. Blockchain technology can also be used to register intellectual property (IP) rights and to catalog and store original work [603]. Blockchains and smart contracts can also help to automate the contracting process by automating the agreement processes between different stakeholders. Moreover, blockchain-powered cloud manufacturing facilitates machine-level connection and data sharing based on blockchain technology [604, 605].

16.4.2 Edge Computing

The rapid growth of the IoT and the provision of numerous cloud services have introduced a new conceptualization, EC, which enables data processing at the network edge. EC can offer significant value not only in the future Industry 5.0 but also in the transition to Industry 4.0. EC is capable of meeting expectations related to latency costs, battery life constraints, response time requirements, data protection, and privacy [71, 606]. EC minimizes communication overhead and guarantees that applications are productive in remote areas. Additionally, EC has the ability to process data without passing it to the public cloud, thus helping to minimize security issues for the significant events in Industry 5.0. EC can perform some useful operations such as data processing, cache coherency, computing offloading, transferring, and delivering requests [489]. With all these network operations, the edge must be designed efficiently to ensure security, reliability, and privacy. For Industry 5.0 applications, EC ensures low latency, data security and privacy and delivers efficient services to the end users [71]. EC provides real-time communications for next-generation Industrial 5.0 applications such as UAVs, autonomous vehicles [607], and remote patient monitoring. EC enables Industry 5.0 to use more accessible, standard hardware and software resources to access and exchange information related to their industrial sectors. In order to manage huge data, industries are trying to access data from local servers on a regular basis. One of the challenges of analyzing all these machines is that the amount of raw data is too large to be assessed efficiently. EC enables Industry 5.0 to filter data by minimizing the volume of data sent to a centralized server. In Industry 5.0, EC allows preventive analytics, which enable the preemptive detection of machine failure and mitigates this by enabling the manpower to make wise decisions.

16.4.2.1 Other Enabling Technologies

In addition, some of the existing technologies such as Network Slicing (NS), eXtended Reality (XR), and Private Mobile Network (PMN) play a vital role in enabling Industry 5.0 and its applications.

NS concept allows enabling multiple virtualized networks on top of a single physical network infrastructure. It slices physical network resources across these virtualized networks [213]. Each virtualized network can be optimized and tailored to satisfy the requirements of different vertical applications. In this aspect, NS plays a vital role in enabling different Industry 4.0 applications [10, 292] and will be important in Industry 5.0 as well. Since Industry 5.0 supports a diverse set of applications, one physical infrastructure will not be capable of fulfilling heterogeneous network requirements. NS can offer different virtualized networks cost-effectively. In [608, 609] and [610], authors present a way of using NS for self-organization, flexibility, and optimal network resources utilization for network monitoring in industrial internet of things (IIoT) networks. In the future, advanced slicing techniques such as federated slicing, hierarchical slicing, and zero-touch slice automation can play an effective role in the realization of Industry 5.0 applications [292, 611].

XR is an another emerging technology which is used in many application domains [340]. XR can improve human–machine interactions by combining virtual and physical worlds [59]. XR is representing a mixture of Virtual Reality (VR), Augmented Reality, and Mixed Reality (MR) technologies [612]. XR technologies will play a vital role in enabling different Industry 5.0 applications.

XR technologies are already being used in Industry 5.0-related applications such as remote assistance [484, 613], assembly line monitoring [614], health education/training [615], remote healthcare [616], indoor and localized outdoor navigation [617–619], driver/pilot training [620, 621], maintenance [622], drone/UAV pilot training [623], and education [624]. Zero touch networking, edge computing, high capable devices, enhanced communication technologies, and high precision computation capabilities will be important for the further development of XR technologies toward Industry 5.0 applications [625, 626].

The introduction of the network softwarization concept in 5G has eliminated the requirement of dedicated, expensive, and vendor-specific hardware equipment to build mobile networks. Thus, network softwarization enables the possibility to deploy local or private mobile networks [627]. Contrary to the traditional country-wide Mobile Network Operators (MNOs), PMNs are deployed to deliver localized, use case-specific network services. With 5G, Local 5G Operators (L5GOs) can be used in several Industry 5.0 applications such as factories [628, 629], hospitals [630, 631], schools, and universities [632] to deliver location-specific connectivity solutions. The integration of NS, AI, and blockchain technologies would optimize the deployment of PMNs for Industry 5.0 realization [356, 633]. Moreover, regulation, management, and leasing of spectrum for PMNs should also be studied further for cost-efficient deployment and wide adaptation for Industry 5.0 deployments [634].

17

Wild Applications

The advent of 6G is also expected to give rise to a number of futuristic applications, which are referred to as wild applications in this chapter. After reading this chapter, you should be able to

- Understand the nature of futuristic 6G applications.
- Understand the research challenges in future 6G applications.

17.1 Introduction

Communication is vitally paramount to everyone or things, while they are dispersed. The 6G communication is not just the evolutionary advancement of past generations of communication, but it is setting a big steppingstone for the futuristic applications that will go beyond the imaginations. 6G has a tremendous potential and vision that can push new industries, applications, and businesses through connecting earth-to-space, space-to-space (or deep space network), space-to-deep sea, and so on. Hence, in next two decade or so, 6G will not only provide global seamless connectivity, but it will also establish the vision to connect the whole universe. This chapter explores three visionary applications of 6G, namely, metaverse, space communication, and deep-sea tourism.

17.2 Metaverse

Present developments of mobile communication networks have enabled many mixed reality applications ranging from video conferencing to remote-operations [635]. However, metaverse aims to be much more immersive and interactive to enable users experience their presence in a "metaverse" in real-time. The concept

6G Frontiers: Towards Future Wireless Systems, First Edition.
Chamitha de Alwis, Quoc-Viet Pham, and Madhusanka Liyanage.
© 2023 The Institute of Electrical and Electronics Engineers, Inc. Published 2023 by John Wiley & Sons, Inc.

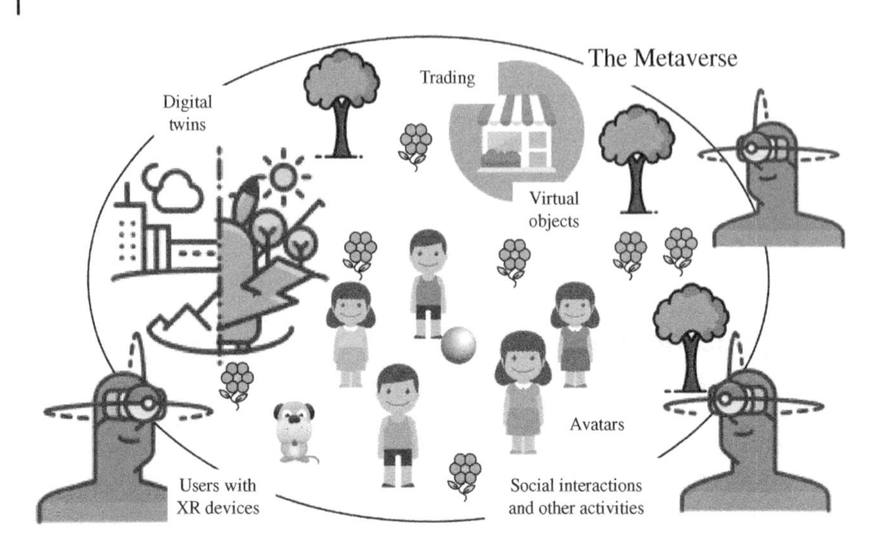

Figure 17.1 The Metaverse Concept.

of the "metaverse" is illustrated in Figure 17.1. Emerging technologies, such as natural language processing, machine vision, blockchain, ultrareliable and ultralow latency communication, digital twins, brain–computer interfaces, and edge artificial intelligence (AI) are considered to be the key enablers of metaverse [636].

The concept of metaverse refers to a universal virtual environment parallel to the physical world to facilitate realistic interactions. This concept was initially discussed in 1992 from the book "Snow Crash" [637]. The metaverse consists of characters named as "Avatars." These avatars can act as real human beings and sometimes go even beyond the capacity of humans in the physical world. In the present context, metaverse is considered to be a blend of real and virtual worlds, enabled through the advents of extended reality (XR), telecommunication, and AI technologies [638].

6G is expected to accelerate the growth of metaverse. The ultrahigh data rates in the Tbps range, extremely low latency, and extreme reliability and availability will enabled through 6G communication networks will facilitate the communication requirements of metaverse applications. Edge AI facilitates powerful processing capabilities with extremely low latency for AI-based XR applications to render graphics with ultrahigh resolution.

These technologies will enable the rendering of realistic 3D environments utilizing large number of sensors connected through the 6G network. Furthermore, the 6G network will handle billions of real-time connections to facilitate a smooth user experience in the virtual world.

The metaverse development is expected to have three development stages [638]:

- Development of digital twins of physical entities and infrastructure, matching the ones in the physical world.
- Moving people in the digital world as digital natives in the form of avatars.
- The dawn of a virtual world that will coexist with the physical world in a sustainable fashion.

The metaverse is expected to be sustainable through having its own ecosystem with many products and services, ownership, trading, and its own economy with metaverse commerce and economic governance [638]. This will pave the path for a new way of interactions between human beings in new dimensions. For instance, XR together with gestures, haptics, and brain–computer interface (BCI) will bring human emotions, sensations, and actions to the virtual world while also facilitating novel means of content creation [639].

17.3 Deep-Sea Explorations

Deep-sea explorations involve the investigating the bed of the ocean for research, commercial, or entertainment purposes. Until now, mankind has amply discovered a small portion of the world's deep oceans. However, ocean depths remain to be largely undiscovered and unexplored, where endless species are also yet to be discovered. Hence, the popularity of exploring the seabeds is on the rise. Furthermore, a new era of deep-sea explorations is dawning where industries are already interested in exploring business opportunities. Few of underwater tour companies (e.g. Blue Marble Private, OceanGate), have started deep dives tour, i.e. the Titanic, to explore the shipwreck [640, 641]. It will open several futuristic use-cases, exploring the twilight zone under the seabed, tracking sea creature, etc. To this end, with the increasing number of activities, there is need for maritime ultra-reliable and high-speed communications to connect devices, ships, and individuals on the water and/or underwater. These devices or vessels would exchange huge amount of digital data including voices, texts, and video streaming, from ship to ship and to shore. Maritime communication still depends on heterogeneous devices such as sensors and drones, and traditional communication technologies, including 2G/3G and satellite networks.

Maritime communication networks utilize Global Maritime Distress and Safety System (GMDSS). This works with high frequency and low bandwidth with limited data rates. Legacy networks also cannot establish communication between underwater nodes and on-shore nodes due to the significant signal attenuation in water. On the other hand, sonar signals transmitted by underwater devices reflect away from the surface without being able to break through. Therefore, existing communication technologies, on their own, cannot provide long-haul and stable channels between off-shore nodes, such as maritime vessels and maritime sensors and on-shore nodes. Satellite communication technologies too are not capable of connecting the growing numbers of maritime communication nodes.

The under-sea network targets on providing Internet facility under the seas and oceans. But it still remains a very controversial topic whether or not it would be a part of 6G networks. This network system involves optical and acoustic communication and radio frequency. But the difficulty remains due to the unpredictability of the underwater environment, challenges that are to be faced, and the risk factors as water exhibits different propagation properties which are different from that of aerial or terrestrial terrains. Therefore, a lot of issues are yet to be resolved for under-water 6G networks.

The recent developments of unmanned aerial vehicles (UAVs) have enabled UAV-assisted mobile relay communication to facilitate maritime communication applications [642]. These systems also utilize UAV-assisted decode-and-forward relay communication for caching to overcome the performance constraints caused by the limited capacity of wireless backhaul links between base stations and UAVs. Furthermore, the optimal UAV placement is calculated to maximize the achievable data rates. In addition, multiaccess edge computing and edge AI technologies are being developed to facilitate ultra-low latency communication and low-computing latency to facilitate AI-based applications through performing computing and caching functions at the edge of the network [643].

In addition, highly accurate communication, positioning, navigation, and timing technologies for deep-sea vehicles, such as deep-sea human-occupied vehicles, remote-operated underwater vehicles, and autonomous underwater vehicles, are also being developed to facilitate deep sea explorations in the 6G era [644].

17.4 Space Tourism

The advancement of science and technologies made humans want to explore what is up on space. This gave rise to the development of rockets and other technologies to communicate and control these space vehicles. Yuri Alekseyevich Gagarin, who was a Soviet pilot became the first ever human to visit space. He traveled one orbit of earth in Vostok 1 capsule in 1961. In 1969, Apollo 11 took humans safely to the moon and back to earth. The legacy continued by putting many men on space and many unmanned rockets on plants. Furthermore, the International Space Station, which is an artificial satellite that humans can occupy, is also placed in space as a joint multinational project between five space agencies, namely, NASA (United States), Roscosmos (Russia), JAXA (Japan), ESA (Europe), and CSA (Canada).

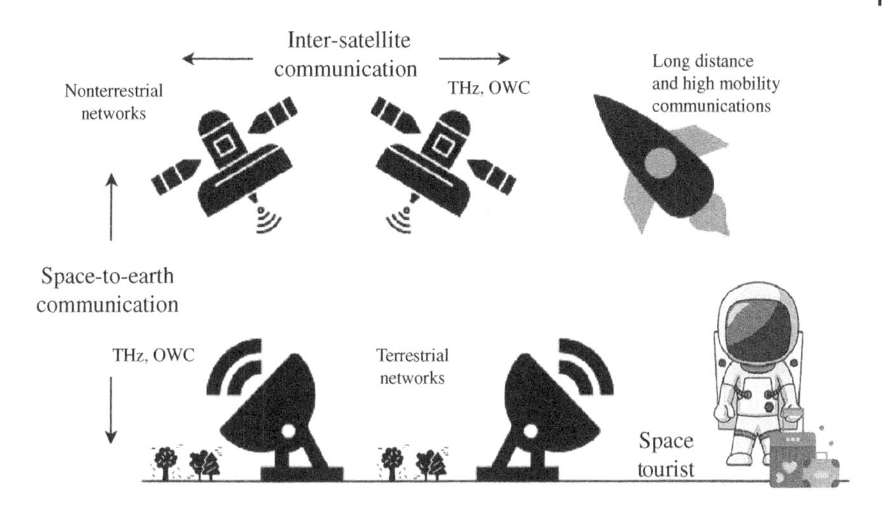

Figure 17.2 6G Communication for Space Tourism.

However, these visits to space were limited to academic and research purposes. Yet, it is envisaged that space tourism will become a blooming industry and enable humans to travel to space in the near future. One such initiative toward space tourism is SpaceX, which is an American aerospace manufacturer that also provides space transportation services [645]. They have already launched Falcon 9 series of rockets over a hundred times and have send space crafts to the International Space Station.

The growth of space tourism will also require communication technologies to evolve to cover extensive ranges to facilitate space communication, as illustrated in Figure 17.2.

In large-dimension networks, Long Distance and High Mobility Communications (LDHMC) are indispensable requirements in 6G to facilitate applications such as space tourism [89]. In 5G, LDHMC services are undeniable as they can support up to 500 km/h. However, future applications, such as space tourism will require seamless mobility and long-distance communication for many thousands of km. However, node mobility is highly challenging in different environments, such as space [90]. Therefore, 6G must require long-distance and high mobility communication (e.g. >1000 km/h) based seamless space communication services. The development of communication technologies are envisaged to enable LDMHC in future networks. For instance, there are several enabling technologies such as accurate channel estimation. Due to severe time and frequency spreading in high mobility wireless communications, channel estimation is very challenging. Filter-based alternative waveforms [94] (alternatives to orthogonal frequency division multiplexing), such as filter-bank multicarrier and universal filtered multicarrier are good candidates for 6G high mobility communications.

Furthermore, recent 6G research also focuses on the growth of Terrestrial Networks (TN) and NTNs. Accordingly, terrestrial-satellites networks are envisaged to play a key role in future mobile networks [121]. One emerging solution, developed by the same SpaceX initiative is known as Starlink [646]. Starlink is a NTN with low-orbit satellite network operated by SpaceX intending to provide satellite Internet access globally and even beyond. Presently, Starlink delivers satellite broadband services with high data rates (>100 Mbps) and low latency (<30 ms) while extending their coverage to unreachable areas with at least two new monthly Starlink launches [121].

6G is expected to operate in both lower-frequency bands and high frequency bands in the THz range. However, higher frequencies experience higher attenuation. This makes it challenging to utilize higher frequencies in high mobility and high attenuation applications, such as space communication. Therefore, massive MIMO and beamforming technologies are expected to evolve toward facilitating space tourism applications. Furthermore, Supplementary Uplinks (SULs), which can associate with a conventional downlink and uplink carrier with a low-frequency SUL carrier can extend the uplink coverage and thereby improve uplink data rates owing to the low-path loss in low-frequency bands. Furthermore, providing reliable line-of-sight links to ground users and stations also require the optimized positioning of NTNs and proper beam management techniques. However, efficient usage of the THz band for space-to-earth and inter-satellite communications requires to be explored further toward facilitating space tourism applications.

Optical Wireless Communication (OWC) is another 6G technology that is envisaged to facilitate space tourism. OWC is capable of providing Tbps capacity with high-energy efficiency compared to RF communication for demanding applications of space tourists. OWC mainly consists of Visible Light Communication (VLC) and Free Space Optics (FSO). Thus, OWC can facilitate intersatellite communication, interspace station communication, and space-to-earth communication. However, many challenges including overcoming the solar background radiation interference and managing line-of-sight in high mobility applications are yet to be solved.

Usage of smart steerable high gain, multiband and multibeam antennas, intelligent surfaces, better error correction mechanisms, and proper path planning for satellites and other space objects are some of the solutions under development. Furthermore, edge AI technologies in 6G and beyond NTN will facilitate the processing of demanding AI-based space applications. Additionally, NTN nodes, satellites, and nano-satellites can be operated and maintained through their digital twins and orchestrated with a multitenant Ground Station Network (GSN), where Ground Stations are the edge nodes of a widespread terrestrial-satellite network. On the other hand, Blockchain can ensure the data security of AI applications that span across distributed 6G nodes and users. However, further research and development work are required to develop these technologies, while ensuring user security and privacy, toward realizing space tourism applications in the 6G era.

Acknowledgement

Dr. Pardeep Kumar with the Department of Computer Science, Swansea University, United Kingdom, has partly contributed for this chapter.

Part V

Conclusion

18

Conclusion

Societal and technological trends envisaged by the 2030 society highlight the limitations of existing 5G mobile networks. Driven by the heterogeneous demands of the hyperconnected 2030 world, a new network paradigm is envisioned to facilitate emerging applications. These applications require 6G networks to provide extreme data rates, extremely low delays, extreme reliability and availability, massive scalability, extreme power efficiency, and extreme mobility. Realizing this 6G vision requires utilizing new network architectures, network intelligence, and other technologies that are envisaged to act as enablers of 6G. These technologies, which are still in the development stage, are required to be fully developed and standardized toward realizing 6G in 2030.

This book provides a visionary insight in to the present and future developments of 6G. The book covers a multitude of 6G aspects, including, what will be the 6G requirements, and how to build different architectures to play together. This is a first book on 6G that will touch upon many exciting technical aspects such as intelligent network softwarization, Open radio access network (RAN), security, privacy and trust, harmonized mobile network, and legal views including standards initiatives. The book provides reference material to design intelligent, smart, and autonomous telecommunication networks. It offers an intelligent insight into the ongoing and future research treads, use-cases, and key technologies toward the 6G networks. The book also discusses 6G applications, including, smart cities and Society 5.0, future healthcare, and Industry 5.0 while it also thrusts on the exciting wild 6G applications such as metaverse, deep-sea explorations, and space tourism.

6G Frontiers: Towards Future Wireless Systems, First Edition.
Chamitha de Alwis, Quoc-Viet Pham, and Madhusanka Liyanage.
© 2023 The Institute of Electrical and Electronics Engineers, Inc. Published 2023 by John Wiley & Sons, Inc.

Bibliography

1 "Focus Group on Technologies for Network 2030," International Telecommunication Union, 2019, [Accessed on 29.03.2021]. [Online]. Available: https://www.itu.int/en/ITU-T/focusgroups/net2030/Pages/default.aspx.
2 "6G Flagship," University of Oulu, 2020, [Accessed on 29.03.2021]. [Online]. Available: https://www.oulu.fi/6gflagship/.
3 M. R. Bhalla and A. V. Bhalla, "Generations of mobile wireless technology: A survey," *International Journal of Computer Applications*, vol. 5, no. 4, pp. 26–32, 2010.
4 V. Pereira and T. Sousa, "Evolution of mobile communications: From 1G to 4G," *Department of Informatics Engineering of the University of Coimbra*, Portugal, 2004.
5 J. R. Churi, T. S. Surendran, S. A. Tigdi, and S. Yewale, "Evolution of networks (2G-5G)," in *International Conference on Advances in Communication and Computing Technologies (ICACACT)*, vol. 51, no. 4, Citeseer, 2012, pp. 8–13.
6 S. Won and S. W. Choi, "Three decades of 3GPP target cell search through 3G, 4G, and 5G," *IEEE Access*, vol. 8, pp. 116 914–116 960, 2020.
7 P. Datta and S. Kaushal, "Exploration and comparison of different 4G technologies implementations: A survey," in *2014 Recent Advances in Engineering and Computational Sciences (RAECS)* IEEE, 2014, pp. 1–6.
8 P. Popovski, K. F. Trillingsgaard, O. Simeone, and G. Durisi, "5G wireless network slicing for eMBB, URLLC, and mMTC: A communication-theoretic view," *IEEE Access*, vol. 6, pp. 55 765–55 779, 2018.
9 M. Liyanage, A. Gurtov, and M. Ylianttila, *Software Defined Mobile Networks (SDMN): Beyond LTE Network Architecture*. John Wiley & Sons, 2015.
10 S. Wijethilaka and M. Liyanage, "Realizing Internet of Things with network slicing: Opportunities and challenges," in *2021 IEEE 18th Annual Consumer Communications & Networking Conference (CCNC)* IEEE, 2021, pp. 1–6.

6G Frontiers: Towards Future Wireless Systems, First Edition.
Chamitha de Alwis, Quoc-Viet Pham, and Madhusanka Liyanage.
© 2023 The Institute of Electrical and Electronics Engineers, Inc. Published 2023 by John Wiley & Sons, Inc.

11 Y. Siriwardhana, C. De Alwis, G. Gür, M. Ylianttila, and M. Liyanage, "The fight against the COVID-19 pandemic with 5G technologies," *IEEE Engineering Management Review*, vol. 48, no. 3, pp. 72–84, 2020.
12 W. Saad, M. Bennis, and M. Chen, "A vision of 6G wireless systems: Applications, trends, technologies, and open research problems," *IEEE Network*, vol. 34, no. 3, pp. 134–142, 2019.
13 F. Fang, Y. Xu, Q.-V. Pham, and Z. Ding, "Energy-efficient design of IRS-NOMA networks," *IEEE Transactions on Vehicular Technology*, vol. 69, no. 11, pp. 14 088–14 092, 2020.
14 Y. Lu and X. Zheng, "6G: A survey on technologies, scenarios, challenges, and the related issues," *Journal of Industrial Information Integration*, vol. 19, p. 100158, 2020.
15 Y. Liu, X. Yuan, Z. Xiong, J. Kang, X. Wang, and D. Niyato, "Federated learning for 6G communications: Challenges, methods, and future directions," *China Communications*, vol. 17, no. 9, pp. 105–118, 2020.
16 K. B. Letaief, W. Chen, Y. Shi, J. Zhang, and Y.-J. A. Zhang, "The roadmap to 6G: AI empowered wireless networks," *IEEE Communications Magazine*, vol. 57, no. 8, pp. 84–90, 2019.
17 M. Giordani, M. Polese, M. Mezzavilla, S. Rangan, and M. Zorzi, "Toward 6G networks: Use cases and technologies," *IEEE Communications Magazine*, vol. 58, no. 3, pp. 55–61, 2020.
18 S. Dang, O. Amin, B. Shihada, and M.-S. Alouini, "What should 6G be?" *Nature Electronics*, vol. 3, no. 1, pp. 20–29, 2020.
19 M. Latva-Aho and K. Leppänen, "Key drivers and research challenges for 6G ubiquitous wireless intelligence (white paper)," Oulu, Finland: *6G Flagship*, 2019.
20 Z. Zhang, Y. Xiao, Z. Ma, M. Xiao, Z. Ding, X. Lei, G. K. Karagiannidis, and P. Fan, "6G wireless networks: Vision, requirements, architecture, and key technologies," *IEEE Vehicular Technology Magazine*, vol. 14, no. 3, pp. 28–41, 2019.
21 L. Mucchi, S. Jayousi, S. Caputo, E. Paoletti, P. Zoppi, S. Geli, and P. Dioniso, "How 6G technology can change the future wireless healthcare," in *2020 2nd 6G Wireless Summit (6G SUMMIT)* IEEE, 2020, pp. 1–6.
22 W. He, D. Goodkind, P. R. Kowal *et al.*, "An Aging World: 2015," 2016.
23 "Global IoT Market will Grow to 24.1 Billion Devices in 2030, Generating $1.5 Trillion Annual Revenue," Transforma Insights research, May, 2020, [Accessed on 29.03.2021]. [Online]. Available: https://transformainsights.com/news/iot-market-24-billion-usd15-trillion-revenue-2030.
24 T.-Y. Chan, Y. Ren, Y.-C. Tseng, and J.-C. Chen, "Multi-slot allocation protocols for massive IoT devices with small-size uploading data," *IEEE Wireless Communications Letters*, vol. 8, no. 2, pp. 448–451, 2018.

25 M. Z. Chowdhury, M. Shahjalal, S. Ahmed, and Y. M. Jang, "6G wireless communication systems: Applications, requirements, technologies, challenges, and research directions," *IEEE Open Journal of the Communications Society*, vol. 1, pp. 957–975, 2020.

26 J. Hu, Q. Wang, and K. Yang, "Energy self-sustainability in full-spectrum 6G," *IEEE Wireless Communications*, vol. 28, pp. 104–111, 2020.

27 S. Wang, X. Zhang, Y. Zhang, L. Wang, J. Yang, and W. Wang, "A survey on mobile edge networks: Convergence of computing, caching and communications," *IEEE Access*, vol. 5, pp. 6757–6779, 2017.

28 A. Bourdoux, A. N. Barreto, B. van Liempd, C. de Lima, D. Dardari, D. Belot, E.-S. Lohan, G. Seco-Granados, H. Sarieddeen, H. Wymeersch *et al.*, "6G White Paper on Localization and Sensing," *arXiv preprint arXiv:2006.01779*, 2020.

29 T. Higashino, A. Uchiyama, S. Saruwatari, H. Yamaguchi, and T. Watanabe, "Context recognition of humans and objects by distributed zero-energy IoT devices," in *2019 IEEE 39th International Conference on Distributed Computing Systems (ICDCS)* IEEE, 2019, pp. 1787–1796.

30 T. Kumar, P. Porambage, I. Ahmad, M. Liyanage, E. Harjula, and M. Ylianttila, "Securing gadget-free digital services," *Computer*, vol. 51, no. 11, pp. 66–77, 2018.

31 I. Ahmad, T. Kumar, M. Liyanage, M. Ylianttila, T. Koskela, T. Braysy, A. Anttonen, V. Pentikinen, J.-P. Soininen, and J. Huusko, "Towards gadget-free internet services: A roadmap of the naked world," *Telematics and Informatics*, vol. 35, no. 1, pp. 82–92, 2018.

32 M. Liyanage, A. Braeken, and M. Ylianttila, "Gadget free authentication," in *IoT Security: Advances in Authentication*, M. Liyanage, A. Braeken, P. Kumar and M. Ylianttila, Eds, 2020, pp. 143–157. https://doi.org/10.1002/9781119527978.ch8.

33 I. Lee and K. Lee, "The Internet of Things (IoT): Applications, investments, and challenges for enterprises," *Business Horizons*, vol. 58, no. 4, pp. 431–440, 2015.

34 M. H. Miraz, M. Ali, P. S. Excell, and R. Picking, "A review on Internet of Things (IoT), internet of everything (IoE) and internet of nano things (IoNT)," in *2015 Internet Technologies and Applications (ITA)* IEEE, 2015, pp. 219–224.

35 N. Janbi, I. Katib, A. Albeshri, and R. Mehmood, "Distributed artificial intelligence-as-a-service (DAIaaS) for smarter IoE and 6G environments," *Sensors*, vol. 20, no. 20, p. 5796, 2020.

36 Q. Qi, X. Chen, C. Zhong, and Z. Zhang, "Integration of energy, computation and communication in 6G cellular Internet of Things," *IEEE Communications Letters*, vol. 24, pp. 1333–1337, 2020.

37 Z. Na, Y. Liu, J. Shi, C. Liu, and Z. Gao, "UAV-supported clustered NOMA for 6G-enabled Internet of Things: Trajectory planning and resource allocation," *IEEE Internet of Things Journal*, vol. 8, no. 20, pp. 15 041–15 048, 2021.

38 B. Sliwa, N. Piatkowski, and C. Wietfeld, "LIMITS: Lightweight machine learning for IoT systems with resource limitations," in *ICC 2020-2020 IEEE International Conference on Communications (ICC)*, IEEE, 2020, pp. 1–7.

39 M. Wang, T. Zhu, T. Zhang, J. Zhang, S. Yu, and W. Zhou, "Security and privacy in 6G networks: New areas and new challenges," *Digital Communications and Networks*, vol. 6, pp. 281–291, 2020.

40 B. Mao, Y. Kawamoto, and N. Kato, "AI-based joint optimization of QoS and security for 6G energy harvesting Internet of Things," *IEEE Internet of Things Journal*, vol. 7, pp. 7032–7042, 2020.

41 R. Sekaran, R. Patan, A. Raveendran, F. Al-Turjman, M. Ramachandran, and L. Mostarda, "Survival study on blockchain based 6G-enabled mobile edge computation for IoT automation," *IEEE Access*, vol. 8, pp. 143 453–143 463, 2020.

42 C. Liu, W. Feng, Y. Chen, C.-X. Wang, and N. Ge, "Cell-free satellite-UAV networks for 6G wide-area Internet of Things," *IEEE Journal on Selected Areas in Communications*, vol. 39, pp. 1116–1131, 2020.

43 S. Sen, "Context-aware energy-efficient communication for IoT sensor nodes," in *2016 53nd ACM/EDAC/IEEE Design Automation Conference (DAC)* IEEE, 2016, pp. 1–6.

44 S. B. Azmy, R. A. Sneineh, N. Zorba, and H. S. Hassanein, "Small data in IoT: An MCS perspective," in *Performability in Internet of Things*. Springer, 2019, pp. 209–229.

45 M. Radhakrishnan, S. Sen, S. Vigneshwaran, A. Misra, and R. Balan, "IoT + Small data: Transforming in-store shopping analytics & services," in *2016 8th international conference on communication systems and networks (COMSNETS)* IEEE, 2016, pp. 1–6.

46 X. Tao and C. Ji, "Clustering massive small data for IoT," in *The 2014 2nd International Conference on Systems and Informatics (ICSAI 2014)* IEEE, 2014, pp. 974–978.

47 B. Liu, C. Liu, and M. Peng, "Resource allocation for energy-efficient MEC in NOMA-enabled massive IoT networks," *IEEE Journal on Selected Areas in Communications*, vol. 39, pp. 1015–1027, 2020.

48 C. de Alwis, H. K. Arachchi, A. Fernando, and A. Kondoz, "Towards minimising the coefficient vector overhead in random linear network coding," in *2013 IEEE International Conference on Acoustics, Speech and Signal Processing* IEEE, 2013, pp. 5127–5131.

49 C. de Alwis, H. K. Arachchi, A. Fernando, and M. Pourazad, "Content and network-aware multicast over wireless networks," in *10th International Conference on Heterogeneous Networking for Quality, Reliability, Security and Robustness* IEEE, 2014, pp. 122–128.

50 L. U. Khan, I. Yaqoob, M. Imran, Z. Han, and C. S. Hong, "6G wireless systems: A vision, architectural elements, and future directions," *IEEE Access*, vol. 8, pp. 147 029–147 044, 2020.

51 S. J. Nawaz, S. K. Sharma, S. Wyne, M. N. Patwary, and M. Asaduzzaman, "Quantum machine learning for 6G communication networks: state-of-the-art and vision for the future," *IEEE Access*, vol. 7, pp. 46 317–46 350, 2019.

52 K. K. Munasinghe, M. N. Dharmaweera, U. L. Wijewardhana, C. De Alwis, and R. Parthiban, "Joint minimization of spectrum and power in impairment-aware elastic optical networks," *IEEE Access*, vol. 9, pp. 43 349–43 363, 2021.

53 E. Yaacoub and M.-S. Alouini, "A key 6G challenge and opportunity–connecting the base of the pyramid: A survey on rural connectivity," *Proceedings of the IEEE*, vol. 108, no. 4, pp. 533–582, 2020.

54 A. Mihovska and M. Sarkar, "Cooperative human-centric sensing connectivity," in *Internet of Things-Technology, Applications and Standardization* IntechOpen, 2018.

55 Y. Xiao, G. Shi, Y. Li, W. Saad, and H. V. Poor, "Toward self-learning edge intelligence in 6G," *IEEE Communications Magazine*, vol. 58, no. 12, pp. 34–40, 2020.

56 E. Peltonen, M. Bennis, M. Capobianco, M. Debbah, A. Ding, F. Gil-Casti neira, M. Jurmu, T. Karvonen, M. Kelanti, A. Kliks *et al.*, "6G White Paper on Edge Intelligence," *arXiv preprint arXiv:2004.14850*, 2020.

57 V. Ziegler, H. Viswanathan, H. Flinck, M. Hoffmann, V. Räisänen, and K. Hätönen, "6G architecture to connect the worlds," *IEEE Access*, vol. 8, pp. 173 508–173 520, 2020.

58 H. Li, K. Ota, and M. Dong, "Energy cooperation in battery-free wireless communications with radio frequency energy harvesting," *ACM Transactions on Embedded Computing Systems (TECS)*, vol. 17, no. 2, pp. 1–17, 2018.

59 F. Tariq, M. R. Khandaker, K.-K. Wong, M. A. Imran, M. Bennis, and M. Debbah, "A speculative study on 6G," *IEEE Wireless Communications*, vol. 27, no. 4, pp. 118–125, 2020.

60 S. Hu, F. Rusek, and O. Edfors, "Beyond massive MIMO: The potential of data transmission with large intelligent surfaces," *IEEE Transactions on Signal Processing*, vol. 66, no. 10, pp. 2746–2758, 2018.

61 Y. Xing and T. S. Rappaport, "Propagation measurement system and approach at 140 GHz-moving to 6G and above 100 GHz," in *2018 IEEE Global Communications Conference (GLOBECOM)* IEEE, 2018, pp. 1–6.

62 W. U. Khan, F. Jameel, M. A. Jamshed, H. Pervaiz, S. Khan, and J. Liu, "Efficient power allocation for NOMA-enabled IoT networks in 6G era," *Physical Communication*, vol. 39, p. 101043, 2020.

63 A. Taha, M. Alrabeiah, and A. Alkhateeb, "Enabling large intelligent surfaces with compressive sensing and deep learning," *IEEE Access*, vol. 9, pp. 44 304–44 321, 2021.

64 C. de Alwis, H. K. Arachchi, V. De Silva, A. Fernando, and A. Kondoz, "Robust video communication using random linear network coding with pre-coding and interleaving," in *2012 19th IEEE International Conference on Image Processing* IEEE, 2012, pp. 2269–2272.

65 "The Naked Approach," [Accessed on 29.01.2022]. [Online]. Available: http://nakedapproach.fi/.

66 E. C. Strinati, S. Barbarossa, J. L. Gonzalez-Jimenez, D. Ktenas, N. Cassiau, L. Maret, and C. Dehos, "6G: The next frontier: From holographic messaging to artificial intelligence using subterahertz and visible light communication," *IEEE Vehicular Technology Magazine*, vol. 14, no. 3, pp. 42–50, 2019.

67 T. Kumar, A. Braeken, A. D. Jurcut, M. Liyanage, and M. Ylianttila, "AGE: Authentication in gadget-free healthcare environments," *Information Technology and Management*, vol. 21, pp. 95–114, 2019.

68 S. Kekade, C.-H. Hseieh, M. M. Islam, S. Atique, A. M. Khalfan, Y.-C. Li, and S. S. Abdul, "The usefulness and actual use of wearable devices among the elderly population," *Computer Methods and Programs in Biomedicine*, vol. 153, pp. 137–159, 2018.

69 S. Malwade, S. S. Abdul, M. Uddin, A. A. Nursetyo, L. Fernandez-Luque, X. K. Zhu, L. Cilliers, C.-P. Wong, P. Bamidis, and Y.-C. J. Li, "Mobile and wearable technologies in healthcare for the ageing population," *Computer methods and Programs in Biomedicine*, vol. 161, pp. 233–237, 2018.

70 S. Nayak and R. Patgiri, "6G communication technology: A vision on intelligent healthcare," in *Health Informatics: A Computational Perspective in Healthcare*. Springer, 2021, pp. 1–18.

71 Q.-V. Pham, F. Fang, V. N. Ha, M. J. Piran, M. Le, L. B. Le, W.-J. Hwang, and Z. Ding, "A survey of multi-access edge computing in 5G and beyond: Fundamentals, technology integration, and state-of-the-art," *IEEE Access*, vol. 8, pp. 116 974–117 017, 2020.

72 E. Markoval, D. Moltchanov, R. Pirmagomedov, D. Ivanova, Y. Koucheryavy, and K. Samouylov, "Priority-based coexistence of eMBB and URLLC traffic in industrial 5G NR deployments," in *2020 12th International Congress on Ultra Modern Telecommunications and Control Systems and Workshops (ICUMT)* IEEE, 2020, pp. 1–6.

73 H. E. Melcherts, *The Internet of Everything and Beyond*. Wiley Online Library, 2017.

74 M. A. Siddiqi, H. Yu, and J. Joung, "5G ultra-reliable low-latency communication implementation challenges and operational issues with IoT devices," *Electronics*, vol. 8, no. 9, p. 981, 2019.

75 J. Zhao, "A Survey of Intelligent Reflecting Surfaces (IRSs): Towards 6G Wireless Communication Networks with Massive MIMO 2.0," *arXiv preprint arXiv:1907.04789*, 2019.

76 C. De Alwis, Kalla, A., Pham, Q.-V., Kumar, P., Dev, K., Hwang, W.-J., and Liyanage, M., "Survey on 6G frontiers: Trends, applications, requirements, technologies and future research," *IEEE Open Journal of the Communications Society*, vol. 2, pp. 836–886, 2021.

77 F. Jameel, U. Javaid, B. Sikdar, I. Khan, G. Mastorakis, and C. X. Mavromoustakis, "Optimizing Blockchain Networks with Artificial Intelligence: Towards Efficient and Reliable IoT Applications," in *Convergence of Artificial Intelligence and the Internet of Things*. Springer, 2020, pp. 299–321.

78 M. Naresh, "Towards 6G: Wireless Communication," *Tathapi with ISSN 2320-0693 is an UGC CARE Journal*, vol. 19, no. 9, pp. 335–341, 2020.

79 M. Alsenwi, N. H. Tran, M. Bennis, S. R. Pandey, A. K. Bairagi, and C. S. Hong, "Intelligent resource slicing for eMBB and URLLC coexistence in 5G and beyond: A deep reinforcement learning based approach," *IEEE Transactions on Wireless Communications*, vol. 20, pp. 4585–4600, 2021.

80 I. F. Akyildiz, C. Han, and S. Nie, "Combating the distance problem in the millimeter wave and Terahertz frequency bands," *IEEE Communications Magazine*, vol. 56, no. 6, pp. 102–108, 2018.

81 H. Sarieddeen, N. Saeed, T. Y. Al-Naffouri, and M.-S. Alouini, "Next generation terahertz communications: A rendezvous of sensing, imaging, and localization," *IEEE Communications Magazine*, vol. 58, no. 5, pp. 69–75, 2020.

82 G. Gür, "Expansive networks: Exploiting spectrum sharing for capacity boost and 6G vision," *Journal of Communications and Networks*, vol. 22, no. 6, pp. 444–454, 2020.

83 N. Mahmood, A. Munari, F. Clazzer, and H. Bartz, "Critical and massive machine type communication towards 6G," 2020.

84 N. H. Mahmood, S. Böcker, A. Munari, F. Clazzer, I. Moerman, K. Mikhaylov, O. Lopez, O.-S. Park, E. Mercier, H. Bartz *et al.*, "White Paper on Critical and Massive Machine Type Communication towards 6G," *arXiv preprint arXiv:2004.14146*, 2020.

85 K. Mikhaylov, V. Petrov, R. Gupta, M. A. Lema, O. Galinina, S. Andreev, Y. Koucheryavy, M. Valkama, A. Pouttu, and M. Dohler, "Energy efficiency of multi-radio massive machine-type communication (MR-MMTC): Applications, challenges, and solutions," *IEEE Communications Magazine*, vol. 57, no. 6, pp. 100–106, 2019.

86 F. A. De Figueiredo, F. A. Cardoso, I. Moerman, and G. Fraidenraich, "On the application of massive MIMO systems to machine type communications," *IEEE Access*, vol. 7, pp. 2589–2611, 2018.

87 "Robotic Surgery," https://www.womencentre.com.au/robotic-surgery.html., 2019, [Accessed on 29.03.2021].

88 C. Huang, A. Zappone, G. C. Alexandropoulos, M. Debbah, and C. Yuen, "Reconfigurable intelligent surfaces for energy efficiency in wireless communication," *IEEE Transactions on Wireless Communications*, vol. 18, no. 8, pp. 4157–4170, 2019.

89 C.-X. Wang, J. Huang, H. Wang, X. Gao, X. You, and Y. Hao, "6G wireless channel measurements and models: Trends and challenges," *IEEE Vehicular Technology Magazine*, vol. 15, no. 4, pp. 22–32, 2020.

90 H. Yang, A. Alphones, Z. Xiong, D. Niyato, J. Zhao, and K. Wu, "Artificial intelligence-enabled intelligent 6G networks," *IEEE Network*, vol. 34, no. 6, pp. 272–280, 2020.

91 "Space Tourism May Mean One Giant Leap for Researchers," https://www.nytimes.com/2011/03/01/science/space/01orbit.html, [Online; Accessed 25.03.2021].

92 "How to visit the deep sea for around," https://www.telegraph.co.uk/travel/activity-and-adventure/how-to-dive-to-the-deep-sea/, [Online; Accessed 25.03.2021].

93 "high speed rail," https://www.iconfinder.com/icons/2539308/bullet_high_speed_rail_railway_train_icon, [Online; Accessed 25.03.2021].

94 P. Fan, J. Zhao, and I. Chih-Lin, "5G high mobility wireless communications: Challenges and solutions," *China Communications*, vol. 13, no. 2, pp. 1–13, 2016.

95 J. Chen, S. Li, J. Tao, S. Fu, and G. E. Sobelman, "Wireless beam modulation: An energy-and spectrum-efficient communication technology for future massive iot systems," *IEEE Wireless Communications*, vol. 27, no. 5, pp. 60–66, 2020.

96 H. A. B. Salameh, S. Al-Masri, E. Benkhelifa, and J. Lloret, "Spectrum assignment in hardware-constrained cognitive radio iot networks under varying channel-quality conditions," *IEEE Access*, vol. 7, pp. 42 816–42 825, 2019.

97 S. P. Rout, "6G wireless communication: Its vision, viability, application, requirement, technologies, encounters and research," in *2020 11th International Conference on Computing, Communication and Networking Technologies (ICCCNT)* IEEE, 2020, pp. 1–8.

98 S. Han, I. Chih-Lin, T. Xie, S. Wang, Y. Huang, L. Dai, Q. Sun, and C. Cui, "Achieving high spectrum efficiency on high speed train for 5G new radio and beyond," *IEEE Wireless Communications*, vol. 26, no. 5, pp. 62–69, 2019.

99 J. Navarro-Ortiz, P. Romero-Diaz, S. Sendra, P. Ameigeiras, J. J. Ramos-Munoz, and J. M. Lopez-Soler, "A survey on 5G usage scenarios and traffic models," *IEEE Communication Surveys and Tutorials*, vol. 22, no. 2, pp. 905–929, 2020.

100 P. Fernandes and U. Nunes, "Multiplatooning leaders positioning and cooperative behavior algorithms of communicant automated vehicles for high traffic capacity," *IEEE Transactions on Intelligent Transportation Systems*, vol. 16, no. 3, pp. 1172–1187, 2015.

101 M. Ergen, F. Inan, O. Ergen, I. Shayea, M. F. Tuysuz, A. Azizan, N. K. Ure, and M. Nekovee, "Edge on wheels with OMNIBUS networking for 6G technology," *IEEE Access*, vol. 8, pp. 215 928–215 942, 2020.

102 M. S. Elbamby, C. Perfecto, M. Bennis, and K. Doppler, "Toward low-latency and ultra-reliable virtual reality," *IEEE Network*, vol. 32, no. 2, pp. 78–84, 2018.

103 A. Alshahrani, I. A. Elgendy, A. Muthanna, A. M. Alghamdi, and A. Alshamrani, "Efficient multi-player computation offloading for VR edge-cloud computing systems," *Applied Sciences*, vol. 10, no. 16, p. 5515, 2020.

104 T. Park and W. Saad, "Distributed learning for low latency machine type communication in a massive Internet of Things," *IEEE Internet of Things Journal*, vol. 6, no. 3, pp. 5562–5576, 2019.

105 F. Tang, Y. Kawamoto, N. Kato, and J. Liu, "Future intelligent and secure vehicular network toward 6G: Machine-learning approaches," *Proceedings of the IEEE*, vol. 108, no. 2, pp. 292–307, 2020.

106 H. Elayan, O. Amin, R. M. Shubair, and M.-S. Alouini, "Terahertz communication: The opportunities of wireless technology beyond 5G," in *2018 International Conference on Advanced Communication Technologies and Networking (CommNet)* IEEE, 2018, pp. 1–5.

107 C. Han, Y. Wu, Z. Chen, and X. Wang, "Terahertz Communications (TeraCom): Challenges and Impact on 6G Wireless Systems," *arXiv preprint arXiv:1912.06040*, 2019.

108 T. Nagatsuma, "Terahertz technologies: Present and future," *IEICE Electronics Express*, vol. 8, no. 14, pp. 1127–1142, 2011.

109 X. Wu and K. Sengupta, "Dynamic waveform shaping with picosecond time widths," *IEEE Journal of Solid-State Circuits*, vol. 52, no. 2, pp. 389–405, 2016.

110 J. M. Jornet and I. F. Akyildiz, "Graphene-based plasmonic nano-transceiver for terahertz band communication," in *The 8th European conference on antennas and propagation (EuCAP 2014)* IEEE, 2014, pp. 492–496.

111 C. J. Docherty and M. B. Johnston, "Terahertz properties of graphene," *Journal of Infrared, Millimeter, and Terahertz Waves*, vol. 33, no. 8, pp. 797–815, 2012.

112 M. Hasan, S. Arezoomandan, H. Condori, and B. Sensale-Rodriguez, "Graphene terahertz devices for communications applications," *Nano Communication Networks*, vol. 10, pp. 68–78, 2016.

113 I. F. Akyildiz and J. M. Jornet, "Realizing ultra-massive MIMO (1024 × 1024) communication in the (0.06–10) terahertz band," *Nano Communication Networks*, vol. 8, pp. 46–54, 2016.

114 J. M. Jornet and I. F. Akyildiz, "Channel modeling and capacity analysis for electromagnetic wireless nanonetworks in the terahertz band," *IEEE Transactions on Wireless Communications*, vol. 10, no. 10, pp. 3211–3221, 2011.

115 S. Priebe and T. Kurner, "Stochastic modeling of THz indoor radio channels," *IEEE Transactions on Wireless Communications*, vol. 12, no. 9, pp. 4445–4455, 2013.

116 S. Kim and A. Zajić, "Statistical modeling and simulation of short-range device-to-device communication channels at sub-THz frequencies," *IEEE Transactions on Wireless Communications*, vol. 15, no. 9, pp. 6423–6433, 2016.

117 D. He, K. Guan, A. Fricke, B. Ai, R. He, Z. Zhong, A. Kasamatsu, I. Hosako, and T. Kürner, "Stochastic channel modeling for kiosk applications in the terahertz band," *IEEE Transactions on Terahertz Science and Technology*, vol. 7, no. 5, pp. 502–513, 2017.

118 M. A. Khalighi and M. Uysal, "Survey on free space optical communication: A communication theory perspective," *IEEE Communication Surveys and Tutorials*, vol. 16, no. 4, pp. 2231–2258, 2014.

119 S. Mollahasani and E. Onur, "Evaluation of terahertz channel in data centers," in *NOMS 2016-2016 IEEE/IFIP Network Operations and Management Symposium* IEEE, 2016, pp. 727–730.

120 M. Mozaffari, A. T. Z. Kasgari, W. Saad, M. Bennis, and M. Debbah, "Beyond 5G with UAVs: Foundations of a 3D wireless cellular network," *IEEE Transactions on Wireless Communications*, vol. 18, no. 1, pp. 357–372, 2018.

121 M. Giordani and M. Zorzi, "Non-terrestrial networks in the 6G era: Challenges and opportunities," *IEEE Network*, vol. 35, no. 2, pp. 244–251, 2020.

122 N.-N. Dao, Q.-V. Pham, N. H. Tu, T. T. Thanh, V. N. Q. Bao, D. S. Lakew, and S. Cho, "Survey on aerial radio access networks: Toward a comprehensive 6G access infrastructure," *IEEE Communication Surveys and Tutorials*, vol. 23, no. 2, pp. 1193–1225, 2021.

123 "FAA Aerospace Forecast Fiscal Year 201-2038," Federal Aviation Administration, https://www.faa.gov/data_research/aviation/aerospace_

forecasts/media/FY2019-39_FAA_Aerospace_Forecast.pdf [Accessed on 29.03.2021].

124 S. D. Intelligence, "The global UAV payload market 2012–2022," *Strategic Defence Intelligence: White Papers*, 2013.

125 H. Chang, C.-X. Wang, Y. Liu, J. Huang, J. Sun, W. Zhang, and X. Gao, "A novel non-stationary 6G UAV-to-ground wireless channel model with 3D arbitrary trajectory changes," *IEEE Internet of Things Journal*, vol. 8, pp. 9865–9877, 2020.

126 E. C. Strinati, S. Barbarossa, T. Choi, A. Pietrabissa, A. Giuseppi, E. De Santis, J. Vidal, Z. Becvar, T. Haustein, N. Cassiau *et al.*, "6G in the sky: On-demand intelligence at the edge of 3D networks," *ETRI Journal*, vol. 42, no. 5, pp. 643–657, 2020.

127 T. Li, A. K. Sahu, A. Talwalkar, and V. Smith, "Federated learning: Challenges, methods, and future directions," *IEEE Signal Processing Magazine*, vol. 37, no. 3, pp. 50–60, 2020.

128 D. C. Nguyen, M. Ding, Q.-V. Pham, P. N. Pathirana, L. B. Le, A. Seneviratne, J. Li, D. Niyato, and H. V. Poor, "Federated learning meets blockchain in edge computing: Opportunities and challenges," *IEEE Internet of Things Journal*, vol. 8, no. 16, pp. 12 806–12 825, 2021.

129 M. Parimala, R. M. Swarna Priya, Q.-V. Pham, K. Dev, P. K. R. Maddikunta, T. R. Gadekallu, and T. Huynh-The, "Fusion of Federated Learning and Industrial Internet of Things: A Survey," *arXiv preprint arXiv:2101.00798*, 2021.

130 Q. Yang, Y. Liu, T. Chen, and Y. Tong, "Federated machine learning: Concept and applications," *ACM Transactions on Intelligent Systems and Technology*, vol. 10, no. 2, pp. 1–19, 2019.

131 T. Huynh-The, C.-H. Hua, Q.-V. Pham, and D.-S. Kim, "MCNet: An efficient CNN architecture for robust automatic modulation classification," *IEEE Communications Letters*, vol. 24, no. 4, pp. 811–815, 2020.

132 H. Li, K. Ota, and M. Dong, "Learning IoT in edge: Deep learning for the Internet of Things with edge computing," *IEEE Network*, vol. 32, no. 1, pp. 96–101, 2018.

133 Y. Xiao, Y. Li, G. Shi, and H. V. Poor, "Optimizing resource-efficiency for federated edge intelligence in IoT networks," in *2020 International Conference on Wireless Communications and Signal Processing (WCSP)* IEEE, 2020, pp. 86–92.

134 Y. Xiao, G. Shi, and M. Krunz, "Towards Ubiquitous AI in 6G with Federated Learning," *arXiv preprint arXiv:2004.13563*, 2020.

135 N. C. Luong, D. T. Hoang, S. Gong, D. Niyato, P. Wang, Y.-C. Liang, and D. I. Kim, "Applications of deep reinforcement learning in communications and networking: A survey," *IEEE Communication Surveys and Tutorials*, vol. 21, no. 4, pp. 3133–3174, 2019.

136 Y. Sun, M. Peng, Y. Zhou, Y. Huang, and S. Mao, "Application of machine learning in wireless networks: Key techniques and open issues," *IEEE Communication Surveys and Tutorials*, vol. 21, no. 4, pp. 3072–3108, 2019.

137 C. T. Nguyen, N. Van Huynh, N. H. Chu, Y. M. Saputra, D. T. Hoang, D. N. Nguyen, Q.-V. Pham, D. Niyato, E. Dutkiewicz, and W.-J. Hwang, "Transfer Learning for Future Wireless Networks: A Comprehensive Survey," *arXiv preprint arXiv:2102.07572*, 2021.

138 C. Zhang, P. Patras, and H. Haddadi, "Deep learning in mobile and wireless networking: A survey," *IEEE Communication Surveys and Tutorials*, vol. 21, no. 3, pp. 2224–2287, 2019.

139 Q.-V. Pham, D. C. Nguyen, S. Mirjalili, D. T. Hoang, D. N. Nguyen, P. N. Pathirana, and W.-J. Hwang, "Swarm intelligence for next-generation networks: Recent advances and applications," *Journal of Network and Computer Applications*, vol. 191, p. 103141, 2021.

140 L. U. Khan, S. R. Pandey, N. H. Tran, W. Saad, Z. Han, M. N. Nguyen, and C. S. Hong, "Federated learning for edge networks: Resource optimization and incentive mechanism," *IEEE Communications Magazine*, vol. 58, no. 10, pp. 88–93, 2020.

141 J. Kang, Z. Xiong, D. Niyato, S. Xie, and J. Zhang, "Incentive mechanism for reliable federated learning: A joint optimization approach to combining reputation and contract theory," *IEEE Internet of Things Journal*, vol. 6, no. 6, pp. 10 700–10 714, 2019.

142 W. Y. B. Lim, N. C. Luong, D. T. Hoang, Y. Jiao, Y.-C. Liang, Q. Yang, D. Niyato, and C. Miao, "Federated learning in mobile edge networks: A comprehensive survey," *IEEE Communication Surveys and Tutorials*, vol. 22, no. 3, pp. 2031–2063, 2020.

143 H. F. Atlam and G. B. Wills, "Intersections between IoT and distributed ledger," in *Advances in Computers*. Elsevier, 2019, vol. 115, pp. 73–113.

144 Y. Lu, "The blockchain: State-of-the-art and research challenges," *Journal of Industrial Information Integration*, vol. 15, pp. 80–90, 2019.

145 S. Daley, "25 Blockchain Applications & Real-World Use Cases Disrupting the Status Quo," 2020, [Accessed on 29.03.2021]. [Online]. Available: https://builtin.com/blockchain/blockchain-applications.

146 T. Maksymyuk, J. Gazda, M. Volosin, G. Bugar, D. Horvath, M. Klymash, and M. Dohler, "Blockchain-empowered framework for decentralized network managementin 6G," *IEEE Communications Magazine*, vol. 58, no. 9, pp. 86–92, 2020.

147 D. C. Nguyen, P. N. Pathirana, M. Ding, and A. Seneviratne, "Blockchain for 5G and beyond networks: A state of the art survey," *Journal of Network and Computer Applications*, vol. 166, p. 102693, 2020.

148 S. Yrjölä, "How could blockchain transform 6G towards open ecosystemic business models?" in *2020 IEEE International Conference on Communications Workshops (ICC Workshops)* IEEE, 2020, pp. 1–6.

149 T. Hewa, G. Gür, A. Kalla, M. Ylianttila, A. Bracken, and M. Liyanage, "The role of blockchain in 6G: Challenges, opportunities and research directions," in *2020 2nd 6G Wireless Summit (6G SUMMIT)* IEEE, 2020, pp. 1–5.

150 X. Ling, J. Wang, Y. Le, Z. Ding, and X. Gao, "Blockchain radio access network beyond 5G," *IEEE Wireless Communications*, vol. 27, pp. 160–168, 2020.

151 W. Li, Z. Su, R. Li, K. Zhang, and Y. Wang, "Blockchain-based data security for artificial intelligence applications in 6G networks," *IEEE Network*, vol. 34, no. 6, pp. 31–37, 2020.

152 M. Satyanarayanan, P. Bahl, R. Caceres, and N. Davies, "The case for vm-based cloudlets in mobile computing," *IEEE Pervasive Computing*, vol. 8, no. 4, pp. 14–23, 2009.

153 S. Barbarossa, S. Sardellitti, and P. D. Lorenzo, "Communicating while computing: Distributed mobile cloud computing over 5G heterogeneous networks," *IEEE Signal Processing Magazine*, vol. 31, no. 6, pp. 45–55, Nov. 2014.

154 C. Mouradian, D. Naboulsi, S. Yangui, R. H. Glitho, M. J. Morrow, and P. A. Polakos, "A comprehensive survey on fog computing: State-of-the-art and research challenges," *IEEE Communication Surveys and Tutorials*, vol. 20, no. 1, pp. 416–464, 2018.

155 M. Mukherjee, L. Shu, and D. Wang, "Survey of fog computing: Fundamental, network applications, and research challenges," *IEEE Communication Surveys and Tutorials*, vol. 20, pp. 1826–1857, 2018.

156 P. Mach and Z. Becvar, "Mobile edge computing: A survey on architecture and computation offloading," *IEEE Communication Surveys and Tutorials*, vol. 19, no. 3, pp. 1628–1656, 2017.

157 M. Chiang and T. Zhang, "Fog and IoT: An overview of research opportunities," *IEEE Internet of Things Journal*, vol. 3, no. 6, pp. 854–864, 2016.

158 H. F. Atlam, R. J. Walters, and G. B. Wills, "Fog computing and the Internet of Things: A review," *Big Data and Cognitive Computing*, vol. 2, no. 2, p. 10, 2018.

159 F. Bonomi, R. Milito, J. Zhu, and S. Addepalli, "Fog computing and its role in the Internet of Things," in *Proceedings of the 1st Edition of the MCC Workshop on Mobile Cloud Computing*, 2012, pp. 13–16.

160 K. Dolui and S. K. Datta, "Comparison of edge computing implementations: Fog computing, cloudlet and mobile edge computing," in *2017 Global Internet of Things Summit (GIoTS)*, 2017, pp. 1–6.

161 M. Patel, B. Naughton, C. Chan, N. Sprecher, S. Abeta, A. Neal *et al.*, "Mobile-edge computing introductory technical white paper," *White Paper, Mobile-edge Computing (MEC) industry initiative*, Sept. 2014.

162 A. Ahmed and E. Ahmed, "A survey on mobile edge computing," in *2016 10th International Conference on Intelligent Systems and Control (ISCO)*, 2016, pp. 1–8.

163 S. Kekki *et al.*,"MEC in 5G networks," *ETSI White Paper*, no. 28, pp. 1–28, 2018.

164 Y. Qiao, M. Zhang, Y. Zhou, X. Kong, H. Zhang, J. Bi, M. Xu, and J. Wang, "NetEC: Accelerating erasure coding reconstruction with in-network aggregation," *IEEE Transactions on Parallel and Distributed Systems*, vol. 33, pp. 2571–2583, 2022.

165 Q.-V. Pham, R. Ruby, F. Fang, D. C. Nguyen, Z. Yang, M. Le, Z. Ding, and W.-J. Hwang, "Aerial computing: A new computing paradigm, applications, and challenges," *IEEE Internet of Things Journal*, vol. 9, pp. 8339–8363, 2022.

166 R. L. A. *et al.*, "Smart Networks in the Context of NGI," *White Paper, European Technology Platform NetWorld2020*, May 2020.

167 N. Gisin and R. Thew, "Quantum communication," *Nature Photonics*, vol. 1, no. 3, pp. 165–171, 2007.

168 X. Su, M. Wang, Z. Yan, X. Jia, C. Xie, and K. Peng, "Quantum network based on non-classical light," *Science China Information Sciences*, vol. 63, no. 8, pp. 1–12, 2020.

169 T. Brougham, S. M. Barnett, K. T. McCusker, P. G. Kwiat, and D. J. Gauthier, "Security of high-dimensional quantum key distribution protocols using Franson interferometers," *Journal of Physics B: Atomic, Molecular and Optical Physics*, vol. 46, no. 10, p. 104010, 2013.

170 A. Manzalini, "Quantum communications in future networks and services," *Quantum Reports*, vol. 2, no. 1, pp. 221–232, 2020.

171 R. Arul, G. Raja, A. O. Almagrabi, M. S. Alkatheiri, S. H. Chauhdary, and A. K. Bashir, "A quantum-safe key hierarchy and dynamic security association for LTE/SAE in 5G scenario," *IEEE Transactions on Industrial Informatics*, vol. 16, no. 1, pp. 681–690, 2019.

172 A. A. Abd EL-Latif, B. Abd-El-Atty, S. E. Venegas-Andraca, and W. Mazurczyk, "Efficient quantum-based security protocols for information sharing and data protection in 5G networks," *Future Generation Computer Systems*, vol. 100, pp. 893–906, 2019.

173 D. Zavitsanos, A. Ntanos, G. Giannoulis, and H. Avramopoulos, "On the QKD integration in converged fiber/wireless topologies for secured, low-latency 5G/B5G fronthaul," *Applied Sciences*, vol. 10, no. 15, p. 5193, 2020.

174 M. Z. Chowdhury, M. Shahjalal, M. Hasan, Y. M. Jang *et al.*, "The role of optical wireless communication technologies in 5G/6G and iot solutions: Prospects, directions, and challenges," *Applied Sciences*, vol. 9, no. 20, p. 4367, 2019.

175 L. U. Khan, "Visible light communication: Applications, architecture, standardization and research challenges," *Digital Communications and Networks*, vol. 3, no. 2, pp. 78–88, 2017.

176 D. Karunatilaka, F. Zafar, V. Kalavally, and R. Parthiban, "Led based indoor visible light communications: State of the art," *IEEE Communication Surveys and Tutorials*, vol. 17, no. 3, pp. 1649–1678, 2015.

177 A. Jovicic, J. Li, and T. Richardson, "Visible light communication: Opportunities, challenges and the path to market," *IEEE Communications Magazine*, vol. 51, no. 12, pp. 26–32, 2013.

178 D. Tsonev, S. Videv, and H. Haas, "Towards a 100 Gb/s visible light wireless access network," *Optics Express*, vol. 23, no. 2, pp. 1627–1637, 2015.

179 S. Soderi, "Enhancing security in 6G visible light communications," in *2020 2nd 6G Wireless Summit (6G SUMMIT)* IEEE, 2020, pp. 1–5.

180 M. Katz and I. Ahmed, "Opportunities and challenges for visible light communications in 6G," in *2020 2nd 6G Wireless Summit (6G SUMMIT)* IEEE, 2020, pp. 1–5.

181 S. Ariyanti and M. Suryanegara, "Visible light communication (vlc) for 6G technology: The potency and research challenges," in *2020 4th World Conference on Smart Trends in Systems, Security and Sustainability (WorldS4)* IEEE, 2020, pp. 490–493.

182 S. U. Rehman, S. Ullah, P. H. J. Chong, S. Yongchareon, and D. Komosny, "Visible light communication: A system perspective–overview and challenges," *Sensors*, vol. 19, no. 5, p. 1153, 2019.

183 J. Chen and Z. Wang, "Topology control in hybrid VLC/RF vehicular ad-hoc network," *IEEE Transactions on Wireless Communications*, vol. 19, no. 3, pp. 1965–1976, 2019.

184 R. Alghamdi, R. Alhadrami, D. Alhothali, H. Almorad, A. Faisal, S. Helal, R. Shalabi, R. Asfour, N. Hammad, A. Shams *et al.*, "Intelligent surfaces for 6G wireless networks: A survey of optimization and performance analysis techniques," *IEEE Access*, vol. 8, pp. 202795–202818, 2020.

185 M. Di Renzo, K. Ntontin, J. Song, F. H. Danufane, X. Qian, F. Lazarakis, J. De Rosny, D.-T. Phan-Huy, O. Simeone, R. Zhang *et al.*, "Reconfigurable intelligent surfaces vs. relaying: Differences, similarities, and performance comparison," *IEEE Open Journal of the Communications Society*, vol. 1, pp. 798–807, 2020.

186 M. Jung, W. Saad, and G. Kong, "Performance Analysis of Large Intelligent Surfaces (LISs): Uplink Spectral Efficiency and Pilot Training," *arXiv preprint arXiv:1904.00453*, 2019.

187 E. Basar, "Reconfigurable intelligent surface-based index modulation: A new beyond MIMO paradigm for 6G," *IEEE Transactions on Communications*, vol. 68, no. 5, pp. 3187–3196, 2020.

188 C. J. Vaca-Rubio, P. Ramirez-Espinosa, K. Kansanen, Z.-H. Tan, E. de Carvalho, and P. Popovski, "Assessing Wireless Sensing Potential with Large Intelligent Surfaces," *arXiv preprint arXiv:2011.08465*, 2020.

189 S. Stanković, "Compressive sensing: Theory, algorithms and applications," in *2015 4th Mediterranean Conference on Embedded Computing (MECO)* IEEE, 2015, pp. 4–6.

190 Z. Gao, L. Dai, S. Han, I. Chih-Lin, Z. Wang, and L. Hanzo, "Compressive sensing techniques for next-generation wireless communications," *IEEE Wireless Communications*, vol. 25, no. 3, pp. 144–153, 2018.

191 Y. Huo, X. Dong, W. Xu, and M. Yuen, "Enabling multi-functional 5G and beyond user equipment: A survey and tutorial," *IEEE Access*, vol. 7, pp. 116 975–117 008, 2019.

192 M. B. Shahab, R. Abbas, M. Shirvanimoghaddam, and S. J. Johnson, "Grant-free non-orthogonal multiple access for IoT: A survey," *IEEE Communication Surveys and Tutorials*, vol. 22, no. 3, pp. 1805–1838, 2020.

193 "Zero-touch network and service management (ZSM); terminology for concepts in ZSM," August 2019, [Accessed on 29.03.2021]. [Online]. Available: https://www.etsi.org/deliver/etsi_gs/ZSM/001_099/007/01.01.01_60/gs_ZSM007v010101p.pdf.

194 "Zero-touch network and Service Management (ZSM); Requirements based on documented scenarios," Oct. 2019, [Accessed on 29.03.2022]. [Online]. Available: https://www.etsi.org/deliver/etsi_gs/ZSM/001_099/001/01.01.01_60/gs_ZSM001v010101p.pdf.

195 T. Darwish, G. K. Kurt, H. Yanikomeroglu, G. Senarath, and P. Zhu, "A Vision of Self-evolving Network Management for Future Intelligent Vertical HetNet," *arXiv preprint arXiv:2009.02771*, 2020.

196 C. Benzaid and T. Taleb, "ZSM security: Threat surface and best practices," *IEEE Network*, vol. 34, no. 3, pp. 124–133, 2020.

197 N. H. Mahmood, H. Alves, O. A. López, M. Shehab, D. P. M. Osorio, and M. Latva-aho, "Six Key Enablers for Machine Type Communication in 6G," *arXiv preprint arXiv:1903.05406*, 2019.

198 S. Hu, X. Chen, W. Ni, X. Wang, and E. Hossain, "Modeling and analysis of energy harvesting and smart grid-powered wireless communication networks: A contemporary survey," *IEEE Transactions on Green Communications and Networking*, vol. 4, no. 2, pp. 461–496, 2020.

199 A. A. Nasir, X. Zhou, S. Durrani, and R. A. Kennedy, "Relaying protocols for wireless energy harvesting and information processing," *IEEE Transactions on Wireless Communications*, vol. 12, no. 7, pp. 3622–3636, 2013.

200 M. A. Hossain, R. M. Noor, K.-L. A. Yau, I. Ahmedy, and S. S. Anjum, "A survey on simultaneous wireless information and power transfer with cooperative relay and future challenges," *IEEE Access*, vol. 7, pp. 19 166–19 198, 2019.

201 R. Khan, P. Kumar, D. N. K. Jayakody, and M. Liyanage, "A survey on security and privacy of 5G technologies: Potential solutions, recent advancements, and future directions," *IEEE Communication Surveys and Tutorials*, vol. 22, no. 1, pp. 196–248, 2019.

202 P. Porambage and M. Liyanage, "Security in network slicing," *Wiley 5G Ref: The Essential 5G Reference Online*, pp. 1–12, 2019.

203 W. Xia, Y. Wen, C. H. Foh, D. Niyato, and H. Xie, "A survey on software-defined networking," *IEEE Communication Surveys and Tutorials*, vol. 17, no. 1, pp. 27–51, 2014.

204 B. Han, V. Gopalakrishnan, L. Ji, and S. Lee, "Network function virtualization: Challenges and opportunities for innovations," *IEEE Communications Magazine*, vol. 53, no. 2, pp. 90–97, 2015.

205 Y. Dinitz, S. Dolev, S. Frenkel, A. Binun, and D. Khankin, "Network cloudification," in *International Symposium on Cyber Security Cryptography and Machine Learning*. Springer, 2019, pp. 249–259.

206 P. Porambage, J. Okwuibe, M. Liyanage, M. Ylianttila, and T. Taleb, "Survey on multi-access edge computing for Internet of Things realization," *IEEE Communication Surveys and Tutorials*, vol. 20, no. 4, pp. 2961–2991, 2018.

207 T. Huang, W. Yang, J. Wu, J. Ma, X. Zhang, and D. Zhang, "A survey on green 6G network: Architecture and technologies," *IEEE Access*, vol. 7, pp. 175 758–175 768, 2019.

208 Q. Yu, J. Ren, H. Zhou, and W. Zhang, "A cybertwin based network architecture for 6G," in *2020 2nd 6G Wireless Summit (6G SUMMIT)* IEEE, 2020, pp. 1–5.

209 V. W. Wong, R. Schober, D. W. K. Ng, and L.-C. Wang, *Key Technologies for 5G Wireless Systems*. Cambridge University Press, 2017.

210 I. Alam, K. Sharif, F. Li, Z. Latif, M. Karim, S. Biswas, B. Nour, and Y. Wang, "A survey of network virtualization techniques for Internet of Things using SDN and NFV," *ACM Computing Surveys (CSUR)*, vol. 53, no. 2, pp. 1–40, 2020.

211 J. G. Herrera and J. F. Botero, "Resource allocation in NFV: A comprehensive survey," *IEEE Transactions on Network and Service Management*, vol. 13, no. 3, pp. 518–532, 2016.

212 W. Kellerer, P. Kalmbach, A. Blenk, A. Basta, M. Reisslein, and S. Schmid, "Adaptable and data-driven softwarized networks: Review, opportunities, and challenges," *Proceedings of the IEEE*, vol. 107, no. 4, pp. 711–731, 2019.

213 I. Afolabi, T. Taleb, K. Samdanis, A. Ksentini, and H. Flinck, "Network slicing and softwarization: A survey on principles, enabling technologies, and solutions," *IEEE Communication Surveys and Tutorials*, vol. 20, no. 3, pp. 2429–2453, 2018.

214 C. Benzaid and T. Taleb, "AI-driven zero touch network and service management in 5G and beyond: Challenges and research directions," *IEEE Network*, vol. 34, no. 2, pp. 186–194, 2020.

215 M. Bunyakitanon, X. Vasilakos, R. Nejabati, and D. Simeonidou, "End-to-end performance-based autonomous VNF placement with adopted reinforcement learning," *IEEE Transactions on Cognitive Communications and Networking*, vol. 6, no. 2, pp. 534–547, 2020.

216 D. Bega, M. Gramaglia, M. Fiore, A. Banchs, and X. Costa-Perez, "AZTEC: Anticipatory capacity allocation for zero-touch network slicing," in *IEEE INFOCOM 2020-IEEE Conference on Computer Communications* IEEE, 2020, pp. 794–803.

217 "Zero-touch network and Service Management (ZSM); Reference Architecture," [Accessed on 29.03.2021]. [Online]. Available: https://www.etsi.org/deliver/etsi_gs/ZSM/001_099/002/01.01.01_60/gs_ZSM002v010101p.pdf.

218 "Zero-touch network and Service Management (ZSM); End to end management and orchestration of network slicing," [Accessed on 29.03.2021]. [Online]. Available: https://portal.etsi.org/webapp/WorkProgram/Report_WorkItem.asp?WKI_ID=54284.

219 "Zero-touch network and Service Management (ZSM); Landscape," [Accessed on 29.03.2021]. [Online]. Available: https://www.etsi.org/deliver/etsi_gr/ZSM/001_099/004/01.01.01_60/gr_ZSM004v010101p.pdf.

220 "Zero-touch network and Service Management (ZSM); Means of Automation," [Accessed on 29.03.2021]. [Online]. Available: https://www.etsi.org/deliver/etsi_gr/ZSM/001_099/005/01.01.01_60/gr_ZSM005v010101p.pdf.

221 "Zero touch network and Service Management (ZSM); Proof of Concept Framework," [Accessed on 29.03.2021]. [Online]. Available: https://www.etsi.org/deliver/etsi_gs/ZSM/001_099/006/01.01.01_60/gs_ZSM006v010101p.pdf.

222 "Zero-touch network and Service Management (ZSM); Terminology for concepts in ZSM," [Accessed on 29.03.2021]. [Online]. Available:

https://www.etsi.org/deliver/etsi_gs/ZSM/001_099/007/01.01.01_60/gs_ZSM007v010101p.pdf.

223 "Zero-touch network and Service Management (ZSM); Cross-domain E2E service lifecycle management," [Accessed on 29.03.2021]. [Online]. Available: https://portal.etsi.org/webapp/WorkProgram/Report_WorkItem.asp?WKI_ID=56825.

224 "Zero-Touch Network and Service Managment (ZSM) Closed-loop automation: Solutions for automation of E2E service and network management use cases," [Accessed on 29.03.2021]. [Online]. Available: https://portal.etsi.org/webapp/WorkProgram/Report_WorkItem.asp?WKI_ID=58055.

225 "Zero-touch network and Service Management (ZSM); General Security Aspects," [Accessed on 29.03.2021]. [Online]. Available: https://portal.etsi.org/webapp/WorkProgram/Report_WorkItem.asp?WKI_ID=58436.

226 I. Sanchez-Navarro, P. Salva-Garcia, Q. Wang, and J. M. A. Calero, "New immersive interface for zero-touch management in 5G networks," in *2020 IEEE 3rd 5G World Forum (5GWF)* IEEE, 2020, pp. 145–150.

227 A. Oi, R. Sato, Y. Suto, K. Sakata, M. Nakajima, and T. Furukawa, "A study on automation of network maintenance in telecom carriers for zero-touch operations," in *2020 21st Asia-Pacific Network Operations and Management Symposium (APNOMS)* IEEE, 2020, pp. 1–6.

228 F. Rezazadeh, H. Chergui, L. Alonso, and C. Verikoukis, "Continuous multi-objective zero-touch network slicing via twin delayed DDPG and OpenAI Gym," in *GLOBECOM 2020-2020 IEEE Global Communications Conference* IEEE, 2020, pp. 1–6.

229 V. Räisänen, "A framework for capability provisioning in B5G," in *2020 2nd 6G wireless summit (6G SUMMIT)* IEEE, 2020, pp. 1–4.

230 F. Wilhelmi, S. Barrachina-Mu noz, B. Bellalta, C. Cano, A. Jonsson, and V. Ram, "A flexible machine-learning-aware architecture for future wlans," *IEEE Communications Magazine*, vol. 58, no. 3, pp. 25–31, 2020.

231 A. M. Zarca, M. Bagaa, J. B. Bernabe, T. Taleb, and A. F. Skarmeta, "Semantic-aware security orchestration in SDN/NFV-enabled IoT systems," *Sensors*, vol. 20, no. 13, p. 3622, 2020.

232 J. Prados-Garzon and T. Taleb, "Asynchronous time-sensitive networking for 5G backhauling," *IEEE Network*, vol. 35, no. 2, pp. 144–151, 2021.

233 C. Benzaid, T. Taleb, and M. Z. Farooqi, "Trust in 5G and beyond networks," *IEEE Network*, vol. 35, pp. 212–222, 2021.

234 M. Bagaa, T. Taleb, J. B. Bernabe, and A. Skarmeta, "QoS and resource-aware security orchestration and life cycle management," *IEEE Transactions on Mobile Computing*, vol. 21, pp. 2978–2993, 2020.

235 I. Vaishnavi and L. Ciavaglia, "Challenges towards automation of live telco network management: Closed control loops," in *2020 16th International Conference on Network and Service Management (CNSM)* IEEE, 2020, pp. 1–5.

236 H. Hantouti, N. Benamar, and T. Taleb, "Service function chaining in 5G & beyond networks: Challenges and open research issues," *IEEE Network*, vol. 34, no. 4, pp. 320–327, 2020.

237 R. Rokui, H. Yu, L. Deng, D. Allabaugh, M. Hemmati, and C. Janz, "A standards-based, model-driven solution for 5G transport slice automation and assurance," in *2020 6th IEEE Conference on Network Softwarization (NetSoft)* IEEE, 2020, pp. 106–113.

238 I. Afolabi, M. Bagaa, W. Boumezer, and T. Taleb, "Toward a real deployment of network services orchestration and configuration convergence framework for 5G network slices," *IEEE Network*, vol. 35, pp. 242–250, 2020.

239 E. G. ZSM, "Zero touch network and service management (ZSM) landscape, version 1.1. 1," *ETSI: Sophia Antipolis, France*, 2020.

240 Q. Duan, "Intelligent and autonomous management in cloud-native future networks–a survey on related standards from an architectural perspective," *Future Internet*, vol. 13, no. 2, p. 42, 2021.

241 A. Boudi, M. Bagaa, P. Pöyhönen, T. Taleb, and H. Flinck, "AI-based resource management in beyond 5G cloud native environment," *IEEE Network*, vol. 35, no. 2, pp. 128–135, 2021.

242 A. Muhammad, T. A. Khan, K. Abbass, and W.-C. Song, "An end-to-end intelligent network resource allocation in iov: A machine learning approach," in *2020 IEEE 92nd Vehicular Technology Conference (VTC2020-Fall)* IEEE, pp. 1–5.

243 K. Samdanis and T. Taleb, "The road beyond 5G: A vision and insight of the key technologies," *IEEE Network*, vol. 34, no. 2, pp. 135–141, 2020.

244 J. Baranda, J. Mangues-Bafalluy, E. Zeydan, L. Vettori, R. Martínez, X. Li, A. Garcia-Saavedra, C. Chiasserini, C. Casetti, K. Tomakh *et al.*, "On the integration of AI/ML-based scaling operations in the 5Growth platform," in *2020 IEEE Conference on Network Function Virtualization and Software Defined Networks (NFV-SDN)* IEEE, 2020, pp. 105–109.

245 M. Chahbar, G. Diaz, A. Dandoush, C. Cérin, and K. Ghoumid, "A comprehensive survey on the E2E 5G network slicing model," *IEEE Transactions on Network and Service Management*, vol. 18, pp. 49–62, 2020.

246 O. Hassane, S. Mustafiz, F. Khendek, and M. Toeroe, "A model traceability framework for network service management," in *Proceedings of the 12th System Analysis and Modelling Conference*, 2020, pp. 64–73.

247 S. Zhang and D. Zhu, "Towards artificial intelligence enabled 6G: State of the art, challenges, and opportunities,"*Computer Networks*, vol. 183, p. 107556, 2020.

248 D. Bega, M. Gramaglia, R. Perez, M. Fiore, A. Banchs, and X. Costa-Perez, "AI-based autonomous control, management, and orchestration in 5G: From standards to algorithms," *IEEE Network*, vol. 34, no. 6, pp. 14–20, 2020.

249 M. Xie, J. S. Pujol-Roig, F. Michelinakis, T. Dreibholz, C. Guerrero, A. G. Sanchez, W. Y. Poe, Y. Wang, and A. M. Elmokashfi, "AI-driven closed-loop service assurance with service exposures," in *2020 European Conference on Networks and Communications (EuCNC)* IEEE, 2020, pp. 265–270.

250 N. Blefari-Melazzi, S. Bartoletti, L. Chiaraviglio, F. Morselli, E. Baena, G. Bernini, D. Giustiniano, M. Hunukumbure, G. Solmaz, and K. Tsagkaris, "LOCUS: Localization and analytics on-demand embedded in the 5G ecosystem," in *2020 European Conference on Networks and Communications (EuCNC)* IEEE, 2020, pp. 170–175.

251 M. McClellan, C. Cervelló-Pastor, and S. Sallent, "Deep learning at the mobile edge: Opportunities for 5G networks," *Applied Sciences*, vol. 10, no. 14, p. 4735, 2020.

252 K. Dev, R. K. Poluru, L. Kumar, P. K. R. Maddikunta, and S. A. Khowaja, "Optimal radius for enhanced lifetime in IoT using hybridization of rider and grey wolf optimization," *IEEE Transactions on Green Communications and Networking*, vol. 5, pp. 635–644, 2021.

253 P. K. R. Maddikunta, T. R. Gadekallu, R. Kaluri, G. Srivastava, R. M. Parizi, and M. S. Khan, "Green communication in IoT networks using a hybrid optimization algorithm," *Computer Communications*, vol. 159, pp. 97–107, 2020.

254 A. Osseiran, F. Boccardi, V. Braun, K. Kusume, P. Marsch, M. Maternia, O. Queseth, M. Schellmann, H. Schotten, H. Taoka *et al.*, "Scenarios for 5G mobile and wireless communications: The vision of the METIS project," *IEEE Communications Magazine*, vol. 52, no. 5, pp. 26–35, 2014.

255 X. Foukas, G. Patounas, A. Elmokashfi, and M. K. Marina, "Network slicing in 5G: Survey and challenges," *IEEE Communications Magazine*, vol. 55, no. 5, pp. 94–100, 2017.

256 B. Holfeld, D. Wieruch, T. Wirth, L. Thiele, S. A. Ashraf, J. Huschke, I. Aktas, and J. Ansari, "Wireless communication for factory automation: An opportunity for LTE and 5G systems," *IEEE Communications Magazine*, vol. 54, no. 6, pp. 36–43, 2016.

257 "Zero-touch network and Service Management (ZSM); end-to-end architectural framework for network and service automation," [Accessed on 29.03.2021]. [Online]. Available: https://www.etsi.org/committee?id=1673.

258 N. Mohan, L. Corneo, A. Zavodovski, S. Bayhan, W. Wong, and J. Kangasharju, "Pruning edge research with latency shears," in *Proceedings* *of the 19th ACM Workshop on Hot Topics in Networks*, ser. HotNets '20, 2020, p. 182–189.

259 E. Peltonen *et al.*, "6G white paper on edge intelligence," *6G RESEARCH VISIONS, NO. 8*, 2020.

260 Q.-V. Pham, M. Le, T. Huynh-The, Z. Han, and W.-J. Hwang, "Energy-efficient federated learning over UAV-enabled wireless powered communications," *IEEE Transactions on Vehicular Technology*, vol. 71, pp. 4977–4990, 2022. [Online]. Available: https://ieeexplore.ieee.org/document/9709639.

261 Joe, "New AI solution to shield drones from advanced cyber attacks," Commercial Drone Professional, January, 2020. [Online]. Available: https://www.commercialdroneprofessional.com/new-ai-solution-to-shield-drones-from-advanced-cyber-attacks/.

262 S. Greengard, "Ai on edge," *Communications of the ACM*, vol. 63, no. 9, pp. 18–20, 2020.

263 Z. Zhou, X. Chen, E. Li, L. Zeng, K. Luo, and J. Zhang, "Edge intelligence: Paving the last mile of artificial intelligence with edge computing," *Proceedings of the IEEE*, vol. 107, no. 8, pp. 1738–1762, 2019.

264 T. R. Gadekallu, Q.-V. Pham, D. C. Nguyen, P. K. R. Maddikunta, N. Deepa, B. Prabadevi, P. N. Pathirana, J. Zhao, and W.-J. Hwang, "Blockchain for edge of things: Applications, opportunities, and challenges," *IEEE Internet of Things Journal*, vol. 9, no. 2, pp. 964–988, 2022.

265 D. C. Nguyen, Q.-V. Pham, P. N. Pathirana, M. Ding, A. Seneviratne, Z. Lin, O. Dobre, and W.-J. Hwang, "Federated learning for smart healthcare: A survey," *ACM Computing Surveys (CSUR)*, vol. 55, no. 3, pp. 1–37, 2022.

266 T. Huynh-The, Q.-V. Pham, T.-V. Nguyen, and D.-S. Kim, "Deep learning for coexistence radar-communication waveform recognition," in *2021 International Conference on Information and Communication Technology Convergence (ICTC)* IEEE, 2021, pp. 1725–1727.

267 J. Shao and J. Zhang, "Communication-computation trade-off in resource-constrained edge inference," *IEEE Communications Magazine*, vol. 58, no. 12, pp. 20–26, 2020.

268 Y. Shi, K. Yang, T. Jiang, J. Zhang, and K. B. Letaief, "Communication-efficient edge AI: Algorithms and systems," *IEEE Communication Surveys and Tutorials*, vol. 22, no. 4, pp. 2167–2191, 2020.

269 Q.-V. Pham, N. T. Nguyen, T. Huynh-The, L. B. Le, K. Lee, and W.-J. Hwang, "Intelligent radio signal processing: A survey," *IEEE Access*, vol. 9, pp. 83 818–83 850, 2021.

270 S. Maghsudi and M. Davy, "Computational models of human decision-making with application to the internet of everything," *IEEE Wireless Communications*, vol. 28, no. 1, pp. 152–159, 2021.

271 Q. Ye and Y. Zhang, "Participation behavior and social welfare in repeated task allocations," in *2016 IEEE International Conference on Agents (ICA)*, Sept. 2016, pp. 94–97.

272 M. Matinmikko-Blue, S. Aalto, M. I. Asghar, H. Berndt, Y. Chen, S. Dixit, R. Jurva, P. Karppinen, M. Kekkonen, M. Kinnula *et al.*, "White Paper on 6G Drivers and the UN SDGs," *arXiv preprint arXiv:2004.14695*, 2020.

273 S. Maghsudi and M. van der Schaar, "Distributed task management in cyber-physical systems: How to cooperate under uncertainty?" *IEEE Transactions on Cognitive Communications and Networking*, vol. 5, no. 1, pp. 165–180, 2019.

274 M. L. Rahman, J. A. Zhang, K. Wu, X. Huang, Y. J. Guo, S. Chen, and J. Yuan, "Enabling Joint Communication and Radio Sensing in Mobile Networks - A Survey," *ArXiv*, vol. abs/2006.07559, 2020.

275 M. Liyanage, I. Ahmed, J. Okwuibe, M. Ylianttila, H. Kabir, J. L. Santos, R. Kantola, O. L. Perez, M. U. Itzazelaia, and E. M. De Oca, "Enhancing security of software defined mobile networks," *IEEE Access*, vol. 5, pp. 9422–9438, 2017.

276 M. W. Akhtar, S. A. Hassan, R. Ghaffar, H. Jung, S. Garg, and M. S. Hossain, "The shift to 6G communications: Vision and requirements," *Human-centric Computing and Information Sciences*, vol. 10, no. 1, pp. 1–27, 2020.

277 G. Mirjalily and Z. Luo, "Optimal network function virtualization and service function chaining: A survey," *Chinese Journal of Electronics*, vol. 27, no. 4, pp. 704–717, 2018.

278 H. Cao, J. Du, H. Zhao, D. X. Luo, N. Kumar, L. Yang, and F. R. Yu, "Resource-ability assisted service function chain embedding and scheduling for 6G networks with virtualization," *IEEE Transactions on Vehicular Technology*, vol. 70, no. 4, pp. 3846–3859, 2021.

279 "Service Function Chaining (SFC) Architecture," [Accessed on 14.03.2022]. [Online]. Available: https://datatracker.ietf.org/doc/html/rfc7665.

280 O. Hireche, C. Benzaïd, and T. Taleb, "Deep data plane programming and ai for zero-trust self-driven networking in beyond 5G," *Computer Networks*, vol. 203, p. 108668, 2022.

281 P. Bosshart, D. Daly, G. Gibb, M. Izzard, N. McKeown, J. Rexford, C. Schlesinger, D. Talayco, A. Vahdat, G. Varghese *et al.*, "P4: Programming protocol-independent packet processors," *ACM SIGCOMM Computer Communication Review*, vol. 44, no. 3, pp. 87–95, 2014.

282 N. Hu, Z. Tian, X. Du, and M. Guizani, "An energy-efficient in-network computing paradigm for 6G," *IEEE Transactions on Green Communications and Networking*, vol. 5, no. 4, pp. 1722–1733, 2021.

283 X. Jin, X. Li, H. Zhang, R. Soulé, J. Lee, N. Foster, C. Kim, and I. Stoica, "Netcache: Balancing key-value stores with fast in-network caching," in

Proceedings of the 26th Symposium on Operating Systems Principles, 2017, pp. 121–136.

284 A. Sapio, I. Abdelaziz, A. Aldilaijan, M. Canini, and P. Kalnis, "In-network computation is a dumb idea whose time has come," in *Proceedings of the 16th ACM Workshop on Hot Topics in Networks*, 2017, pp. 150–156.

285 F. Yang, Z. Wang, X. Ma, G. Yuan, and X. An, "SwitchAgg: A further step towards in-network computation," *arXiv preprint arXiv:1904.04024*, 2019.

286 D. Sanvito, G. Siracusano, and R. Bifulco, "Can the network be the ai accelerator?" in *Proceedings of the 2018 Morning Workshop on In-Network Computing*, 2018, pp. 20–25.

287 A. A. Zaidi, R. Baldemair, V. Molés-Cases, N. He, K. Werner, and A. Cedergren, "OFDM numerology design for 5G new radio to support IoT, eMBB, and MBSFN," *IEEE Communications Standards Magazine*, vol. 2, no. 2, pp. 78–83, 2018.

288 N. Chi, Y. Zhou, Y. Wei, and F. Hu, "Visible light communication in 6G: Advances, challenges, and prospects," *IEEE Vehicular Technology Magazine*, vol. 15, no. 4, pp. 93–102, 2020.

289 Q.-V. Pham, T. Huynh-The, M. Alazab, J. Zhao, and W.-J. Hwang, "Sum-rate maximization for UAV-assisted visible light communications using NOMA: Swarm intelligence meets machine learning," *IEEE Internet of Things Journal*, vol. 7, no. 10, pp. 10 375–10 387, 2020.

290 X. Shen, J. Gao, W. Wu, M. Li, C. Zhou, and W. Zhuang, "Holistic network virtualization and pervasive network intelligence for 6G," *IEEE Communication Surveys and Tutorials*, vol. 24, pp. 1–30, 2022.

291 D. C. Nguyen, M. Ding, P. N. Pathirana, A. Seneviratne, J. Li, D. Niyato, O. Dobre, and H. V. Poor, "6G Internet of Things: A comprehensive survey," *IEEE Internet of Things Journal*, vol. 9, pp. 359–383, 2021.

292 S. Wijethilaka and M. Liyanage, "Survey on network slicing for Internet of Things realization in 5G networks," *IEEE Communication Surveys and Tutorials*, vol. 23, no. 2, pp. 957–994, 2021.

293 W. Shi, H. Zhou, J. Li, W. Xu, N. Zhang, and X. Shen, "Drone assisted vehicular networks: Architecture, challenges and opportunities," *IEEE Network*, vol. 32, no. 3, pp. 130–137, 2018.

294 X. Huang, J. A. Zhang, R. P. Liu, Y. J. Guo, and L. Hanzo, "Airplane-aided integrated networking for 6G wireless: Will it work?" *IEEE Vehicular Technology Magazine*, vol. 14, no. 3, pp. 84–91, 2019.

295 D. S. Lakew, U. Sa'ad, N.-N. Dao, W. Na, and S. Cho, "Routing in flying ad hoc networks: A comprehensive survey," *IEEE Communications Survey and Tutorials*, vol. 22, no. 2, pp. 1071–1120, 2020.

296 Z. Lin, M. Lin, J.-B. Wang, T. de Cola, and J. Wang, "Joint beamforming and power allocation for satellite-terrestrial integrated networks with non-orthogonal multiple access," *IEEE Journal of Selected Topics in Signal Processing*, vol. 13, no. 3, pp. 657–670, 2019.

297 L. Song, Z. Han, and B. Di, "Aerial access networks for 6G: From UAV, HAP, to satellite communication networks," in *IEEE International Conference on Communications (ICC)*, Virtual Program, 2020, pp. 1–1.

298 United States. Department of the Army, "'Eyes of the army': U.S. army roadmap for unmanned systems, 2010–2035," Apr. 2010.

299 M. Kishk, A. Bader, and M.-S. Alouini, "Aerial base station deployment in 6G cellular networks using tethered drones: The mobility and endurance tradeoff," *IEEE Vehicular Technology Magazine*, vol. 15, no. 4, pp. 103–111, 2020.

300 S. C. Arum, D. Grace, and P. D. Mitchell, "A review of wireless communication using high-altitude platforms for extended coverage and capacity," *Computer Communications*, vol. 157, no. 1, pp. 232–256, 2020.

301 Y. Su, Y. Liu, Y. Zhou, J. Yuan, H. Cao, and J. Shi, "Broadband LEO satellite communications: Architectures and key technologies," *IEEE Wireless Communications*, vol. 26, no. 2, pp. 55–61, 2019.

302 F. Davoli, C. Kourogiorgas, M. Marchese, A. Panagopoulos, and F. Patrone, "Small satellites and CubeSats: Survey of structures, architectures, and protocols," *International Journal of Satellite Communications and Networking*, vol. 37, no. 4, pp. 343–359, 2019.

303 T. O'Shea and J. Hoydis, "An introduction to deep learning for the physical layer," *IEEE Transactions on Cognitive Communications and Networking*, vol. 3, no. 4, pp. 563–575, 2017.

304 D. H. Nguyen, "Neural network-optimized channel estimator and training signal design for MIMO systems with few-bit ADCs," *IEEE Signal Processing Letters*, vol. 27, pp. 1370–1374, 2020.

305 N. T. Nguyen and K. Lee, "Deep learning-aided tabu search detection for large MIMO systems," *IEEE Transaction on Wireless Communication*, vol. 19, no. 6, pp. 4262–4275, 2020.

306 A. Garcia-Saavedra and X. Costa-Perez, "O-RAN: Disrupting the virtualized ran ecosystem," *IEEE Communications Standards Magazine*, vol. 5, 96–103, 2021.

307 A. S. Abdalla, P. S. Upadhyaya, V. K. Shah, and V. Marojevic, "Toward Next Generation open Radio Access Network–What O-RAN Can and Cannot Do!" *arXiv preprint arXiv:2111.13754*, 2021.

308 L. Bonati, S. D'Oro, M. Polese, S. Basagni, and T. Melodia, "Intelligence and learning in O-RAN for data-driven NextG cellular networks," *IEEE Communications Magazine*, vol. 59, no. 10, pp. 21–27, 2021.

309 G. Gui, M. Liu, F. Tang, N. Kato, and F. Adachi, "6G: Opening new horizons for integration of comfort, security and intelligence," *IEEE Wireless Communications*, vol. 27, pp. 126–132, 2020.

310 P. Porambage, G. Gür, D. P. M. Osorio, M. Liyanage, and M. Ylianttila, "6G security challenges and potential solutions," in *2021 Joint European Conference on Networks and Communications (EuCNC) and 6G Summit* IEEE, 2021, pp. 1–6.

311 M. Liyanage, I. Ahmad, A. B. Abro, A. Gurtov, and M. Ylianttila, *A Comprehensive Guide to 5G Security*. John Wiley & Sons, 2018.

312 M. Liyanage, A. B. Abro, M. Ylianttila, and A. Gurtov, "Opportunities and challenges of software-defined mobile networks in network security," *IEEE Security & Privacy*, vol. 14, no. 4, pp. 34–44, 2016.

313 B. Schneier, "Artificial intelligence and the attack/defense balance," *IEEE Security & Privacy*, vol. 16, no. 2, pp. 96–96, 2018.

314 A. Pouttu, F. Burkhardt, C. Patachia, L. Mendes, G. R. Brazil, S. Pirttikangas, E. Jou, P. Kuvaja, F. T. Finland, M. Heikkilä *et al.*, "6G White Paper on Validation and Trials for Verticals Towards 2030's." [Online]. Available: https://www.6gchannel.com/wp-content/uploads/2020/04/6g-white-paper-validation-trials.pdf.

315 United Nations (UN) #Envision2030 Sustainable Development Goals, "United Nations (UN)." [Online]. Available: https://sdgs.un.org/goals.

316 V. Ziegler and S. Yrjola, "6G indicators of value and performance," in *2020 2nd 6G wireless summit (6G SUMMIT)* IEEE, 2020, pp. 1–5.

317 K. B. Letaief, W. Chen, Y. Shi, J. Zhang, and Y. A. Zhang, "The roadmap to 6G: AI empowered wireless networks," *IEEE Communications Magazine*, vol. 57, no. 8, pp. 84–90, 2019.

318 M. Yao, M. Sohul, V. Marojevic, and J. H. Reed, "Artificial intelligence defined 5G radio access networks," *IEEE Communications Magazine*, vol. 57, no. 3, pp. 14–20, 2019.

319 H. Viswanathan and P. E. Mogensen, "Communications in the 6G era," *IEEE Access*, vol. 8, pp. 57 063–57 074, 2020.

320 S. Deng, H. Zhao, W. Fang, J. Yin, S. Dustdar, and A. Y. Zomaya, "Edge intelligence: The confluence of edge computing and artificial intelligence," *IEEE Internet of Things Journal*, vol. 7, no. 8, pp. 7457–7469, 2020.

321 G. Plastiras, M. Terzi, C. Kyrkou, and T. Theocharidcs, "Edge intelligence: Challenges and opportunities of near-sensor machine learning applications," in *2018 IEEE 29th International Conference on Application-specific Systems, Architectures and Processors (ASAP)*, 2018, pp. 1–7.

322 S. Xu, Y. Qian, and R. Q. Hu, "Edge intelligence assisted gateway defense in cyber security," *IEEE Network*, vol. 34, no. 4, pp. 14–19, 2020.

323 M. Mukherjee, R. Matam, C. X. Mavromoustakis, H. Jiang, G. Mastorakis, and M. Guo, "Intelligent edge computing: Security and privacy challenges," *IEEE Communications Magazine*, vol. 58, no. 9, pp. 26–31, 2020.

324 R. Kalaiprasath, R. Elankavi, and R. Udayakumar, "Cloud security and compliance-a semantic approach in end to end security."*International Journal on Smart Sensing & Intelligent Systems*, vol. 10, pp. 482–494, 2017.

325 G. ETSI, "004, Zero-touch network and service management (ZSM)," *Reference Architecture*, 2020. [Online]. Available: https://www.etsi.org/deliver/etsi_gs/ZSM/001_099/002/01.01.01_60/gs_ZSM002v010101p.pdf.

326 J. Ortiz, R. Sanchez-Iborra, J. B. Bernabe, A. Skarmeta, C. Benzaid, T. Taleb, P. Alemany, R. Mu noz, R. Vilalta, C. Gaber *et al.*, "Inspire-5Gplus: Intelligent security and pervasive trust for 5G and beyond networks," in *Proceedings of the 15th International Conference on Availability, Reliability and Security*, 2020, pp. 1–10.

327 Y. Siriwardhana, P. Porambage, M. Liyanage, and M. Ylianttila, "AI and 6G security: Opportunities and challenges," in *2021 Joint European Conference on Networks and Communications (EuCNC) and 6G Summit. IEEE*, 2021, pp. 1–6.

328 M. Hyder and M. Ismail, "INMTD: Intent-based moving target defense framework using software defined networks," *Engineering, Technology & Applied Science Research*, vol. 10, no. 1, pp. 5142–5147, 2020.

329 Y. Han, J. Li, D. Hoang, J.-H. Yoo, and J. W.-K. Hong, "An intent-based network virtualization platform for SDN," in *2016 12th International Conference on Network and Service Management (CNSM)* IEEE, 2016, pp. 353–358.

330 Y. Wei, M. Peng, and Y. Liu, "Intent-based networks for 6G: Insights and challenges," *Digital Communications and Networks*, vol. 6, no. 3, pp. 270–280, 2020.

331 M. Ylianttila, R. Kantola, A. Gurtov, L. Mucchi, I. Oppermann, Z. Yan, T. H. Nguyen, F. Liu, T. Hewa, M. Liyanage *et al.*, "6G White Paper: Research Challenges for Trust, Security and Privacy," *arXiv preprint arXiv:2004.11665*, 2020.

332 R. Yasmin, J. Petäjäjärvi, K. Mikhaylov, and A. Pouttu, "On the integration of lorawan with the 5G test network," in *2017 IEEE 28th Annual International Symposium on Personal, Indoor, and Mobile Radio Communications (PIMRC)* IEEE, 2017, pp. 1–6.

333 S. Chen, Y.-C. Liang, S. Sun, S. Kang, W. Cheng, and M. Peng, "Vision, requirements, and technology trend of 6G: How to tackle the challenges of system coverage, capacity, user data-rate and movement speed," *IEEE Wireless Communications*, vol. 27, no. 2, pp. 218–228, 2020.

334 B. Deebak and F. Al-Turjman, "Drone of IoT in 6G wireless communications: Technology, challenges, and future aspects," in *Unmanned Aerial Vehicles in Smart Cities*. Springer, 2020, pp. 153–165.

335 H. Menouar, I. Guvenc, K. Akkaya, A. S. Uluagac, A. Kadri, and A. Tuncer, "UAV-enabled intelligent transportation systems for the smart city: Applications and challenges," *IEEE Communications Magazine*, vol. 55, no. 3, pp. 22–28, 2017.

336 P. B. Johnston and A. K. Sarbahi, "The impact of US drone strikes on terrorism in Pakistan," *International Studies Quarterly*, vol. 60, no. 2, pp. 203–219, 2016.

337 J. O'Malley, "The no drone zone," *Engineering & Technology*, vol. 14, no. 2, pp. 34–38, 2019.

338 I. Petrov and T. Janevski, "5G mobile technologies and early 6G viewpoints," *European Journal of Engineering Research and Science*, vol. 5, no. 10, pp. 1240–1246, 2020.

339 K. Lebeck, K. Ruth, T. Kohno, and F. Roesner, "Towards security and privacy for multi-user augmented reality: Foundations with end users," in *2018 IEEE Symposium on Security and Privacy (SP)* IEEE, 2018, pp. 392–408.

340 Y. Siriwardhana, P. Porambage, M. Liyanage, and M. Ylinattila, "A survey on mobile augmented reality with 5G mobile edge computing: Architectures, applications and technical aspects," *IEEE Communication Surveys and Tutorials*, vol. 23, pp. 1160–1192, 2021.

341 C. Insights, "40+ corporations working on autonomous vehicles," 2019.

342 J. He, K. Yang, and H.-H. Chen, "6G Cellular Networks and Connected Autonomous Vehicles," *arXiv preprint arXiv:2010.00972*, 2020.

343 H. Shahinzadeh, J. Moradi, G. B. Gharehpetian, H. Nafisi, and M. Abedi, "Internet of energy (IoE) in smart power systems," in *2019 5th Conference on Knowledge Based Engineering and Innovation (KBEI)* IEEE, 2019, pp. 627–636.

344 I. Andrea, C. Chrysostomou, and G. Hadjichristofi, "Internet of things: Security vulnerabilities and challenges," in *Proceedings - IEEE Symposium on Computers and Communications*, vol. 2016-February, no. August 2017, pp. 180–187, 2016.

345 D. Pliatsios, P. Sarigiannidis, T. Lagkas, and A. G. Sarigiannidis, "A survey on SCADA systems: Secure protocols, incidents, threats and tactics," *IEEE Communication Surveys and Tutorials*, vol. 22, no. 3, pp. 1942–1976, 2020.

346 E. Hossain, I. Khan, F. Un-Noor, S. S. Sikander, and M. S. H. Sunny, "Application of big data and machine learning in smart grid, and associated security concerns: A review," *IEEE Access*, vol. 7, no. c, pp. 13 960–13 988, 2019.

347 Z. Li, J. Kang, R. Yu, D. Ye, Q. Deng, and Y. Zhang, "Consortium blockchain for secure energy trading in industrial Internet of Things," *IEEE Transactions on Industrial Informatics*, vol. 14, no. 8, pp. 3690–3700, 2018.

348 A. Miglani, N. Kumar, V. Chamola, and S. Zeadally, "Blockchain for internet of energy management: Review, solutions, and challenges," *Computer Communications*, vol. 151, pp. 395–418, 2020.

349 S. Nahavandi, "Industry 5.0–A human-centric solution," *Sustainability*, vol. 11, no. 16, p. 4371, 2019.

350 X. Xu, "From cloud computing to cloud manufacturing," *Robotics and Computer-integrated Manufacturing*, vol. 28, no. 1, pp. 75–86, 2012.

351 M. Grieves and J. Vickers, "Digital twin: Mitigating unpredictable, undesirable emergent behavior in complex systems," in *Transdisciplinary Perspectives on Complex Systems*. Springer, 2017, pp. 85–113.

352 M. W. Grieves, "Virtually intelligent product systems: digital and physical twins," 2019.

353 T. Hewa, M. Ylianttila, and M. Liyanage, "Survey on blockchain based smart contracts: Applications, opportunities and challenges," *Journal of Network and Computer Applications*, vol. 177, p. 102857, 2020.

354 M. Barreno, B. Nelson, R. Sears, A. D. Joseph, and J. D. Tygar, "Can machine learning be secure?" in *Proceedings of the 2006 ACM Symposium on Information, computer and communications security*, 2006, pp. 16–25.

355 H. Kim, J. Park, M. Bennis, and S.-L. Kim, "On-device Federated Learning Via Blockchain and its Latency Analysis," *arXiv preprint arXiv:1808.03949*, 2018.

356 N. Weerasinghe, T. Hewa, M. Liyanage, S. S. Kanhere, and M. Ylianttila, "A novel blockchain-as-a-service (BaaS) platform for local 5G operators," *IEEE Open Journal of the Communications Society*, vol. 2, pp. 575–601, 2021.

357 J. Liu and Z. Liu, "A survey on security verification of blockchain smart contracts," *IEEE Access*, vol. 7, pp. 77 894–77 904, 2019.

358 T. M. Hewa, Y. Hu, M. Liyanage, S. Kanhare, and M. Ylianttila, "Survey on blockchain based smart contracts: Technical aspects and future research," *IEEE Access*, vol. 9, pp. 87643–87662, 2021.

359 S. Dey, "Securing majority-attack in blockchain using machine learning and algorithmic game theory: A proof of work," in *2018 10th Computer Science and Electronic Engineering (CEEC)* IEEE, 2018, pp. 7–10.

360 J. Moubarak, E. Filiol, and M. Chamoun, "On blockchain security and relevant attacks," in *2018 IEEE Middle East and North Africa Communications Conference (MENACOMM)* IEEE, 2018, pp. 1–6.

361 D. Efanov and P. Roschin, "The all-pervasiveness of the blockchain technology," *Procedia Computer Science*, vol. 123, pp. 116–121, 2018.

362 U. W. Chohan, "The double spending problem and cryptocurrencies," *Available at SSRN 3090174*, 2017.

363 S. Zhang and J.-H. Lee, "Double-spending with a sybil attack in the bitcoin decentralized network," *IEEE Transactions on Industrial Informatics*, vol. 15, no. 10, pp. 5715–5722, 2019.

364 M. I. Mehar, C. L. Shier, A. Giambattista, E. Gong, G. Fletcher, R. Sanayhie, H. M. Kim, and M. Laskowski, "Understanding a revolutionary and flawed grand experiment in blockchain: The dao attack," *Journal of Cases on Information Technology (JCIT)*, vol. 21, no. 1, pp. 19–32, 2019.

365 P. Otte, M. de Vos, and J. Pouwelse, "Trustchain: A sybil-resistant scalable blockchain," *Future Generation Computer Systems*, vol. 107, pp. 770–780, 2020.

366 Y. Cai and D. Zhu, "Fraud detections for online businesses: A perspective from blockchain technology," *Financial Innovation*, vol. 2, no. 1, p. 20, 2016.

367 Q. Feng, D. He, S. Zeadally, M. K. Khan, and N. Kumar, "A survey on privacy protection in blockchain system," *Journal of Network and Computer Applications*, vol. 126, pp. 45–58, 2019.

368 B. Bünz, S. Agrawal, M. Zamani, and D. Boneh, "Zether: Towards privacy in a smart contract world," in *International Conference on Financial Cryptography and Data Security* Springer, 2020, pp. 423–443.

369 A. Dorri, M. Steger, S. S. Kanhere, and R. Jurdak, "Blockchain: A distributed solution to automotive security and privacy," *IEEE Communications Magazine*, vol. 55, no. 12, pp. 119–125, 2017.

370 Z. Bao, Q. Wang, W. Shi, L. Wang, H. Lei, and B. Chen, "When blockchain meets SGX: An overview, challenges, and open issues," *IEEE Access*, vol. 8, pp. 170404–170420, 2020.

371 A. Groce, J. Feist, G. Grieco, and M. Colburn, "What are the actual flaws in important smart contracts (and how can we find them)?" in *International Conference on Financial Cryptography and Data Security* Springer, 2020, pp. 634–653.

372 C. Liu, J. Gao, Y. Li, H. Wang, and Z. Chen, "Studying gas exceptions in blockchain-based cloud applications," *Journal of Cloud Computing*, vol. 9, no. 1, pp. 1–25, 2020.

373 X. Li, P. Jiang, T. Chen, X. Luo, and Q. Wen, "A survey on the security of blockchain systems," *Future Generation Computer Systems*, vol. 107, pp. 841–853, 2020.

374 K. Chatterjee, A. K. Goharshady, and A. Pourdamghani, "Probabilistic smart contracts: Secure randomness on the blockchain," in *2019 IEEE International Conference on Blockchain and Cryptocurrency (ICBC)* IEEE, 2019, pp. 403–412.

375 C. G. Harris, "The risks and challenges of implementing ethereum smart contracts," in *2019 IEEE International Conference on Blockchain and Cryptocurrency (ICBC)* IEEE, 2019, pp. 104–107.

376 S. Kim and S. Ryu, "Analysis of blockchain smart contracts: Techniques and insights," in *2020 IEEE Secure Development (SecDev)* IEEE, 2020, pp. 65–73.

377 H. Poston, "Mapping the owasp top ten to blockchain," *Procedia Computer Science*, vol. 177, pp. 613–617, 2020.

378 G. Karame and S. Capkun, "Blockchain security and privacy," *IEEE Security & Privacy*, vol. 16, no. 04, pp. 11–12, 2018.

379 F. H. Pohrmen, R. K. Das, and G. Saha, "Blockchain-based security aspects in heterogeneous internet-of-things networks: A survey," *Transactions on Emerging Telecommunications Technologies*, vol. 30, no. 10, p. e3741, 2019.

380 A. Singh, R. M. Parizi, Q. Zhang, K.-K. R. Choo, and A. Dehghantanha, "Blockchain smart contracts formalization: Approaches and challenges to address vulnerabilities," *Computers & Security*, vol. 88, p. 101654, 2020.

381 Y. Zhang, S. Ma, J. Li, K. Li, S. Nepal, and D. Gu, "SMARTSHIELD: Automatic smart contract protection made easy," in *2020 IEEE 27th International Conference on Software Analysis, Evolution and Reengineering (SANER)* IEEE, 2020, pp. 23–34.

382 N. Atzei, M. Bartoletti, and T. Cimoli, "A Survey of attacks on ethereum smart contracts (SOK)," in *International Conference on Principles of Security and Trust* Springer, 2017, pp. 164–186.

383 M. Wohrer and U. Zdun, "Smart contracts: Security patterns in the ethereum ecosystem and solidity," in *2018 International Workshop on Blockchain Oriented Software Engineering (IWBOSE)*, March 2018, pp. 2–8.

384 C. Liu, H. Liu, Z. Cao, Z. Chen, B. Chen, and B. Roscoe, "ReGuard: Finding reentrancy bugs in smart contracts," in *Proceedings of the 40th International Conference on Software Engineering: Companion Proceeedings* ACM, 2018, pp. 65–68.

385 B. Jiang, Y. Liu, and W. Chan, "Contractfuzzer: Fuzzing smart contracts for vulnerability detection," in *Proceedings of the 33rd ACM/IEEE International Conference on Automated Software Engineering* ACM, 2018, pp. 259–269.

386 L. Brent, A. Jurisevic, M. Kong, E. Liu, F. Gauthier, V. Gramoli, R. Holz, and B. Scholz, "Vandal: A Scalable Security Analysis Framework for Smart Contracts," *arXiv preprint arXiv:1809.03981*, 2018.

387 K. Bhargavan, A. Delignat-Lavaud, C. Fournet, A. Gollamudi, G. Gonthier, N. Kobeissi, A. Rastogi, T. Sibut-Pinote, N. Swamy, and S. Zanella-Béguelin, "Short paper: Formal verification of smart contracts," in *Proceedings of the 11th ACM Workshop on Programming Languages and Analysis for Security (PLAS), in Conjunction with ACM CCS*, 2016, pp. 91–96.

388 T. Abdellatif and K.-L. Brousmiche, "Formal verification of smart contracts based on users and blockchain behaviors models," in *2018 9th IFIP International Conference on New Technologies, Mobility and Security (NTMS)* IEEE, 2018, pp. 1–5.

389 Z. Nehai, P.-Y. Piriou, and F. Daumas, "Model-checking of smart contracts," in *IEEE International Conference on Blockchain*, 2018, pp. 980–987.

390 E. Albert, P. Gordillo, B. Livshits, A. Rubio, and I. Sergey, "EthIR: A framework for high-level analysis of ethereum bytecode," in *International Symposium on Automated Technology for Verification and Analysis* Springer, 2018, pp. 513–520.

391 P. Schaar, "Privacy by design," *Identity in the Information Society*, vol. 3, no. 2, pp. 267–274, 2010.

392 R. Cheng, F. Zhang, J. Kos, W. He, N. Hynes, N. Johnson, A. Juels, A. Miller, and D. Song, "Ekiden: A Platform for Confidentiality-Preserving, Trustworthy, and Performant Smart Contracts," 1804.

393 R. Yuan, Y.-B. Xia, H.-B. Chen, B.-Y. Zang, and J. Xie, "Shadoweth: private smart contract on public blockchain," *Journal of Computer Science and Technology*, vol. 33, no. 3, pp. 542–556, 2018.

394 R. Gupta, S. Tanwar, F. Al-Turjman, P. Italiya, A. Nauman, and S. W. Kim, "Smart contract privacy protection using AI in cyber-physical systems: Tools, techniques and challenges," *IEEE Access*, vol. 8, pp. 24 746–24 772, 2020.

395 N. Kapsoulis, A. Psychas, G. Palaiokrassas, A. Marinakis, A. Litke, and T. Varvarigou, "Know your customer (KYC) implementation with smart contracts on a privacy-oriented decentralized architecture," *Future Internet*, vol. 12, no. 2, p. 41, 2020.

396 M. Niranjanamurthy, B. Nithya, and S. Jagannatha, "Analysis of Blockchain technology: pros, cons and SWOT," *Cluster Computing*, vol. 22, no. 6, pp. 14 743–14 757, 2019.

397 M. Roetteler, M. Naehrig, K. M. Svore, and K. Lauter, "Quantum resource estimates for computing elliptic curve discrete logarithms," in *International Conference on the Theory and Application of Cryptology and Information Security* Springer, 2017, pp. 241–270.

398 D. J. Bernstein and T. Lange, "Post-quantum cryptography," *Nature*, vol. 549, no. 7671, pp. 188–194, 2017.

399 S. Tarantino, B. Da Lio, D. Cozzolino, and D. Bacco, "Feasibility of quantum communications in aquatic scenarios," *Optik*, vol. 216, p. 164639, 2020.

400 F. Bouchard, R. Fickler, R. W. Boyd, and E. Karimi, "High-dimensional quantum cloning and applications to quantum hacking," *Science advances*, vol. 3, no. 2, e1601915, 2017.

401 A. Lohachab, A. Lohachab, and A. Jangra, "A comprehensive survey of prominent cryptographic aspects for securing communication in post-quantum iot networks," *Internet of Things*, vol. 9, p. 100174, 2020.

402 T. Saito, K. Xagawa, and T. Yamakawa, "Tightly-secure key-encapsulation mechanism in the quantum random oracle model," in *Annual International Conference on the Theory and Applications of Cryptographic Techniques* Springer, 2018, pp. 520–551.

403 ENISA, "Artificial intelligence cybersecurity challenges," ENISA, Tech. Rep., December 2020.

404 R. S. S. Kumar, D. O. Brien, K. Albert, S. Viljöen, and J. Snover, "Failure Modes in Machine Learning Systems," *arXiv e-prints*, arXiv:1911.11034, Nov. 2019.

405 M. S. Jere, T. Farnan, and F. Koushanfar, "A taxonomy of attacks on federated learning," *IEEE Security & Privacy*, vol. 19, no. 2, pp. 20–28, 2021.

406 N. Khurana, S. Mittal, A. Piplai, and A. Joshi, "Preventing poisoning attacks on ai based threat intelligence systems," in *2019 IEEE 29th International Workshop on Machine Learning for Signal Processing (MLSP)* IEEE, 2019, pp. 1–6.

407 M. Pawlicki, M. Choraś, and R. Kozik, "Defending network intrusion detection systems against adversarial evasion attacks," *Future Generation Computer Systems*, vol. 110, pp. 148–154, 2020.

408 H. Xiao, B. Biggio, G. Brown, G. Fumera, C. Eckert, and F. Roli, "Is feature selection secure against training data poisoning?" in *Proceedings of the 32nd International Conference on International Conference on Machine Learning - Volume 37*, ser. ICML'15 JMLR.org, 2015, pp. 1689–1698.

409 A. Kurakin, D. Boneh, F. Tramèr, I. Goodfellow, N. Papernot, and P. McDaniel, "Ensemble adversarial training: Attacks and defenses," 2018. [Online]. Available: https://openreview.net/pdf?id=rkZvSe-RZ.

410 M. Soll, T. Hinz, S. Magg, and S. Wermter, "Evaluating defensive distillation for defending text processing neural networks against adversarial examples," in *Artificial Neural Networks and Machine Learning – ICANN 2019: Image Processing*, I. V. Tetko, V. Kůrková, P. Karpov, and F. Theis, Eds Cham: Springer International Publishing, 2019, pp. 685–696.

411 A. Roy, A. Chhabra, C. A. Kamhoua, and P. Mohapatra, "A moving target defense against adversarial machine learning," in *Proceedings of the 4th ACM/IEEE Symposium on Edge Computing*, ser. SEC '19 New York, NY, USA: Association for Computing Machinery, 2019, pp. 383–388. [Online]. Available: https://doi.org/10.1145/3318216.3363338.

412 S. Sengupta, T. Chakraborti, and S. Kambhampati, "Mtdeep: Boosting the security of deep neural nets against adversarial attacks with moving target defense," in *International Conference on Decision and Game Theory for Security* Springer, 2019, pp. 479–491.

413 J. Liu, L. Chen, A. Miné, and J. Wang, "Input Validation for Neural Networks Via Runtime Local Robustness Verification," *CoRR*, vol. abs/2002.03339, 2020. [Online]. Available: https://arxiv.org/abs/2002.03339.

414 B. Li, C. Chen, W. Wang, and L. Carin, "Certified Adversarial Robustness with Additive Noise," *arXiv e-prints*, p. arXiv:1809.03113, Sep. 2018.

415 J. Ma, R. Shrestha, J. Adelberg, C.-Y. Yeh, E. K. Zahed Hossain, J. M. Jornet, and D. M. Mittleman, "Security and eavesdropping in terahertz wireless links," *Nature*, vol. 563, no. 8, pp. 89–93, 2018.

416 V. Petrov, D. Moltchanov, J. M. Jornet, and Y. Koucheryavy, "Exploiting multipath terahertz communications for physical layer security in beyond 5G networks," in *IEEE INFOCOM 2019 - IEEE Conference on Computer Communications Workshops (INFOCOM WKSHPS)*, 2019, pp. 865–872.

417 M. M. U. Rahman, Q. H. Abbasi, N. Chopra, K. Qaraqe, and A. Alomainy, "Physical layer authentication in nano networks at terahertz frequencies for biomedical applications," *IEEE Access*, vol. 5, pp. 7808–7815, 2017.

418 M. S. Saud, H. Chowdhury, and M. Katz, "Heterogeneous software-defined networks: Implementation of a hybrid radio-optical wireless network," in *2017 IEEE Wireless Communications and Networking Conference (WCNC)*, 2017, pp. 1–6.

419 M. A. Arfaoui, M. D. Soltani, I. Tavakkolnia, A. Ghrayeb, M. Safari, C. M. Assi, and H. Haas, "Physical layer security for visible light communication systems: A survey," *IEEE Communication Surveys and Tutorials*, vol. 22, no. 3, pp. 1887–1908, 2020.

420 M. A. Arfaoui, A. Ghrayeb, and C. M. Assi, "Secrecy performance of the MIMO VLC wiretap channel with randomly located eavesdropper," *IEEE Transactions on Wireless Communications*, vol. 19, no. 1, pp. 265–278, 2020.

421 J. Chen and T. Shu, "Statistical modeling and analysis on the confidentiality of indoor VLC systems," *IEEE Transactions on Wireless Communications*, vol. 19, no. 7, pp. 4744–4757, 2020.

422 S. Soderi, "Enhancing security in 6G visible light communications," in *2020 2nd 6G Wireless Summit (6G SUMMIT)*, 2020, pp. 1–5.

423 Q. Wu and R. Zhang, "Intelligent reflecting surface enhanced wireless network via joint active and passive beamforming," *IEEE Transactions on Wireless Communications*, vol. 18, no. 11, pp. 5394–5409, 2019.

424 M. Cui, G. Zhang, and R. Zhang, "Secure wireless communication via intelligent reflecting surface," *IEEE Wireless Communications Letters*, vol. 8, no. 5, pp. 1410–1414, 2019.

425 Z. Ji, P. L. Yeoh, D. Zhang, G. Chen, Y. Zhang, Z. He, H. Yin, and Y. Li, "Secret key generation for intelligent reflecting surface assisted wireless communication networks," *IEEE Transactions on Vehicular Technology*, vol. 70, no. 1, pp. 1030–1034, 2021.

426 T. Nakano, Y. Okaie, S. Kobayashi, T. Hara, Y. Hiraoka, and T. Haraguchi, "Methods and applications of mobile molecular communication," *Proceedings of the IEEE*, vol. 107, no. 7, pp. 1442–1456, 2019.

427 F. Dressler and F. Kargl, "Towards security in nano-communication: Challenges and opportunities," *Nano Communication Networks*, vol. 3, no. 3, pp. 151–160, 2012.

428 L. Mucchi, A. Martinelli, S. Jayousi, S. Caputo, and M. Pierobon, "Secrecy capacity and secure distance for diffusion-based molecular communication systems," *IEEE Access*, vol. 7, pp. 110 687–110 697, 2019.

429 Y. Qu, L. Gao, T. H. Luan, Y. Xiang, S. Yu, B. Li, and G. Zheng, "Decentralized privacy using blockchain-enabled federated learning in fog computing," *IEEE Internet of Things Journal*, vol. 7, no. 6, pp. 5171–5183, 2020.

430 C. Dwork, A. Roth *et al.*, "The algorithmic foundations of differential privacy." *Foundations and Trends in Theoretical Computer Science*, vol. 9, no. 3–4, pp. 211–407, 2014.

431 S. Li, S. Zhao, G. Min, L. Qi, and G. Liu, "Lightweight privacy-preserving scheme using homomorphic encryption in industrial Internet of Things," *IEEE Internet of Things Journal*, vol. 9, pp. 14542–14550, 2021.

432 T. Nguyen, N. Tran, L. Loven, J. Partala, M.-T. Kechadi, and S. Pirttikangas, "Privacy-aware blockchain innovation for 6G: Challenges and opportunities," in *2020 2nd 6G Wireless Summit (6G SUMMIT)* IEEE, 2020, pp. 1–5.

433 Y. Sun, J. Liu, J. Wang, Y. Cao, and N. Kato, "When machine learning meets privacy in 6G: A survey," *IEEE Communication Surveys and Tutorials*, vol. 22, no. 4, pp. 2694–2724, 2020.

434 A. A. Abd EL-Latif, B. Abd-El-Atty, E. M. Abou-Nassar, and S. E. Venegas-Andraca, "Controlled alternate quantum walks based privacy preserving healthcare images in Internet of Things," *Optics & Laser Technology*, vol. 124, p. 105942, 2020.

435 D. Dharminder and D. Mishra, "LCPPA: Lattice-based conditional privacy preserving authentication in vehicular communication," *Transactions on Emerging Telecommunications Technologies*, vol. 31, no. 2, p. e3810, 2020.

436 Y. Du, M.-H. Hsieh, T. Liu, D. Tao, and N. Liu, "Quantum Noise Protects Quantum Classifiers Against Adversaries," *arXiv preprint arXiv:2003.09416*, 2020.

437 M. Diaz, H. Wang, F. P. Calmon, and L. Sankar, "On the robustness of information-theoretic privacy measures and mechanisms," *IEEE Transactions on Information Theory*, vol. 66, no. 4, pp. 1949–1978, 2020.

438 M. Bloch, O. Günlü, A. Yener, F. Oggier, H. V. Poor, L. Sankar, and R. F. Schaefer, "An overview of information-theoretic security and privacy: Metrics, limits and applications," *IEEE Journal on Selected Areas in Information Theory*, vol. 2, no. 1, pp. 5–22, 2021.

439 Q. Huang, M. Lin, W.-P. Zhu, J. Cheng, and M.-S. Alouini, "Uplink massive access in mixed RF/FSO satellite-aerial-terrestrial networks," *IEEE Transactions on Communications*, vol. 69, no. 4, pp. 2413–2426, 2021.

440 X. Wu, M. D. Soltani, L. Zhou, M. Safari, and H. Haas, "Hybrid LiFi and WiFi networks: A survey," *IEEE Communication Surveys and Tutorials*, vol. 23, no. 2, pp. 1398–1420, 2021.

441 S. Szott, K. Kosek-Szott, P. Gawłowicz, J. T. Gómez, B. Bellalta, A. Zubow, and F. Dressler, "WiFi Meets ML: A Survey on Improving IEEE 802.11 Performance with Machine Learning," *arXiv preprint arXiv:2109.04786*, 2021.

442 L. T. Tan and R. Q. Hu, "Mobility-aware edge caching and computing in vehicle networks: A deep reinforcement learning," *IEEE Transactions on Vehicular Technology*, vol. 67, no. 11, pp. 10 190–10 203, 2018.

443 L. Huang, S. Bi, and Y.-J. A. Zhang, "Deep reinforcement learning for online computation offloading in wireless powered mobile-edge computing networks," *IEEE Transactions on Mobile Computing*, vol. 19, no. 11, pp. 2581–2593, 2020.

444 Y. He, F. R. Yu, N. Zhao, and H. Yin, "Secure social networks in 5G systems with mobile edge computing, caching, and device-to-device communications," *IEEE Wireless Communications*, vol. 25, no. 3, pp. 103–109, 2018.

445 Z. Liang, H. Chen, Y. Liu, and F. Chen, "Data sensing and offloading in edge computing networks: TDMA or NOMA?" *IEEE Transactions on Wireless Communications*, vol. 21, pp. 4497–4508, 2022.

446 A. Bordetsky, C. Glose, S. Mullins, and E. Bourakov, "Machine Learning of Semi-Autonomous Intelligent Mesh Networks Operation Expertise," in *Proceedings of the 52nd Hawaii International Conference on System Sciences*, 2019.

447 I. F. Akyildiz, A. Kak, and S. Nie, "6G and beyond: The future of wireless communications systems," *IEEE Access*, vol. 8, pp. 133 995–134 030, 2020.

448 S. Zhang, J. Liu, H. Guo, M. Qi, and N. Kato, "Envisioning device-to-device communications in 6G," *IEEE Network*, vol. 34, no. 3, pp. 86–91, 2020.

449 R. Xie, Q. Tang, S. Qiao, H. Zhu, F. R. Yu, and T. Huang, "When serverless computing meets edge computing: Architecture, challenges, and open issues," *IEEE Wireless Communications*, vol. 28, no. 5, pp. 126–133, 2021.

450 I. Philbeck, "Connecting the unconnected: Working together to achieve connect 2020 agenda targets," *ITU White Paper*, 2017.

451 M. S. Alam, G. K. Kurt, H. Yanikomeroglu, P. Zhu, and N. D. Dào, "High altitude platform station based super macro base station constellations," *IEEE Communications Magazine*, vol. 59, no. 1, pp. 103–109, 2021.

452 G. K. Kurt, M. G. Khoshkholgh, S. Alfattani, A. Ibrahim, T. S. Darwish, M. S. Alam, H. Yanikomeroglu, and A. Yongacoglu, "A vision and framework for the high altitude platform station (HAPS) networks of the future," *IEEE Communication Surveys and Tutorials*, vol. 23, no. 2, pp. 729–779, 2021.

453 P. K. R. Maddikunta, S. Hakak, M. Alazab, S. Bhattacharya, T. R. Gadekallu, W. Z. Khan, and Q.-V. Pham, "Unmanned aerial vehicles in smart agriculture: Applications, requirements, and challenges," *IEEE Sensors Journal*, vol. 21, no. 16, pp. 17 608–17 619, 2021.

454 V. Sharma, M. Bennis, and R. Kumar, "UAV-assisted heterogeneous networks for capacity enhancement," *IEEE Communications Letters*, vol. 20, no. 6, pp. 1207–1210, 2016.

455 U. Siddique, H. Tabassum, E. Hossain, and D. I. Kim, "Wireless backhauling of 5G small cells: Challenges and solution approaches," *IEEE Wireless Communications*, vol. 22, no. 5, pp. 22–31, 2015.

456 V. Jamali, H. Ajam, M. Najafi, B. Schmauss, R. Schober, and H. V. Poor, "Intelligent reflecting surface assisted free-space optical communications," *IEEE Communications Magazine*, vol. 59, no. 10, pp. 57–63, 2021.

457 L. Wei, S. Zhao, O. F. Bourahla, X. Li, F. Wu, Y. Zhuang, J. Han, and M. Xu, "End-to-end video saliency detection via a deep contextual spatiotemporal network," *IEEE Transactions on Neural Networks and Learning Systems*, vol. 32, no. 4, pp. 1691–1702, 2021.

458 M. Mozaffari, W. Saad, M. Bennis, Y.-H. Nam, and M. Debbah, "A tutorial on UAVs for wireless networks: applications, challenges, and open problems," *IEEE Communication Surveys and Tutorials*, vol. 21, no. 3, pp. 2334–2360, 2019.

459 S. Wang, M. A. Qureshi, L. Miralles-Pechuaán, T. Huynh-The, T. R. Gadekallu, and M. Liyanage, "Explainable AI for B5G/6G: Technical aspects, use cases, and research challenges," *arXiv preprint arXiv:2112.04698*, 2021.

460 W. Guo, "Explainable artificial intelligence for 6G: Improving trust between human and machine," *IEEE Communications Magazine*, vol. 58, no. 6, pp. 39–45, 2020.

461 "General Data Protection Regulation (GDPR)," [Accessed on 02.02.2022]. [Online]. Available: https://www.privacy-regulation.eu/en/recital-71-GDPR.htm.

462 "French Digital Republic Act," [Accessed on 02.02.2022]. [Online]. Available: https://www.alstonprivacy.com/french-digital-republic-act-new-powers-french-data-protection-authority-enhanced-rights-individuals/.

463 "Equal Credit Opportunity Act (Regulation B)," [Accessed on 02.02.2022]. [Online]. Available: https://www.ecfr.gov/current/title-12/chapter-X/part-1002.

464 "Algorithmic Accountability Act of 2019," [Accessed on 02.02.2022]. [Online]. Available: https://www.congress.gov/bill/116th-congress/house-bill/2231/text.

465 "China passes new personal data privacy law," [Accessed on 02.02.2022]. [Online]. Available: https://www.reuters.com/world/china/china-passes-new-personal-data-privacy-law-take-effect-nov-1-2021-08-20/.

466 "Federal Law of 27 July 2006 N 152-FZ ON PERSONAL DATA," [Accessed on 02.02.2022]. [Online]. Available: https://pd.rkn.gov.ru/authority/p146/p164/.

467 "European Telecommunications Standards Institute," [Accessed on 29.03.2021]. [Online]. Available: https://www.etsi.org/.

468 "The Next Generation Mobile Networks," [Accessed on 29.03.2021]. [Online]. Available: https://www.ngmn.org/.

469 "Alliance for Telecommunications Industry Solutions," [Accessed on 29.03.2021]. [Online]. Available: https://www.atis.org/.

470 "Next G Alliance," [Accessed on 29.03.2021]. [Online]. Available: https://nextgalliance.org/.

471 "5G Automotive Association," [Accessed on 29.03.2021]. [Online]. Available: https://www.5gaa.org/.

472 "Association of Radio Industries and Businesses," [Accessed on 29.03.2021]. [Online]. Available: https://www.arib.or.jp/.

473 "5G Alliance for Connected Industries and Automation," [Accessed on 29.03.2021]. [Online]. Available: https://www.5g-acia.org/.

474 "3rd Generation Partnership Project," [Accessed on 29.03.2021]. [Online]. Available: https://www.3gpp.org/.

475 "International Telecommunication Union - Telecommunication (ITU-T)," [Accessed on 29.03.2021]. [Online]. Available: https://www.itu.int/.

476 "Institute of Electrical and Electronics Engineers (IEEE)," [Accessed on 29.03.2021]. [Online]. Available: https://www.ieee.org.

477 "Inter-American Telecommunication Commission (CITEL)," [Accessed on 29.03.2021]. [Online]. Available: https://www.citel.oas.org/.

478 "Canadian Communication Systems Alliance (CCSA)," [Accessed on 29.03.2021]. [Online]. Available: https://www.ccsaonline.ca/.

479 "Telecommunications Standards Development Society, India (TSDSI)," [Accessed on 29.03.2021]. [Online]. Available: https://www.tsdsi.in.

480 "Telecommunications Technology Association (TTA)," [Accessed on 29.03.2021]. [Online]. Available: http://www.tta.or.kr/eng/.

481 "Telecommunication Technology Committee (TTC)," [Accessed on 29.03.2021]. [Online]. Available: https://www.ttc.or.jp.

482 A. Mahajan, G. Pottie, and W. Kaiser, "Transformation in healthcare by wearable devices for diagnostics and guidance of treatment," *ACM Transactions on Computing for Healthcare*, vol. 1, no. 1, pp. 1–12, 2020.

483 3GPP, "Study on Communication Services for Critical Medical Applications," Technical Report, November 2018. [Online]. Available: https://www.3gpp.org/ftp/Specs/archive/22_series/22.826/.

484 3GPP, "Study on Communication for Automation in Vertical Domains (CAV)," Technical Report, December 2018. [Online]. Available: https://portal.3gpp.org/desktopmodules/Specifications/SpecificationDetails.aspx?specificationId=3187.

485 X. Ge, R. Zhou, and Q. Li, "5G NFV-based tactile internet for mission-critical IoT services," *IEEE Internet of Things Journal*, vol. 7, no. 7, pp. 6150–6163, 2020.

486 A. Valkanis, P. Nicopolitidis, G. Papadimitriou, D. Kallergis, C. Douligeris, and P. D. Bamidis, "Efficient resource allocation in tactile-capable ethernet passive optical healthcare LANs," *IEEE Access*, vol. 8, pp. 52 981–52 995, 2020.

487 K. Paranjape, M. Schinkel, and P. Nanayakkara, "Short keynote paper: Mainstreaming personalized healthcare–transforming healthcare through new era of artificial intelligence," *IEEE Journal of Biomedical and Health Informatics*, vol. 24, no. 7, pp. 1860–1863, 2020.

488 Y. Chen, X. Qin, J. Wang, C. Yu, and W. Gao, "FedHealth: A federated transfer learning framework for wearable healthcare," *IEEE Intelligent Systems*, vol. 35, no. 4, pp. 83–93, 2020.

489 N. Deepa, Q.-V. Pham, D. C. Nguyen, S. Bhattacharya, B. Prabadevi, T. R. Gadekallu, P. K. R. Maddikunta, F. Fang, and P. N. Pathirana, "A survey on blockchain for big data: Approaches, opportunities, and future directions," *Future Generation Computer Systems*, vol. 131, pp. 209–226, 2022.

490 D. P. Isravel, S. Silas, and E. B. Rajsingh, "SDN-based traffic management for personalized ambient assisted living healthcare system," in *Intelligence in Big Data Technologies–Beyond the Hype*. Springer, 2020, pp. 379–388.

491 S. Movassaghi, M. Abolhasan, J. Lipman, D. Smith, and A. Jamalipour, "Wireless body area networks: A survey," *IEEE Communication surveys and tutorials*, vol. 16, no. 3, pp. 1658–1686, 2014.

492 I. F. Akyildiz, M. Pierobon, S. Balasubramaniam, and Y. Koucheryavy, "The internet of bio-nano things," *IEEE Communications Magazine*, vol. 53, no. 3, pp. 32–40, 2015.

493 M. Šiškins, M. Lee, D. Wehenkel, R. van Rijn, T. W. de Jong, J. R. Renshof, B. C. Hopman, W. S. Peters, D. Davidovikj, H. S. van der Zant *et al.*, "Sensitive capacitive pressure sensors based on graphene membrane arrays," *Microsystems & Nanoengineering*, vol. 6, no. 1, pp. 1–9, 2020.

494 N. A. Abbasi and O. B. Akan, "An information theoretical analysis of human insulin-glucose system toward the internet of bio-nano things," *IEEE Transactions on Nanobioscience*, vol. 16, no. 8, pp. 783–791, 2017.

495 J. Wang, M. Peng, Y. Liu, X. Liu, and M. Daneshmand, "Performance analysis of signal detection for amplify-and-forward relay in diffusion-based molecular communication systems," *IEEE Internet of Things Journal*, vol. 7, no. 2, pp. 1401–1412, 2020.

496 M. Kuscu, E. Dinc, B. A. Bilgin, H. Ramezani, and O. B. Akan, "Transmitter and receiver architectures for molecular communications: A survey on physical design with modulation, coding, and detection techniques," *Proceedings of the IEEE*, vol. 107, no. 7, pp. 1302–1341, 2019.

497 T. Hewa, A. Kalla, A. Nag, M. Ylianttila, and M. Liyanage, "Blockchain for 5G and IoT: Opportunities and challenges."

498 N. M. Kumar and P. K. Mallick, "Blockchain technology for security issues and challenges in iot," *Procedia Computer Science*, vol. 132, pp. 1815–1823, 2018.

499 EC The High-Level Expert Group on Artificial Intelligence (AI HLEG), "Ethics guidelines for trustworthy ai," 2019. [Online]. Available: https://digital-strategy.ec.europa.eu/en/library/ethics-guidelines-trustworthy-ai.

500 A. Gharaibeh, M. A. Salahuddin, S. J. Hussini, A. Khreishah, I. Khalil, M. Guizani, and A. Al-Fuqaha, "Smart cities: A survey on data management, security, and enabling technologies," *IEEE Communication Surveys and Tutorials*, vol. 19, no. 4, pp. 2456–2501, 2017.

501 P. Kumar, R. Kumar, G. Srivastava, G. P. Gupta, R. Tripathi, T. R. Gadekallu, and N. Xiong, "PPSF: A privacy-preserving and secure framework using blockchain-based machine-learning for IoT-driven smart cities," *IEEE Transactions on Network Science and Engineering*, vol. 8, no. 3, pp. 2326–2341, 2021.

502 F. Cirillo, D. Gómez, L. Diez, I. E. Maestro, T. B. J. Gilbert, and R. Akhavan, "Smart city IoT services creation through large-scale collaboration," *IEEE Internet of Things Journal*, vol. 7, no. 6, pp. 5267–5275, 2020.

503 Z. Ullah, F. Al-Turjman, L. Mostarda, and R. Gagliardi, "Applications of artificial intelligence and machine learning in smart cities," *Computer Communications*, vol. 154, pp. 313–323, 2020.

504 A. Gohar and G. Nencioni, "The role of 5G technologies in a smart city: The case for intelligent transportation system," *Sustainability*, vol. 13, no. 9, p. 5188, 2021.

505 T. R. Gadekallu, Q.-V. Pham, T. Huynh-The, S. Bhattacharya, P. K. R. Maddikunta, and M. Liyanage, "Federated Learning for Big Data: A Survey on Opportunities, Applications, and Future Directions," *arXiv preprint arXiv:2110.04160*, 2021.

506 D. Jiang, "The construction of smart city information system based on the Internet of Things and cloud computing," *Computer Communications*, vol. 150, pp. 158–166, 2020.

507 I. Ahmad, T. Kumar, M. Liyanage, J. Okwuibe, M. Ylianttila, and A. Gurtov, "Overview of 5G security challenges and solutions," *IEEE Communications Standards Magazine*, vol. 2, no. 1, pp. 36–43, 2018.

508 T. Han, X. Ge, L. Wang, K. S. Kwak, Y. Han, and X. Liu, "5G converged cell-less communications in smart cities," *IEEE Communications Magazine*, vol. 55, no. 3, pp. 44–50, 2017.

509 M. Dalla Cia, F. Mason, D. Peron, F. Chiariotti, M. Polese, T. Mahmoodi, M. Zorzi, and A. Zanella, "Using smart city data in 5G self-organizing networks," *IEEE Internet of Things Journal*, vol. 5, no. 2, pp. 645–654, 2017.

510 F. Qi, X. Zhu, G. Mang, M. Kadoch, and W. Li, "UAV network and iot in the sky for future smart cities," *IEEE Network*, vol. 33, no. 2, pp. 96–101, 2019.

511 X. Cheng, Z. Huang, and S. Chen, "Vehicular communication channel measurement, modelling, and application for beyond 5G and 6G," *IET Communications*, vol. 14, no. 19, pp. 3303–3311, 2020.

512 S. Yrjölä, P. Ahokangas, and M. Matinmikko-Blue, "Sustainability as a challenge and driver for novel ecosystemic 6G business scenarios," *Sustainability*, vol. 12, no. 21, p. 8951, 2020.

513 C. Chen, B. Liu, S. Wan, P. Qiao, and Q. Pei, "An edge traffic flow detection scheme based on deep learning in an intelligent transportation system," *IEEE Transactions on Intelligent Transportation Systems*, vol. 22, no. 3, pp. 1840–1852, 2020.

514 D. Zhao, H. Qin, B. Song, Y. Zhang, X. Du, and M. Guizani, "A reinforcement learning method for joint mode selection and power adaptation in the V2V communication network in 5G," *IEEE Transactions on Cognitive Communications and Networking*, vol. 6, no. 2, pp. 452–463, 2020.

515 X. Wang, Y. Liu, and K.-K. R. Choo, "Fault-tolerant multisubset aggregation scheme for smart grid," *IEEE Transactions on Industrial Informatics*, vol. 17, no. 6, pp. 4065–4072, 2020.

516 S. A. A. Abir, A. Anwar, J. Choi, and A. Kayes, "IoT-enabled smart energy grid: Applications and challenges," *IEEE Access*, vol. 9, pp. 50 961–50 981, 2021.

517 Y. Liu, X. Yang, W. Wen, and M. Xia, "Smarter grid in the 5G era: Integrating power Internet of Things with cyber physical system," *Frontiers in Communications and Networks*, vol. 2, p. 23, 2021.

518 T. Dragičević, P. Siano, S. Prabaharan *et al.*, "Future generation 5G wireless networks for smart grid: A comprehensive review," *Energies*, vol. 12, no. 11, p. 2140, 2019.

519 M. Tariq, M. Ali, F. Naeem, and H. V. Poor, "Vulnerability assessment of 6G-enabled smart grid cyber–physical systems," *IEEE Internet of Things Journal*, vol. 8, no. 7, pp. 5468–5475, 2020.

520 M. Alazab, S. Khan, S. S. R. Krishnan, Q.-V. Pham, M. P. K. Reddy, and T. R. Gadekallu, "A multidirectional LSTM model for predicting the stability of a smart grid," *IEEE Access*, vol. 8, pp. 85 454–85 463, 2020.

521 A. K. Singh, R. Singh, and B. C. Pal, "Stability analysis of networked control in smart grids," *IEEE Transactions on Smart Grid*, vol. 6, no. 1, pp. 381–390, 2014.

522 A. K. Bashir, S. Khan, B. Prabadevi, N. Deepa, W. S. Alnumay, T. R. Gadekallu, and P. K. R. Maddikunta, "Comparative analysis of machine learning algorithms for prediction of smart grid stability," *International Transactions on Electrical Energy Systems*, vol. 31, p. e12706, 2021.

523 Z. Yan and H. Wen, "Electricity theft detection base on extreme gradient boosting in AMI," *IEEE Transactions on Instrumentation and Measurement*, vol. 70, pp. 1–9, 2021.

524 M. Ismail, M. F. Shaaban, M. Naidu, and E. Serpedin, "Deep learning detection of electricity theft cyber-attacks in renewable distributed generation," *IEEE Transactions on Smart Grid*, vol. 11, no. 4, pp. 3428–3437, 2020.

525 C. Hu, J. Yan, and X. Liu, "Adaptive feature boosting of multi-sourced deep autoencoders for smart grid intrusion detection," in *2020 IEEE Power & Energy Society General Meeting (PESGM)* IEEE, 2020, pp. 1–5.

526 I. Taboada and H. Shee, "Understanding 5G technology for future supply chain management," *International Journal of Logistics Research and Applications*, vol. 24, no. 4, pp. 392–406, 2021.

527 L. Chettri and R. Bera, "A comprehensive survey on Internet of Things (IoT) toward 5G wireless systems," *IEEE Internet of Things Journal*, vol. 7, no. 1, pp. 16–32, 2019.

528 Z. Allam and D. S. Jones, "Future (post-COVID) digital, smart and sustainable cities in the wake of 6G: Digital twins, immersive realities and new urban economies," *Land Use Policy*, vol. 101, p. 105201, 2021.

529 F. Aslam, W. Aimin, M. Li, and K. Ur Rehman, "Innovation in the era of IoT and industry 5.0: Absolute innovation management (AIM) framework," *Information*, vol. 11, no. 2, p. 124, 2020.

530 Y. Lu, "Industry 4.0: A survey on technologies, applications and open research issues," *Journal of industrial information integration*, vol. 6, pp. 1–10, 2017.

531 S. Echchakoui and N. Barka, "Industry 4.0 and its impact in plastics industry: A literature review," *Journal of Industrial Information Integration*, vol. 20, p. 100172, 2020.

532 O. A. ElFar, C.-K. Chang, H. Y. Leong, A. P. Peter, K. W. Chew, and P. L. Show, "Prospects of industry 5.0 in algae: Customization of production and new advance technology for clean bioenergy generation," *Energy Conversion and Management: X*, vol. 10, p. 100048, 2020.

533 L. D. Xu, "Industry 4.0 – frontiers of fourth industrial revolution," *Systems Research and Behavioral Science*, vol. 37, no. 4, pp. 531–534, 2020.

534 L. D. Xu, "The contribution of systems science to industry 4.0," *Systems Research and Behavioral Science*, vol. 37, no. 4, pp. 618–631, 2020.

535 L. Li, "China's manufacturing locus in 2025: With a comparison of "made-in-china 2025" and "industry 4.0"," *Technological Forecasting and Social Change*, vol. 135, pp. 66–74, 2018.

536 H. Lasi, P. Fettke, H.-G. Kemper, T. Feld, and M. Hoffmann, "Industry 4.0," *Business & Information Systems Engineering*, vol. 6, no. 4, pp. 239–242, 2014.

537 V. Priya, I. S. Thaseen, T. R. Gadekallu, M. K. Aboudaif, and E. A. Nasr, "Robust attack detection approach for IIoT using ensemble classifier," *Computers, Materials & Continua*, vol. 66, no. 3, pp. 2457–2470, 2021.

538 I. de la Pe na Zarzuelo, M. J. F. Soeane, and B. L. Bermúdez, "Industry 4.0 in the port and maritime industry: A literature review," *Journal of Industrial Information Integration*, vol. 20, p. 100173, 2020.

539 M. Azeem, A. Haleem, and M. Javaid, "Symbiotic relationship between machine learning and industry 4.0: A review," *Journal of Industrial Integration and Management*, vol. 6, p. 2130002, 2021.

540 C. Zhang and Y. Chen, "A review of research relevant to the emerging industry trends: Industry 4.0, IoT, blockchain, and business analytics," *Journal of Industrial Integration and Management*, vol. 5, no. 01, pp. 165–180, 2020.

541 K. A. Demir, G. Döven, and B. Sezen, "Industry 5.0 and human-robot co-working," *Procedia Computer Science*, vol. 158, pp. 688–695, 2019.

542 P. K. Sharma, N. Kumar, and J. H. Park, "Blockchain-based distributed framework for automotive industry in a smart city," *IEEE Transactions on Industrial Informatics*, vol. 15, no. 7, pp. 4197–4205, 2018.

543 D. He, M. Ma, S. Zeadally, N. Kumar, and K. Liang, "Certificateless public key authenticated encryption with keyword search for industrial Internet of Things," *IEEE Transactions on Industrial Informatics*, vol. 14, no. 8, pp. 3618–3627, 2017.

544 I. H. Khan and M. Javaid, "Role of Internet of Things (IoT) in adoption of industry 4.0," *Journal of Industrial Integration and Management*, vol. 6, p. 2150006, 2021.

545 J. H. Kim, "A review of cyber-physical system research relevant to the emerging IT trends: Industry 4.0, IoT, big data, and cloud computing,"

Journal of Industrial Integration and Management, vol. 2, no. 03, p. 1750011, 2017.

546 H. Chen, "Theoretical foundations for cyber-physical systems: A literature review," *Journal of Industrial Integration and Management*, vol. 2, no. 03, p. 1750013, 2017.

547 Y. Lu, "Industry 4.0: A survey on technologies, applications and open research issues," *Journal of Industrial Information Integration*, vol. 6, pp. 1–10, 2017.

548 L. D. Xu, E. L. Xu, and L. Li, "Industry 4.0: State of the art and future trends," *International Journal of Production Research*, vol. 56, no. 8, pp. 2941–2962, 2018.

549 M. Rada, "Industry 5.0 definition," May 2020. [Online]. Available: https://michael-rada.medium.com/industry-5-0-definition-6a2f9922dc48.

550 R. Sattiraju, J. Kochems, and H. D. Schotten, "Machine learning based obstacle detection for Automatic Train Pairing," in *2017 IEEE 13th International Workshop on Factory Communication Systems (WFCS)* IEEE, 2017, pp. 1–4.

551 B. Friedman and D. G. Hendry, *Value Sensitive Design: Shaping Technology with Moral Imagination* Mit Press, 2019.

552 P. J. Koch, M. K. van Amstel, P. Debska, M. A. Thormann, A. J. Tetzlaff, S. Bøgh, and D. Chrysostomou, "A skill-based robot co-worker for industrial maintenance tasks," *Procedia Manufacturing*, vol. 11, pp. 83–90, 2017.

553 Y. K. Leong, J. H. Tan, K. W. Chew, and P. L. Show, "Significance of industry 5.0," in *The Prospect of Industry 5.0 in Biomanufacturing*, P. L. Show, K. W. Chew, and T. C. Ling, Eds. CRC Press, 2020, Ch. 2.2, pp. 1–20.

554 M. Sanchez, E. Exposito, and J. Aguilar, "Autonomic computing in manufacturing process coordination in industry 4.0 context," *Journal of Industrial Information Integration*, vol. 19, p. 100159, 2020.

555 A. Majeed, Y. Zhang, S. Ren, J. Lv, T. Peng, S. Waqar, and E. Yin, "A big data-driven framework for sustainable and smart additive manufacturing," *Robotics and Computer-Integrated Manufacturing*, vol. 67, p. 102026, 2020.

556 A. Haleem and M. Javaid, "Additive manufacturing applications in industry 4.0: a review," *Journal of Industrial Integration and Management*, vol. 4, no. 04, p. 1930001, 2019.

557 T. Zonta, C. A. da Costa, R. da Rosa Righi, M. J. de Lima, E. S. da Trindade, and G. P. Li, "Predictive maintenance in the industry 4.0: A systematic literature review," *Computers & Industrial Engineering*, vol. 150, p. 106889, 2020.

558 M. Compare, P. Baraldi, and E. Zio, "Challenges to IoT-enabled predictive maintenance for industry 4.0," *IEEE Internet of Things Journal*, vol. 7, no. 5, pp. 4585–4597, 2019.

559 H. Yetış and M. Karaköse, "Optimization of mass customization process using quantum-inspired evolutionary algorithm in industry 4.0," in *2020 IEEE International Symposium on Systems Engineering (ISSE)* IEEE, 2020, pp. 1–5.

560 Y. Lu, "Cyber physical system (CPS)-based industry 4.0: A survey," *Journal of Industrial Integration and Management*, vol. 2, no. 03, p. 1750014, 2017.

561 L. D. Xu and L. Duan, "Big data for cyber physical systems in industry 4.0: A survey," *Enterprise Information Systems*, vol. 13, no. 2, pp. 148–169, 2019.

562 C. S. de Oliveira, C. Sanin, and E. Szczerbicki, "Visual content representation and retrieval for cognitive cyber physical systems," *Procedia Computer Science*, vol. 159, pp. 2249–2257, 2019.

563 O. A. Topal, M. O. Demir, Z. Liang, A. E. Pusane, G. Dartmann, G. Ascheid, and G. K. Kur, "A physical layer security framework for cognitive cyber-physical systems," *IEEE Wireless Communications*, vol. 27, no. 4, pp. 32–39, 2020.

564 S. Wang, H. Wang, J. Li, H. Wang, J. Chaudhry, M. Alazab, and H. Song, "A fast cp-abe system for cyber-physical security and privacy in mobile healthcare network," *IEEE Transactions on Industry Applications*, vol. 56, no. 4, pp. 4467–4477, 2020.

565 X. Chen, M. A. Eder, and A. Shihavuddin, "A concept for human-cyber-physical systems of future wind turbines towards industry 5.0," 2020. [Online]. Available: 10.36227/techrxiv.13106108.v1.

566 H. Akbaripour, M. Houshmand, T. Van Woensel, and N. Mutlu, "Cloud manufacturing service selection optimization and scheduling with transportation considerations: mixed-integer programming models," *The International Journal of Advanced Manufacturing Technology*, vol. 95, no. 1-4, pp. 43–70, 2018.

567 Y. Liu, X. Xu, L. Zhang, and F. Tao, "An extensible model for multitask-oriented service composition and scheduling in cloud manufacturing," *Journal of Computing and Information Science in Engineering*, vol. 16, no. 4, p. 041009, 2016.

568 P. Helo, D. Phuong, and Y. Hao, "Cloud manufacturing–scheduling as a service for sheet metal manufacturing," *Computers & Operations Research*, vol. 110, pp. 208–219, 2019.

569 F. Tao, Y. Zuo, L. Da Xu, and L. Zhang, "IoT-based intelligent perception and access of manufacturing resource toward cloud manufacturing," *IEEE Transactions on Industrial Informatics*, vol. 10, no. 2, pp. 1547–1557, 2014.

570 B.-H. Li, L. Zhang, S.-L. Wang, F. Tao, J. Cao, X. Jiang, X. Song, and X. Chai, "Cloud manufacturing: A new service-oriented networked manufacturing model," *Computer Integrated Manufacturing Systems*, vol. 16, no. 1, pp. 1–7, 2010.

571 F. Tao, L. Zhang, V. Venkatesh, Y. Luo, and Y. Cheng, "Cloud manufacturing: A computing and service-oriented manufacturing model," *Proceedings of the Institution of Mechanical Engineers, Part B: Journal of Engineering Manufacture*, vol. 225, no. 10, pp. 1969–1976, 2011.

572 Y. Lu, C. Liu, I. Kevin, K. Wang, H. Huang, and X. Xu, "Digital twin-driven smart manufacturing: Connotation, reference model, applications and research issues," *Robotics and Computer-Integrated Manufacturing*, vol. 61, p. 101837, 2020.

573 Z. Jiang, Y. Guo, and Z. Wang, "Digital twin to improve the virtual-real integration of industrial IoT," *Journal of Industrial Information Integration*, vol. 22, p. 100196, 2021.

574 F. Tao, J. Cheng, Q. Qi, M. Zhang, H. Zhang, and F. Sui, "Digital twin-driven product design, manufacturing and service with big data," *The International Journal of Advanced Manufacturing Technology*, vol. 94, no. 9–12, pp. 3563–3576, 2018.

575 S. Y. Teng, M. Touš, W. D. Leong, B. S. How, H. L. Lam, and V. Máša, "Recent advances on industrial data-driven energy savings: Digital twins and infrastructures," *Renewable and Sustainable Energy Reviews*, vol. 135, p. 110208, 2021.

576 J. Van, *Mechanical Advantage*, December 11 1996, https://www.chicagotribune.com/news/ct-xpm-1996-12-11-9612110101-story.html.

577 A. C. Sim oes, A. L. Soares, and A. C. Barros, "Factors influencing the intention of managers to adopt collaborative robots (cobots) in manufacturing organizations," *Journal of Engineering and Technology Management*, vol. 57, p. 101574, 2020.

578 K. Sowa, A. Przegalinska, and L. Ciechanowski, "Cobots in knowledge work: Human–AI collaboration in managerial professions," *Journal of Business Research*, vol. 125, pp. 135–142, 2020.

579 L. Li, "Education supply chain in the era of industry 4.0," *Systems Research and Behavioral Science*, vol. 37, no. 4, pp. 579–592, 2020.

580 J. A. Marmolejo-Saucedo, M. Hurtado-Hernandez, and R. Suarez-Valdes, "Digital twins in supply chain management: a brief literature review," in *International Conference on Intelligent Computing & Optimization* Springer, 2019, pp. 653–661.

581 D. Ivanov and A. Dolgui, "New disruption risk management perspectives in supply chains: Digital twins, the ripple effect, and resileanness," *IFAC-PapersOnLine*, vol. 52, no. 13, pp. 337–342, 2019.

582 N. Simchenko, S. Tsohla, and P. Chyvatkin, "IoT & digital twins concept integration effects on supply chain strategy: Challenges and effect," *International Journal of Supply Chain Management*, vol. 8, no. 6, pp. 803–808, 2019.

583 M. D. Kent and P. Kopacek, "Do we need synchronization of the human and robotics to make industry 5.0 a success story?" in *The International Symposium for Production Research* Springer, 2020, pp. 302–311.

584 M. Yli-Ojanperä, S. Sierla, N. Papakonstantinou, and V. Vyatkin, "Adapting an agile manufacturing concept to the reference architecture model industry 4.0: A survey and case study," *Journal of industrial information integration*, vol. 15, pp. 147–160, 2019.

585 C. Huang, S. Hu, G. C. Alexandropoulos, A. Zappone, C. Yuen, R. Zhang, M. Di Renzo, and M. Debbah, "Holographic MIMO surfaces for 6G wireless networks: Opportunities, challenges, and trends," *IEEE Wireless Communications*, vol. 27, no. 5, pp. 118–125, 2020.

586 X. Li and L. Da Xu, "A review of Internet of Things – resource allocation," *IEEE Internet of Things Journal*, vol. 8, no. 11, pp. 8657–8666, 2021.

587 S. P. RM, S. Bhattacharya, P. K. R. Maddikunta, S. R. K. Somayaji, K. Lakshmanna, R. Kaluri, A. Hussien, and T. R. Gadekallu, "Load balancing of energy cloud using wind driven and firefly algorithms in internet of everything," *Journal of Parallel and Distributed Computing*, vol. 142, pp. 16–26, 2020.

588 S. Higginbotham, "What 5G hype gets wrong - [Internet of everything]," *IEEE Spectrum*, vol. 57, no. 3, pp. 22–22, 2020.

589 Y. Cheng, K. Chen, H. Sun, Y. Zhang, and F. Tao, "Data and knowledge mining with big data towards smart production," *Journal of Industrial Information Integration*, vol. 9, pp. 1–13, 2018.

590 G. T. Reddy, M. P. K. Reddy, K. Lakshmanna, R. Kaluri, D. S. Rajput, G. Srivastava, and T. Baker, "Analysis of dimensionality reduction techniques on big data," *IEEE Access*, vol. 8, pp. 54 776–54 788, 2020.

591 M. Javaid, A. Haleem, R. P. Singh, and R. Suman, "Significant applications of big data in industry 4.0," *Journal of Industrial Integration and Management*, vol. 6, pp. 1–19, 2021.

592 E. Hämäläinen and T. Inkinen, "Industrial applications of big data in disruptive innovations supporting environmental reporting," *Journal of Industrial Information Integration*, vol. 16, p. 100105, 2019.

593 A. Mitra, "On the capabilities of cellular automata-based mapreduce model in industry 4.0," *Journal of Industrial Information Integration*, vol. 21, p. 100195, 2021.

594 K. Fukuda, "Science, technology and innovation ecosystem transformation toward society 5.0," *International Journal of Production Economics*, vol. 220, p. 107460, 2020.

595 A. Majeed, Y. Zhang, S. Ren, J. Lv, T. Peng, S. Waqar, and E. Yin, "A big data-driven framework for sustainable and smart additive manufacturing," *Robotics and Computer-Integrated Manufacturing*, vol. 67, p. 102026, 2021.

596 W. Viriyasitavat and D. Hoonsopon, "Blockchain characteristics and consensus in modern business processes," *Journal of Industrial Information Integration*, vol. 13, pp. 32–39, 2019.

597 W. Viriyasitavat, L. Da Xu, Z. Bi, and A. Sapsomboon, "Blockchain-based business process management (BPM) framework for service composition in industry 4.0," *Journal of Intelligent Manufacturing*, vol. 31, pp. 1737–1748, 2018.

598 B. Prabadevi, N. Deepa, Q.-V. Pham, D. C. Nguyen, M. Praveen Kumar Reddy, G. Thippa Reddy, P. N. Pathirana, and O. Dobre, "Toward blockchain for edge-of-things: A new paradigm, opportunities, and future directions," *IEEE Internet of Things Magazine*, vol. 4, no. 2, pp. 102–108, 2021.

599 S. He, W. Ren, T. Zhu, and K.-K. R. Choo, "Bosmos: A blockchain-based status monitoring system for defending against unauthorized software updating in industrial Internet of Things," *IEEE Internet of Things Journal*, vol. 7, no. 2, pp. 948–959, 2019.

600 N. Mohamed and J. Al-Jaroodi, "Applying blockchain in industry 4.0 applications," in *2019 IEEE 9th Annual Computing and Communication Workshop and Conference (CCWC)* IEEE, 2019, pp. 0852–0858.

601 L. Da Xu, Y. Lu, and L. Li,"Embedding blockchain technology into IoT for security: A survey," *IEEE Internet of Things Journal*, vol. 8, pp. 10452–10473, 2021.

602 A. V. Barenji, Z. Li, and W. M. Wang, "Blockchain cloud manufacturing: Shop floor and machine level," in *Smart SysTech 2018; European Conference on Smart Objects, Systems and Technologies* VDE, 2018, pp. 1–6.

603 A. Mushtaq and I. U. Haq, "Implications of blockchain in industry 4.0," in *2019 International Conference on Engineering and Emerging Technologies (ICEET)* IEEE, 2019, pp. 1–5.

604 Y. Zhang, X. Xu, A. Liu, Q. Lu, L. Xu, and F. Tao, "Blockchain-based trust mechanism for IoT-based smart manufacturing system," *IEEE Transactions on Computational Social Systems*, vol. 6, no. 6, pp. 1386–1394, 2019.

605 W. Wang, H. Xu, M. Alazab, T. R. Gadekallu, Z. Han, and C. Su, "Blockchain-based reliable and efficient certificateless signature for IIoT devices," *IEEE Transactions on Industrial Informatics*, vol. 18, pp. 7059–7067, 2021.

606 W. Shi, J. Cao, Q. Zhang, Y. Li, and L. Xu, "Edge computing: Vision and challenges," *IEEE Internet of Things Journal*, vol. 3, no. 5, pp. 637–646, 2016.

607 M. Abdirad, K. Krishnan, and D. Gupta, "A two-stage metaheuristic algorithm for the dynamic vehicle routing problem in industry 4.0 approach," *Journal of Management Analytics*, vol. 8, no. 1, pp. 69–83, 2021.

608 S. Wijethilaka and M. Liyanage, "Survey on network slicing for Internet of Things realization in 5G networks," *Transport*, vol. 23, no. 2, pp. 957–994, 2021.

609 H. Wu, I. A. Tsokalo, D. Kuss, H. Salah, L. Pingel, and F. H. Fitzek, "Demonstration of network slicing for flexible conditional monitoring in industrial IoT networks," in *2019 16th IEEE Annual Consumer Communications & Networking Conference (CCNC)* IEEE, 2019, pp. 1–2.

610 M. Baddeley, R. Nejabati, G. Oikonomou, S. Gormus, M. Sooriyabandara, and D. Simeonidou, "Isolating SDN control traffic with layer-2 slicing in 6TiSCH industrial IoT networks," in *2017 IEEE Conference on Network Function Virtualization and Software Defined Networks (NFV-SDN)* IEEE, 2017, pp. 247–251.

611 E. N. Tominaga, H. Alves, O. L. A. López, R. D. Souza, J. L. Rebelatto, and M. Latva-Aho, "Network slicing for eMBB and mMTC with NOMA and space diversity reception," in *IEEE 93rd Vehicular Technology Conference (VTC2021-Spring)* IEEE, 2021, pp. 1–6.

612 S. H.-W. Chuah, "Why and who will adopt extended reality technology? Literature review, synthesis, and future research agenda," *Literature Review, Synthesis, and Future Research Agenda (December 13, 2018)*, 2018.

613 R. Masoni, F. Ferrise, M. Bordegoni, M. Gattullo, A. E. Uva, M. Fiorentino, E. Carrabba, and M. Di Donato, "Supporting remote maintenance in industry 4.0 through augmented reality," *Procedia Manufacturing*, vol. 11, pp. 1296–1302, 2017.

614 X. Wang, S. K. Ong, and A. Y. Nee, "A comprehensive survey of augmented reality assembly research," *Advances in Manufacturing*, vol. 4, no. 1, pp. 1–22, 2016.

615 ImmersiveTouch, "Comprehensive Surgical Training using the power of Augmented and Virtual Reality," 2020. [Online]. Available: https://www.immersivetouch.com/immersivesim-training.

616 S. Roy and C. Chowdhury, "Remote health monitoring protocols for IoT-enabled healthcare infrastructure," in *Healthcare Paradigms in the Internet of Things Ecosystem* Elsevier, 2021, pp. 163–188.

617 A. Damala, P. Cubaud, A. Bationo, P. Houlier, and I. Marchal, "Bridging the gap between the digital and the physical: design and evaluation of a mobile augmented reality guide for the museum visit," in *Proceedings of the 3rd International Conference on Digital Interactive Media in Entertainment and Arts* ACM, 2008, pp. 120–127.

618 M. Ding, "Augmented reality in museums," *Museums & Augmented Reality–A Collection of Essays from the Arts Management and Technology Laboratory*, pp. 1–15, 2017.

619 P. Föckler, T. Zeidler, B. Brombach, E. Bruns, and O. Bimber, "PhoneGuide: museum guidance supported by on-device object recognition on mobile phones," in *Proceedings of the 4th international conference on Mobile and ubiquitous multimedia* ACM, 2005, pp. 3–10.

620 D. Sportillo, A. Paljic, and L. Ojeda,"On-road evaluation of autonomous driving training," in *2019 14th ACM/IEEE International Conference on Human-Robot Interaction (HRI)* IEEE, 2019, pp. 182–190.

621 L3HARRIS, "Blue Boxer Extended Reality (BBXR) Training System," 2020. [Online]. Available: https://www.l3t.com/link/assets/uploads/pdf/datasheets/L3Harris_Collateral_BBXR_SellSheet_0719.pdf.

622 F. De Crescenzio, M. Fantini, F. Persiani, L. Di Stefano, P. Azzari, and S. Salti, "Augmented reality for aircraft maintenance training and operations support," *IEEE Computer Graphics and Applications*, vol. 31, no. 1, pp. 96–101, 2010.

623 "AR Based Drone Pilot Training." [Online]. Available: https://dronoss.com/.

624 M. Zikky, K. Fathoni, and M. Firdaus, "Interactive distance media learning collaborative based on virtual reality with solar system subject," in *2018 19th IEEE/ACIS International Conference on Software Engineering, Artificial Intelligence, Networking and Parallel/Distributed Computing (SNPD)* IEEE, 2018, pp. 4–9.

625 J. G. Tromp, D.-N. Le, and C. Van Le, *Emerging Extended Reality Technologies for Industry 4.0: Early Experiences with Conception, Design, Implementation, Evaluation and Deployment*. John Wiley & Sons, 2020.

626 P. Skobelev and S. Y. Borovik, "On the way from industry 4.0 to industry 5.0: From digital manufacturing to digital society," *Industry 4.0*, vol. 2, no. 6, pp. 307–311, 2017.

627 A. Prasad, Z. Li, S. Holtmanns, and M. A. Uusitalo, "5G micro-operator networks–a key enabler for new verticals and markets," in *2017 25th Telecommunication Forum (TELFOR)* IEEE, 2017, pp. 1–4.

628 P. Ahokangas, M. Matinmikko-Blue, S. Yrjölä, V. Seppänen, H. Hämmäinen, R. Jurva, and M. Latva-aho, "Business models for local 5G micro operators," *IEEE Transactions on Cognitive Communications and Networking*, vol. 5, no. 3, pp. 730–740, 2019.

629 Y. Siriwardhana, P. Porambage, M. Ylianttila, and M. Liyanage, "Performance analysis of local 5G operator architectures for industrial internet," *IEEE Internet of Things Journal*, vol. 7, no. 12, pp. 11 559–11 575, 2020.

630 B. Barua, M. Matinmikko-Blue, and M. Latva-aho, "On emerging contractual relationships for local 5G micro operator networks," in *2019 16th International Symposium on Wireless Communication Systems (ISWCS)* IEEE, 2019, pp. 703–708.

631 R. De Silva, Y. Siriwardhana, T. Samarasinghe, M. Ylianttila, and M. Liyanage, "Local 5G operator architecture for delay critical telehealth applications," in *2020 IEEE 3rd 5G World Forum (5GWF)* IEEE, 2020, pp. 257–262.

632 M. Latva-aho, "Micro operators for vertical specifc service deliver in 5G," 2017.

633 J. Backman, S. Yrjölä, K. Valtanen, and O. Mämmelä, "Blockchain network slice broker in 5G: Slice leasing in factory of the future use case," in *2017 Internet of Things Business Models, Users, and Networks* IEEE, 2017, pp. 1–8.

634 M. Matinmikko, M. Latva-aho, P. Ahokangas, and V. Seppänen, "On regulations for 5G: Micro licensing for locally operated networks," *Telecommunications Policy*, vol. 42, no. 8, pp. 622–635, 2018.

635 K. J. Nevelsteen, "Virtual world, defined from a technological perspective and applied to video games, mixed reality, and the metaverse," *Computer Animation and Virtual Worlds*, vol. 29, no. 1, p. e1752, 2018.

636 T. Huynh-The, Q.-V. Pham, X.-Q. Pham, T. T. Nguyen, Z. Han, and D.-S. Kim, "Artificial Intelligence for the Metaverse: A Survey," *arXiv preprint arXiv:2202.10336*, 2022.

637 N. Stephenson, *Snow Crash: A Novel*. Spectra, 2003.

638 L.-H. Lee, T. Braud, P. Zhou, L. Wang, D. Xu, Z. Lin, A. Kumar, C. Bermejo, and P. Hui, "All One Needs to Know About Metaverse: A Complete Survey on Technological Singularity, Virtual Ecosystem, and Research Agenda," *arXiv preprint arXiv:2110.05352*, 2021.

639 L.-H. Lee, Z. Lin, R. Hu, Z. Gong, A. Kumar, T. Li, S. Li, and P. Hui, "When Creators Meet the Metaverse: A Survey on Computational Arts," *arXiv preprint arXiv:2111.13486*, 2021.

640 "Blue Marble Private," 2022. [Online]. Available: https://bluemarbleprivate.com/.

641 "Ocean Gate," 2022. [Online]. Available: https://oceangate.com/.

642 J. Zhang, F. Liang, B. Li, Z. Yang, Y. Wu, and H. Zhu, "Placement optimization of caching uav-assisted mobile relay maritime communication," *China Communications*, vol. 17, no. 8, pp. 209–219, 2020.

643 J. Zeng, J. Sun, B. Wu, and X. Su, "Mobile edge communications, computing, and caching (MEC3) technology in the maritime communication network," *China Communications*, vol. 17, no. 5, pp. 223–234, 2020.

644 T. Zhang, G. Han, C. Lin, N. Guizani, H. Li, and L. Shu, "Integration of communication, positioning, navigation and timing for deep-sea vehicles," *IEEE Network*, vol. 34, no. 2, pp. 121–127, 2020.

645 "SpaceX," 2022. [Online]. Available: https://www.spacex.com/.

646 "Starlink," [Accessed on 22.01.2022]. [Online]. Available: https://www.starlink.com/.

Index

6G Frontiers: Towards Future Wireless Systems, First Edition.
Chamitha de Alwis, Quoc-Viet Pham, and Madhusanka Liyanage.
© 2023 The Institute of Electrical and Electronics Engineers, Inc. Published 2023 by John Wiley & Sons, Inc.

t

u

Printed and bound by CPI Group (UK) Ltd, Croydon, CR0 4YY

16/04/2025

14658596-0003